HBJ
Dictionary
of
Mathematics

CONSULTANT EDITORS

LINDSAY GRIMISON

Senior Lecturer in Mathematics,
Catholic College of Education, Sydney
&
Visiting Scholar,
Shell Centre for Mathematical Education,
University of Nottingham

DAPHNE KERSLAKE

Principal Lecturer,
Education Department,
Bristol Polytechnic

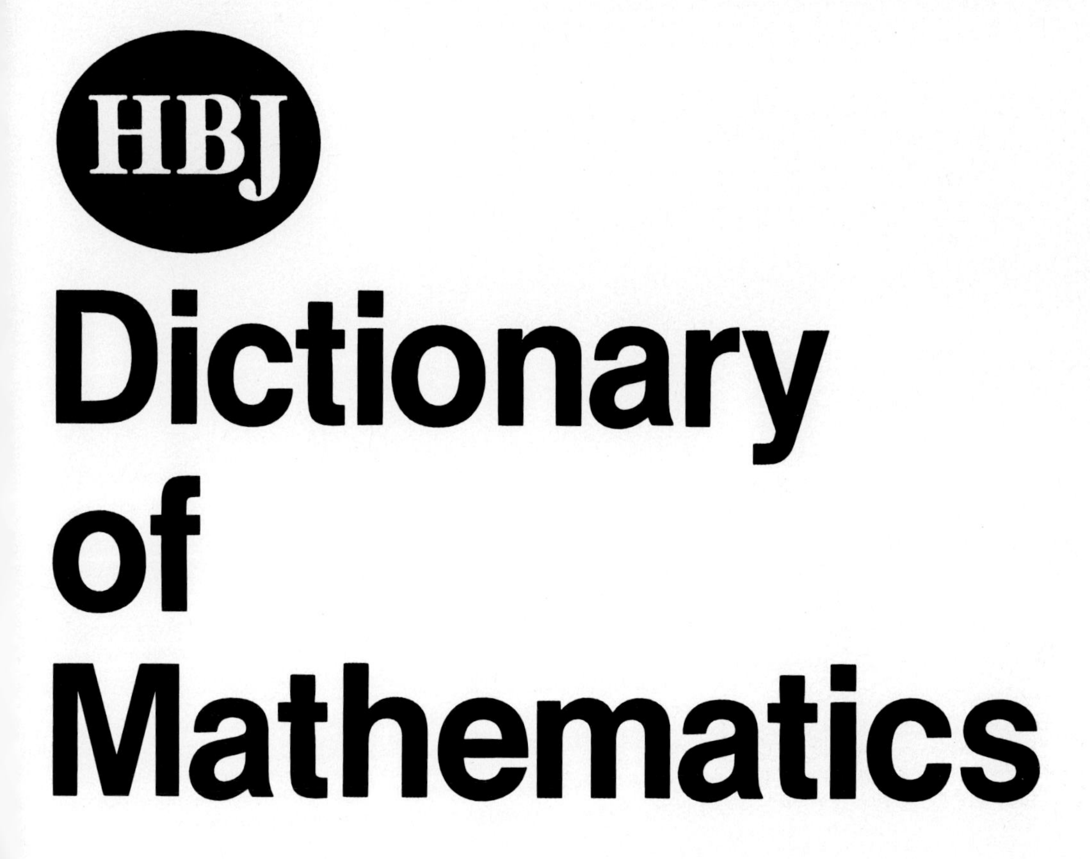

Dictionary of Mathematics

ILLUSTRATED

Compiled by
KEN KLAEBE

HARCOURT BRACE JOVANOVICH, PUBLISHERS
Sydney London Orlando Toronto

Harcourt Brace Jovanovich Group (Australia) Pty Limited
30–52 Smidmore Street, Marrickville, NSW 2204

Harcourt Brace Jovanovich Limited
24/28 Oval Road, London NW1 7DX

Harcourt Brace Jovanovich, Inc.
Orlando, Florida, 32887

Printed in Australia

National Library of Australia Cataloguing-in-Publication Data

---

Dictionary of mathematics
ISBN 0 7295 0224 4.

1. Mathematics - Dictionaries. I. Klaebe, Ken E.
II. Grimison, Lindsay.

510'.3'21

CONTENTS

To L.R.H., whose discoveries in the field of education provided the inspiration for this dictionary.

PUBLISHER'S NOTE
Many people have assisted in the production of this reference work. The publisher would like to thank, first, Ken Klaebe and Lindsay Grimison for their patience and good humour during the production process, Valda Brook, Ken Roussell and John Cleasby for the illustrations, Daphne Kerslake for attention to detail, and the following staff members: Jeremy Fisher (senior editor) James Jackson (production manager), Mary Fuller (production assistant), Grant Walker (HBJ London), and Desney Jackson (senior editor).

HOW TO USE THIS BOOK

Each definition may contain words (or symbols) defined elsewhere in this dictionary. Such words are in **bold type**, and if you are not sure of their meaning, look them up. Most definitions contain examples of their usage. To fully understand a definition be sure you think up your own examples! Sometimes, there are cross references to other entries in SMALL CAPITALS.

The origin of the word is given last in a standard form: for example, LATIN, *numerus*, number. This means it comes from the Latin word *numerus*, meaning number.

ACKNOWLEDGEMENTS

Grateful acknowledgement is expressed to the editors and publishers of *Shorter Oxford English Dictionary* for permission to use origins of selected words. Special thanks to all those who contributed to making this book a reality: Lindsay Grimison (consulting editor); Tony Webster and staff at Computer Reference Guide; my son Simon; the illustrators; and the staff of Harcourt Brace Jovanovich.

SIGNS AND SYMBOLS

Arithmetic, Algebra

Sign/Symbol	*Meanings*	*Examples*
$+$	Plus	$3 + 4 = 7$
$-$	Minus	$5 - 2 = 3$
$\pm$	Plus or minus	if $y^2 = 4$, then $y = \pm 2$
$\times$	Multiplied by	$3 \times 4 = 12$
$\div$	Divided by	$12 \div 2 = 6$
$=$	Equal to	$2 + 1 = 3$
$\neq$	Not equal to	$3 + 4 \neq 5$
$\equiv$	(1) Identical to (2) See also Geometry	$3(a + b) \equiv 3a + 3b$
$\approx$	Approximately equal to	$\pi \approx 22/7$
$\doteq$	Approximately equal to	$\pi \doteq 22/7$
$>$	Greater than	$5 > 3$
$<$	Less than	$3 < 4$
$\geqslant$	Greater than or equal to	$a \geqslant 4$
$\leqslant$	Less than or equal to	$b \leqslant 2$
$:$	Is to; ratio (see also p. xi)	$a:b = 1:2$ (can also be written $\frac{a}{b} = \frac{1}{2}$)
$::$	As, proportion	$a:b :: c:d$
α	Varies as, is proportional to	$a \propto b$ can be written $a = kb$, where k is a constant.
Δ	1. Delta, a small difference	Δx represents a small change in x (an increase or decrease)
	2. Discriminant	$\Delta = b^2 - 4ac$, referring to $ax^2 + bx + c$
	3. Determinant 4. Triangle	If $A = \begin{pmatrix} a & c \\ b & d \end{pmatrix}$, $\Delta = ad - bc$
$\sqrt{\ }$	Square root	$\sqrt{9} = 3$, $\sqrt{5}$ (which can not be written as a ratio of whole numbers)
$\sqrt[3]{\ }$	Cube root	$\sqrt[3]{64} = 4$, since $4 \times 4 \times 4 = 64$

Symbol	Meaning	Example
$\sqrt[4]{\ }$	Fourth root	$\sqrt[4]{16} = 2$, since $2 \times 2 \times 2 \times 2 = 16$
$\cdot$	Multiplied by	$4 \cdot 3 = 12$
.	Decimal point	$4.3 = 4\frac{3}{10}$
mod	Modulus (or modulo)	In using a 12 hour clock, $7 + 6 = 1 \bmod 12$ (adding 6 hours to 7 hours gives 1 o'clock)
( )	Parentheses (groups things together)	$(a + b)^2 = (a + b) \times (a + b)$ $2(3 + 7) = 2 \times 10 = 20$
[ ]	Brackets (groups things together)	$3(1 + [2 + 5]) = 3(1 + 7) = 3 \times 8 = 24$
{ }	(1) Braces (groups things together) (2) Groups a set of things together	$3(1 + 2 + 3\{4 + 1\}) = 3(1 + 2 + 3 \times 5) = 3(1 + 2 + 15) = 3(1 + 17) = 3 \times 18 = 54$
$\overline{\quad}$	Vinculum (groups things together)	$\overline{4 + 3} = 7$
${}^nC_r$, $\binom{n}{r}$ (also C^n_r)	The number of combinations of r things selected (in any order) from n things ${}^nC_r = C^n_r = \frac{n!}{r!(n - r)!} = \binom{n}{r}$	${}^4C_2 = \frac{4!}{2!(4 - 2)!} = \frac{4 \times 3 \times 2 \times 1}{2 \times 1 \times 2 \times 1} = \frac{12}{2} = 6$ i.e. there are 6 ways 2 things can be selected from 4 things. *A B C D* *AB AC AD BC BD CD*
${}^nP_r$ (also P^n_r)	The number of permutations (groups of things in a specific order) of r things which can be selected from n things. ${}^nP_r = \frac{n!}{(n - r)!} = P^n_r$	${}^4P_2 = \frac{4!}{(4 - 2)!} = 12$ *A B C D* *DA BD DB CA AC AD* *AB BA BC CB CD DC*
$\frac{a}{b}$	a divided by b	$\frac{1}{2}, \frac{3}{4}$
a/b	a divided by b	½, ¼
$\dot{}$	(1) Repeating (2) Differential coefficient (with respect to t = time)	$3.\dot{1} = 3.1111\ldots$ $\dot{x} = \frac{dx}{dt}$ (which is a way of writing speed)
a^2	$a \times a$	$2^2 = 2 \times 2 = 4$
a^3	$a \times a \times a$	$4^3 = 4 \times 4 \times 4 = 64$
$\| \ \|$	Absolute value of	$\|-2\| = 2$, $\|+3.6\| = 3.6$

Σ	Sum	$\sum_{i=1}^{4} i = 1 + 2 + 3 + 4$
!, $\llcorner$	Factorial (i.e. the product of all the positive whole numbers up to the number in front of the !)	$6! = 6 \times 5 \times 4 \times 3 \times 2 \times 1$ $\llcorner 6 = 6 \times 5 \times 4 \times 3 \times 2 \times 1$
∞	Infinity, without limit	As $x \to 0, \frac{1}{x} \to \infty$
$\to$	(1) Approaches towards (2) Vector, as in $\vec{PQ}$	$\vec{PQ}$ is a vector
π	Pi (Greek letter). Represents the ratio of the circumference of a circle to its diameter. It is constant for any circle, and is approx. equal to 3·141592...	$= \frac{C}{d}$ 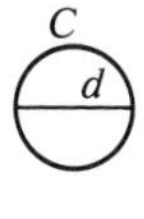
$\therefore$	Therefore	Since $a^2 - b^2 = (a - b)(a + b)$ $\therefore 2^2 - 1^2 = (2-1)(2+1) = 3$
:	Connecting one symbol with another or others	m: $y = -1$ (m is a line described by the equation $y = -1$)
$'$	Transpose of (a matrix)	If $A = \begin{pmatrix} 1 & 2 \\ 3 & 4 \\ 5 & 6 \end{pmatrix}$, then $A' = \begin{pmatrix} 1 & 3 & 5 \\ 2 & 4 & 6 \end{pmatrix}$
$\because$	Because	$\because a^2 + 2a + 1 = (a + 1)^2$ then $(4 + 1)^2 = 4^2 + 2 \times 4 + 1$
$f(x)$	Function of x (i.e. for every value of x, there is a corresponding value of $f(x)$)	$f(y)$, $F(z)$
f^{-1}	Inverse of ($f^{-1}(x)$ is a reflection of $f(x)$ about the line $y = x$) If A is a square matrix, then A^{-1} is the inverse ($AA^{-1} = I$) I is the identity matrix	If $f(x) = \log x$ then $f^{-1}(x) = e^x$ 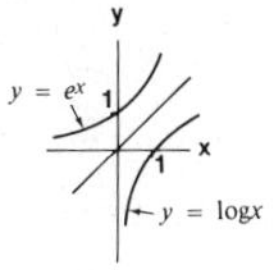
AP	Arithmetic Progression	1, 3, 5, 7, ... is an AP
GP	Geometric Progression	$1, x, x^2, x^3, \ldots$ is a GP
i	$\sqrt{-1}$ (square root of -1) 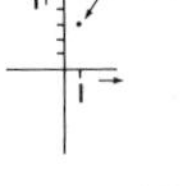	$1 + 3i$ can be represented on a plane
—	Bar, meaning minus (used in logarithms)	$\bar{2}$ means -2 in logarithmic procedures

Calculus

Sign/Symbol	Meanings	Examples
dx	(1) Change in x (2) with respect to x	$\frac{dy}{dx}$ represents the rate of change of y (assumed to be a function of x only) with respect to x.
$\frac{d(\ )}{dx}$	The rate of change of ( ) with respect to x.	
$\int$	Integral	$\int y\, dx$ is the integral of y with respect to x.
∂	Partial differential	$\frac{\partial y}{\partial x}$ represents the rate of change of y (assumed to be a function of x and other variables) with respect to x only.
$\frac{d^2(\ )}{dx^2}$	The rate of change of $\frac{d(\ )}{dx}$ with respect to x	$\frac{d^2y}{dx^2}$ is the second derivative* of y with respect to x. (* See definition)
δ	A small part	δx represents a small increase in x. $\frac{\delta y}{\delta x}$ is approximated by $\frac{dy}{dx}$ at the point of x.
$\rightarrow$	Approaches	As $x \rightarrow 2$, $x^2 \rightarrow 4$
lim	Limit	If $f(x) = x^3 - 1$, $\lim_{x \rightarrow 2} f(x) = 7$ (as x approaches the limit of 2, $f(x)$ approaches 7)
$f(\ )$	Function of	$f(x) = x^2 - 3x + 1$ is a function of x. $f(g(x))$ = function of $g(x)$ (which itself is a function of x)
$'$	First derivative	If $y = f(x)$, then $f'(x) = \frac{dy}{dx}$ (another way of writing the first derivative)
$''$	Second derivative	$f''(x) = \frac{d^2y}{dx^2}$
$\dot{}$	First derivative with respect to time	If $y = f(t)$, $\dot{y} = \frac{dy}{dt} = f'(t)$
$\ddot{}$	Second derivative with respect to time	$\ddot{y} = \frac{d^2y}{dt^2} = f''(t)$

Geometry

Sign/Symbol	*Meanings*	*Examples*
$\angle$	Angle	$\angle ABC$ is 30°
⊾ (also ∟)	Right Angle	$\angle ABC$ = ∟
$\triangle$	Triangle	$\triangle ABC$
$\parallel$	Parallel	$AB \parallel CD$
$\perp$	Perpendicular, at right angles to	$AB \perp XY$
□	Square	$ABCD$ is a □
°	Degree ($\frac{1}{360\text{th}}$ of a circle)	$\angle ABC$ is 30°
′	Minute ($\frac{1}{60\text{th}}$ of a degree)	A is E 31°30′ N
″	Second ($\frac{1}{60\text{th}}$ of a minute)	B is N 25°20′31″ W
$\equiv$	Congruent (identical) to. (equal in all respects, for example corresponding sides equal, corresponding angles equal)	$\triangle ABC \equiv \triangle PQR$

Sets

Sign/Symbol	*Meanings*	*Examples*
{ }	Set (group of things)	{1, 2, 3, 4, . . . 10} is the set of all natural numbers up to and including 10.
$\varnothing$	Null Set (also called the Empty Set)	$\varnothing$ = { }, i.e. there are no members in the set.
;	Such that	$\{x;\ x \leqslant 1\}$ This is a shorthand way of writing a set of values such that each value is less than or equal to 1.
$\exists$	There exists	$\exists\ \{x;\ x \leqslant 2\}$
$\cup$	Union (Pronounced 'cup')	If A = {1, 2, 3, . . . 10} and B = {7, 8, . . . 13} then $A \cup B$ = {1, 2, 3, . . . 13} The Union is the combination of all the different members of each set.

$\cap$	Intersection (Pronounced 'cap')	If $A = \{1, 2, 3, \ldots 10\}$ and $B = \{7, 8, \ldots 13\}$ then $A \cap B = \{7, 8, 9, 10\}$ The intersection is the set of only those elements common to both of the intersecting sets.
$\subset$	Subset, is contained in	If $A = \{1, 2, 3, 4\}$ and $B = \{1, 2, 3, \ldots 10\}$ then $A \subset B$.
ϵ	Is a member of	If $A = \{1, 2, 3, \ldots 10\}$ then $3 \epsilon A$.
ξ, E (or U)	Universal Set	If $\xi = \{1, 2, 3, \ldots \infty\}$ then $A = \{1, 2, 3, \ldots 10\}$ is a subset of ξ.
$'$	Complement of	If A = ⊘, and U = 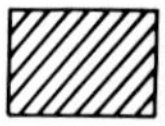 then A' = 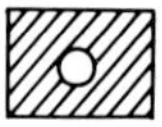

Statistics

Sign/Symbol	*Meanings*	*Examples*
σ	Standard deviation	If σ is low, then the variation of the values in the set of values being considered is low.
$\bar{}$	Mean (or average)	If $x = 2, 3, 2, 2, 3, 2$ then $\bar{x} = \frac{2 + 3 + 2 + 2 + 3 + 2}{6} = \frac{14}{6} = 2\frac{1}{3}$
χ^2	Chi squared	If observing the face of a coin thrown 10 times the results were H, H, T, H, T, T, H, T, H, H, and the *expected* results were H, H, H, T, H, H, T, H, H, T (The coin being assumed biased to Heads), then χ^2 = Sum of $\frac{(\text{Actual} - \text{Expected})^2}{\text{Expected}}$ gives a measure of how close a fit the expected results are.

Trigonometry

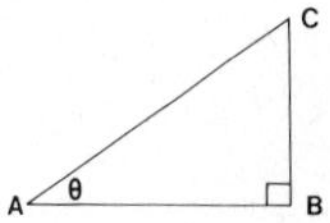

Sign/Symbol	*Meanings*	*Examples*
θ	Theta (Usually represents an angle)	$\angle BAC = \theta$
sin	Sine	$\sin\theta = \frac{BC}{AC}$
cos	Cosine	$\cos\theta = \frac{AB}{AC}$
tan	Tangent	$\tan\theta = \frac{BC}{AB}$
sec	Secant	$\sec\theta = \frac{AC}{AB}$
cosec	Cosecant	$\text{cosec}\,\theta = \frac{AC}{BC}$
cot	Cotangent	$\cot\theta = \frac{AB}{BC}$
$^{-1}$	Inverse of ($^{-1}$ does *not* mean $\frac{1}{(\ )}$ in trigonometry)	$\sin^{-1}x$ is the angle whose sine is x (i.e. if $x = \sin\theta$ then $\theta = \sin^{-1}x$. $\sin^{-1}x$ can also be written as arc sin x (i.e. if $x = \sin\theta$ $\theta = \text{arc sin } x$)
arc sin	Angle whose sine is	arc sin x is the angle whose sine is x. (or $\sin^{-1}x$)
arc cos	Angle whose cosine is	arc cos x is the angle whose cosine is x. (or $\cos^{-1}x$)
arc tan	Angle whose tangent is	arc tan x is the angle whose tangent is x. (Also written as $\tan^{-1}x$)

A θ B C

Aa

A (a) **1.** First letter in the English alphabet. It is often used as a symbol for either a **variable** quantity or a **constant.** **2.** The capital letter is often used to name a point in a geometric figure. **3.** One.
Examples
1. In $a^2 - b^2 = (a+b)(a-b)$, which is true for every value of a and b, a is a variable quantity.
2. In triangle ABC, A represents the vertex.

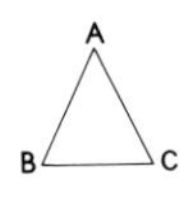

3. A hundred letters were sent (meaning: one hundred letters were sent).
GREEK, α, alpha [first letter in the Greek alphabet]

abacus a device for counting. It is made up of a frame holding a number of rods on which beads or counters slide. Each rod represents a **place value** (such as units, tens or hundreds in a particular counting system, e.g. the decimal system). There are different types of abacuses (or abaci), e.g., those used in east Asia (Japanese or Soroban, Chinese etc.).

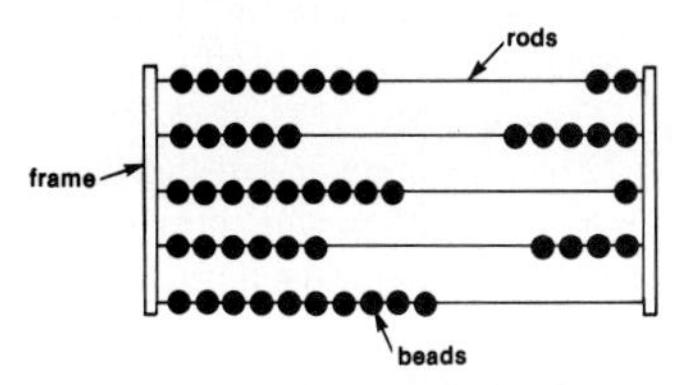

GREEK, *abax*, slab, mathematical table

abscissa (*pl.* ~ e) (**coordinate** geometry). **1.** The distance of a point from the y-axis (the distance from the x-axis to the same point is called the **ordinate**). **2.** The first number in the ordered **pair** of numbers used to locate a point on a plane.
Example The point (3,4) has an abscissa of 3, and an ordinate of 4. The ordered pair is (3,4), and the abscissa is the first number in that pair.

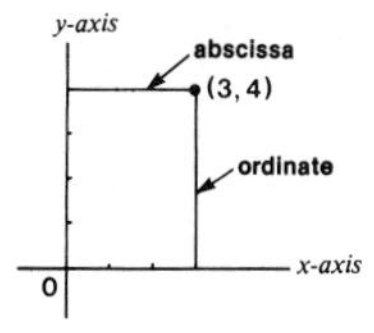

LATIN, *abscissus*, cut off

absolute value **1.** The value of a quantity without using its **sign**. The symbol used is | |. **2.** (of a **complex** number) The square root of the sum of the squares of the real and imaginary parts. **3.** The distance between two points on the number line.
Examples
1. The absolute value of -3 is 3; the absolute value of $+3$ is 3.
2. The absolute value of the complex number $7 + i = |7 + i| = \sqrt{7^2 + 1^2} = \sqrt{50}$

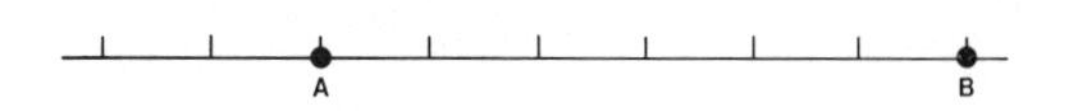

3. The distance between A and B is six units.
LATIN, *absolutus*, completed, unconditional; *valere*, be strong, of value

acceleration the rate of change of velocity.
Examples
1. The car increased its velocity from 40 km/hour to 60 km/hour in 10 seconds moving due South. The acceleration is 20 km/hour in 10 seconds, which is 120 km/hour/minute, or 2km/minute/minute. The latter is sometimes abbreviated to 2km/min^2.
2. The acceleration of a particle moving in accordance with the projection of another particle (moving uniformly around the circumference of a circle) on the circle's diameter is proportional to the distance of the projection from the centre of the circle.

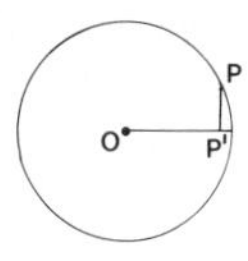

Acceleration of P is proportional to OP'.
LATIN, *accelerare*, to hasten

accumulator (in computing) a part of a computer that counts, or holds information that can be altered.
Example Most computers have many accumulators which are used in adding and subtracting, multiplying and dividing. The meter which measures distance travelled in a car is an example of a simple accumulator, as each kilometre is added to the previous total.
LATIN, *ad*, in addition; *cumulare*, to pile up

accuracy exactness; it is measured usually by the number of figures in an answer or result, or by the last figure.
Examples What is the accuracy of the following lengths? 10.3 cm, 25.8 cm, 8.4 cm. They are all

accurate to the nearest one tenth of a centimetre (cm). 10.3 cm means a length measured between 10.25 and 10.35 cm. Similarly 8.4 represents a length between 8.35 and 8.45 cm. 10.3 and 25.8 cm are said to be accurate to 3 **significant** figures, 8.4 cm to 2 significant figures, which is another way of describing the accuracy.

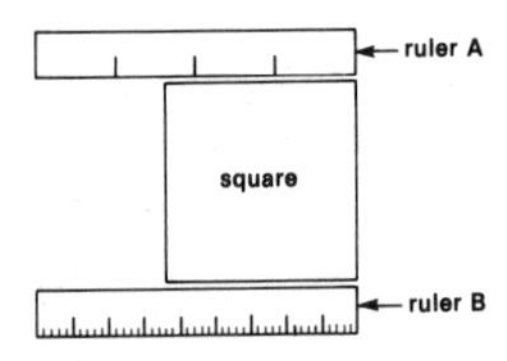

Ruler B has more measuring marks on it than ruler A. It will therefore measure with more accuracy the side of the square.
LATIN, *accuratus*, done with care

accurate exact; precise; correct.
Example I have some very accurate results from my experiment here. They are accurate to the nearest thousandth of a centimetre. The first result is 9.437 cm.
LATIN, *accuratus*, done with care

acute sharp; in geometry an acute angle is a sharply pointed angle whose size is between 0° and 90°.

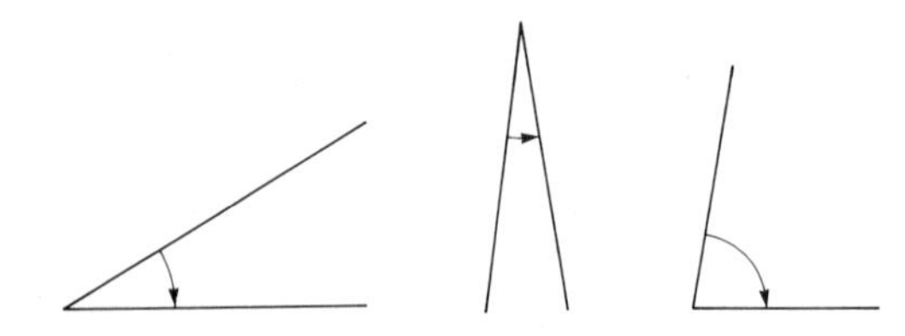

All three angles shown are acute.
LATIN, *acutus*, sharp

add to combine two or more quantities to form another quantity called their sum. The **sign** or **symbol** showing this combination is + (called plus).
Example The sentence "Add 5 and 2 together" can be written as 5 + 2, which equals 7.

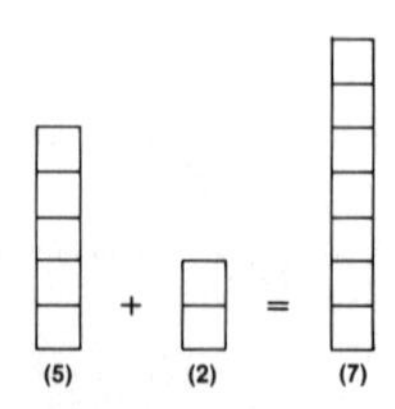

LATIN, *addere*, add, put to

addition the **operation** of combining two or more quantities to obtain another quantity called their sum. The **sign** or **symbol** indicating the addition operation is + (called plus). Addition as an operation in arithmetic follows the laws of addition, namely it is **commutative** (i.e. $2 + 7 = 7 + 2$), **associative** $(1 + (3 + 6)) = ((1 + 3) + 6)$, and **closed** (any 2 numbers added together to give another number).
Example The addition of 9 and 2 gives 11. This can be written as $9 + 2 = 11$.

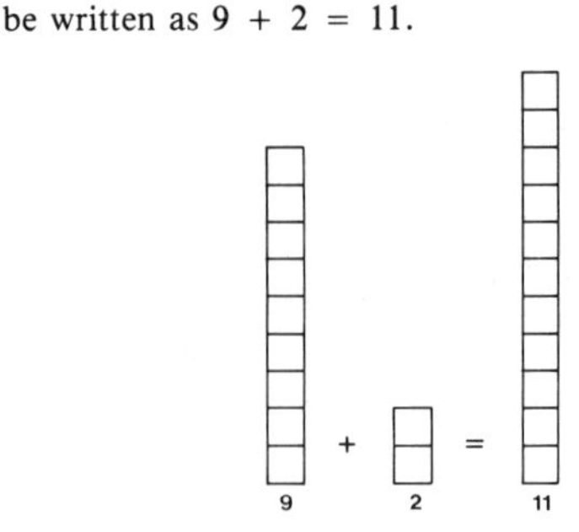

LATIN, *addere*, add, put to

additive inverse the number which, when added to the given number, gives 0.
Example The additive inverse of 4 is −4, since $4 + -4 = 0$.
LATIN, *addere*, add, put to; *inversus*, inside out, upside down

address (in computing) a place where a quantity or value is stored.
Example The number 1273 was found to be at address 0022 in the computer's memory. A post office box compares closely to the idea of a computer address in the computer's memory.

Post box number

1	2	3	4
5	6	7	8
9	10	11	12
13	14	15	16

LATIN, *ad*, towards, to; *directus*, direct

adjacent close to, lying next to.
Examples

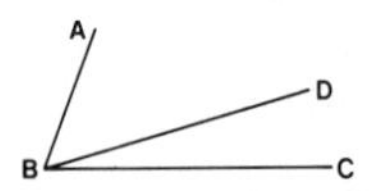

$\angle ABD$ and $\angle DBC$ are adjacent angles, sharing BD as a common arm, and B as a common **vertex**.

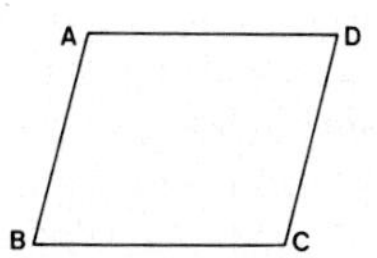

AB and BC are adjacent sides, sharing B as a common vertex.
LATIN, *adjacere*, to lie near

affine transformation a **transformation** which always **maps** (or transforms) figures so that lines originally parallel remain parallel, and the ratio between any three collinear points remains constant. Projecting a figure on to a plane which is then tilted is another way of describing affine transformations.
Examples **translations**, **reflections**, and **enlargements** are examples of affine transformations.

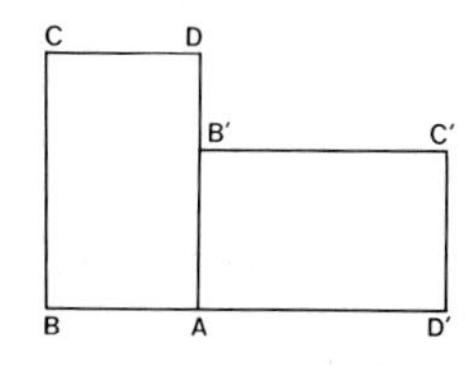

$ABCD$ is rotated to $AB'C'D'$ (about point A).

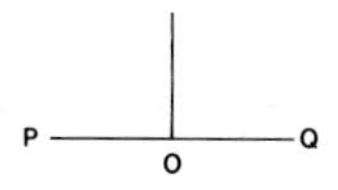

OP is reflected to OQ.

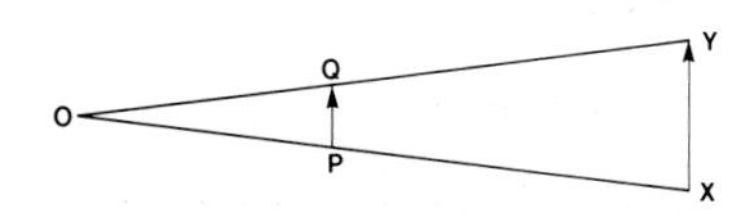

PQ is enlarged to XY.
LATIN, *affinis*, bordering on, related by marriage; *trans*, across; *forma*, a shape

Agnesi, Maria (1718–1799) Italian mathematician appointed as first woman professor of mathematics in 1750 at the University of Bologna. She studied **differential** and **integral calculus** and is especially remembered for her study of the curve,

$$xy^2 = a^2(a-x)$$

—the "Witch of Agnesi".

algebra the part of mathematics which deals with the properties and **relations** of numbers, sets, groups, etc., using general symbols such as letters of the alphabet. Also known as universal or generalised arithmetic. Algebra uses letters in this instance representing **unknown** quantities as well as numbers (hence *algebraic)*.
Examples
$x + y = 10$ is an equation in algebra (x and y are symbols), or algebraic equation.
$x^2 + x - 9 = 0$ is another type of algebraic equation.
ARABIC, *al-jebra*, reunion of broken parts

algorithm, algorism a method or series of steps for working out calculations. There are many kinds of algorithms used in computers to work out difficult problems.
Example Multiplication of 476×62 can be considered as $476 \times 60 + 476 \times 2$, written as follows:

	476	
×	62	
	952	(476 × 2)
	28 560	(476 × 60)
	29 512	(952 + 28 560)

ARABIC, after *al-Khowarazmi*, 9th century mathematician, the translation of whose works first introduced arabic numerals to Europe.

allied angles (also **co-interior angles**) **angles** formed (on the same side) between two **parallel** lines by a straight line cutting across them. Allied angles are supplementary. Note: the straight line cutting across the parallel lines is called a **transversal.**
Example

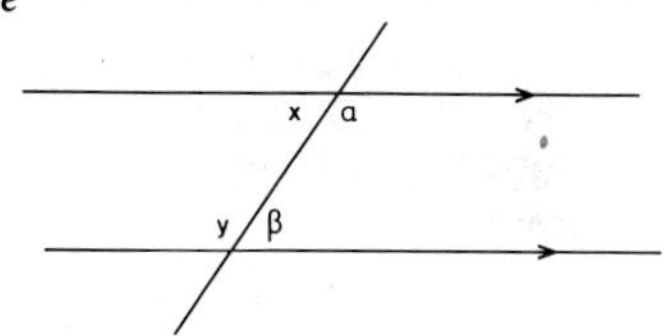

α, β are allied (co-interior angles). x, y are also allied (co-interior) angles.
$\alpha + \beta = 180°$. $x + y = 180°$.
LATIN, *alligare*, to bind; *angulus*, corner

alternate angles a pair of equal angles formed by a straight line cutting a pair of **parallel lines. Each angle** lies on opposite sides of the line. Alternate angles are equal. Note: If the lines are not parallel, then the angles can still be called alternate, but they will not be equal.
Example

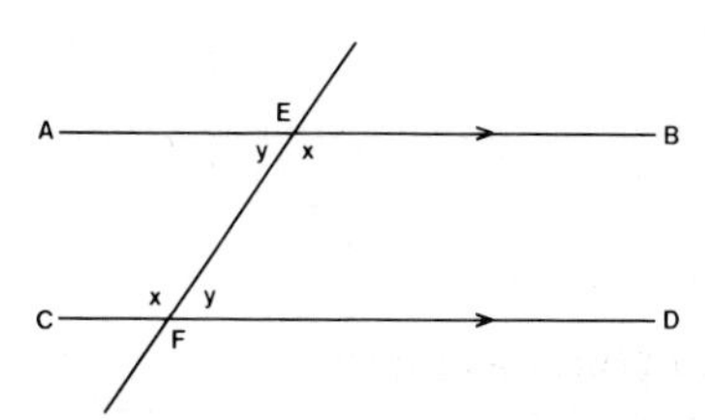

$AEF = EFD = y$ (alternate angles)
$BEF = CFE = x$ (alternate angles)
LATIN, *alternare*, to do one thing after the other

altitude (in **geometry**) the perpendicular distanc from the **base** of a figure to its **vertex** or top point.

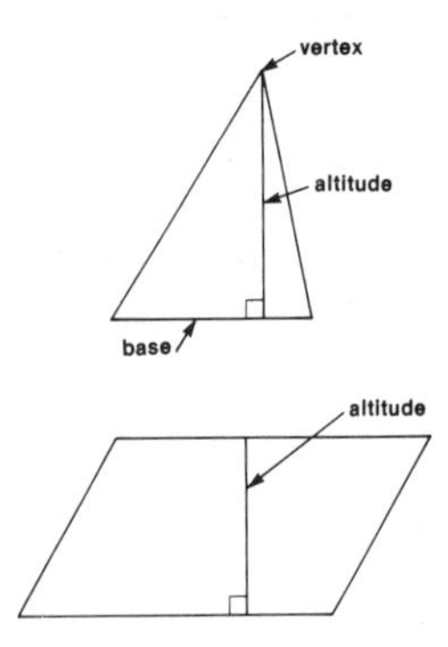

LATIN, *altus*, high.

analog computer a calculating device which uses physical quantities which can vary widely (such as electrical signals) to represent numbers, in contrast to a **digital computer** which uses separate pulses to count with.
Example In an analog computer using electrical signals, a quantity 10.41 to be added to 7.58 can be represented, say, by two voltages 10.41 volts and 7.58 volts which when combined give you 17.99 volts. Therefore the answer to the problem is 17.99.
GREEK, *analogos*, in proportion

analysis **1.** (in general) The action of separating something complex (made up of many different parts) into the parts which make it up. **2.** (referring to the phrase Modern Analysis) The resolving of problems by changing them into equations. **3.** (logic) The tracing of things to their sources.
Examples

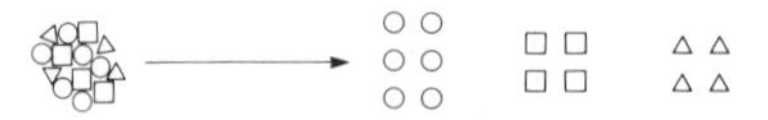

1. In mathematics we use the word analysis in many different areas. The **analysis of variance** is a mathematical method which separates the variation due to one cause from the variation due to another or other causes.
2. The modern analysis of **geometry** means that **points, lines** and **planes** can be represented by **equations**.
3. Analysis is used to find out causes from their effects.
GREEK, *analusis*, a loosening, undoing

analysis of variance (statistics) a mathematical way of separating out the variation due to one group of causes from the variation due to other groups.
Example Suppose we are given three sets of values, each value representing the height of an adult. If we are told the three sets come from the same town we can check using analysis of variance to see how likely that is. We may find that it is very unlikely, which may mean one or more of the sets come from another town.

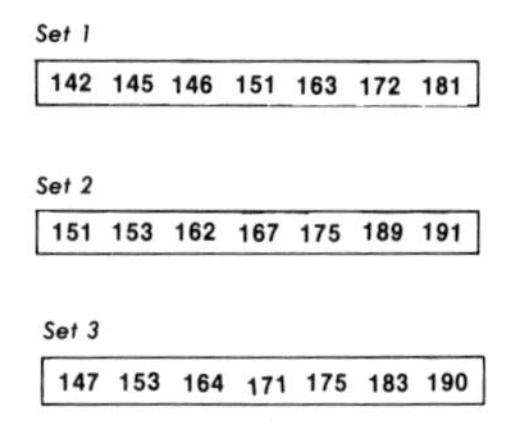

Set 1

142	145	146	151	163	172	181

Set 2

151	153	162	167	175	189	191

Set 3

147	153	164	171	175	183	190

see ANALYSIS; VARIANCE

analytical geometry (also known as **coordinate geometry**) The representation of **points, lines, curves**, shapes by means of **algebra**. The basic idea is to locate points on a plane by referring them to a system of two intersecting lines, or **axes** (three intersecting lines are used to locate points in space for solids).
Examples
A straight line can be represented by the general equation $ax + by + c = 0$, where a, b, c are constants.

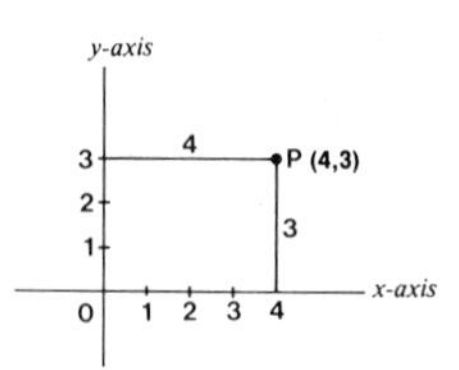

The point P is represented by the two numbers (4,3), which shows how far it is from the y-axis and x-axis respectively.
A **parabola** can be represented by the equation $x^2 = 4ay$, where a is a constant.
GREEK, *analusis*, a loosening, undoing; *geometria*, measuring earth or land.

Anaxagoras (499–427 BC) A Greek mathematician concerned with solving the third famous problem of antiquity—the quadrature (squaring) of the circle. The problem was to construct, with the aid of a straight edge and compass only, a square equal in area to the area of a given circle. Anaxagoras is supposed to have worked on the squaring problem whilst in prison.

angle an angle is the amount of turning between two rays sharing the same common starting point.
Example

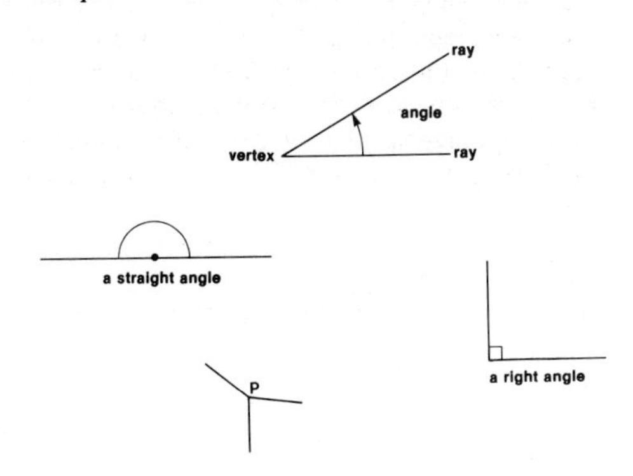

Angles are measured in **degrees** (°) or in **radians** (rad). There are 360 degrees (or 2π radians) in a circle. The sum of the angles at any point (e.g. P) is four right angles (360°).
LATIN, *angulus*, corner or angle

angle of projection The angle made by a line (or plane) with its **projection**.
Example

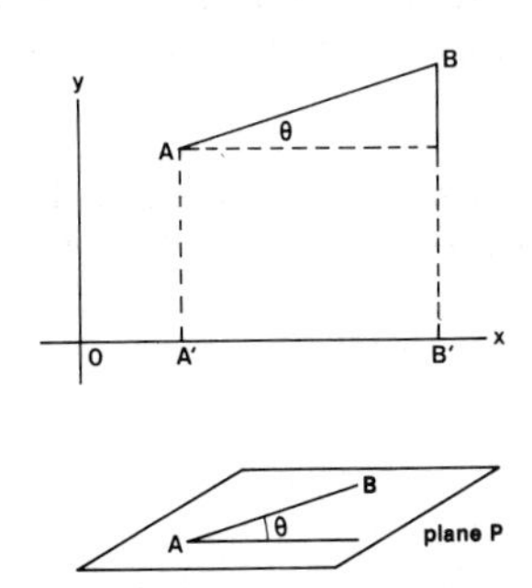

AB is projected on to the x axis at $A'B'$. AB makes $\theta°$ with its projection $A'B'$ as shown.
LATIN, *angulus*, corner or angle; *proicere*, throw forth

annulus a ring-shaped space formed by two concentric circles. The area of the ring space is $\pi(R^2-r^2)$ where R is the radius of the bigger circle, and r is the radius of the smaller.
Example An annulus is formed when an eclipse of the sun occurs and the centres of the sun and moon circular shapes coincide.

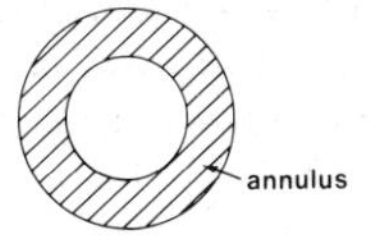

LATIN, *annulus*, little ring

answer **1.** The result one gets upon working out a problem (also **solution**). **2.** (verb). To work out a problem.
Examples
1a. Write the answer for the following:
26 + 17 − 31 + 14
Answer = 26
b. Find the solution to the following equation. Give your answer to two decimal places.
$x^2 + 3x = 6$ (see **completing the square** for solution).
2. Answer the following: 42 + 93 = ?; 61 − 29 = ?; 45 × 27 = ?
OLD ENGLISH, *andswaru*, swear against

anti-clockwise the direction opposite to the direction taken by clock hands (as in grandfather clocks).
Example

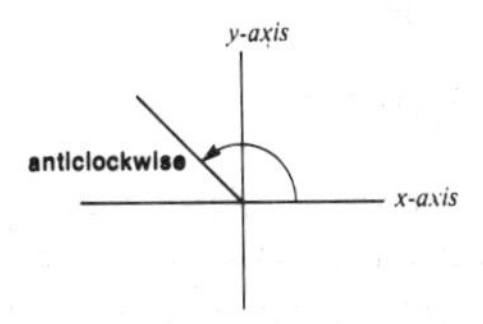

GREEK, *anti*, opposite, against; LATIN, *clocca*, bell [used to tell the time of day in the Roman world]

anti-derivative *see* INTEGRAL, INDEFINITE; PRIMITIVE.

AP *see* ARITHMETIC PROGRESSION

apex the point of a plane figure or solid figure furthest above or below a fixed **base line** or **plane**.

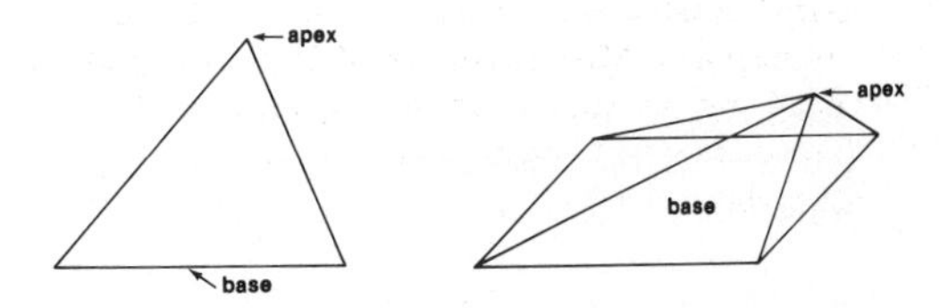

LATIN, *apex*, peak, summit

Apollonius (262–190 BC) A famous Greek mathematician who learned his mathematics from Euclid's successors. Apollonius is known as the "great geometer" and was particularly interested in the study of the **conic** sections. His research summarised all the work on conics completed previously and he further extended his knowledge of the geometry of the conics.

arabic numerals The figures 1, 2, 3, 4, 5, 6, 7, 8, 9, 0.
Example Arabic numerals were called arabic as they were considered to be introduced into Europe by

Arabic mathematicians and scholars many centuries ago.
Arabia, country of the Arabs; LATIN, *numerus*, a number.

arbitrary relating to a **number** or **variable** selected at will.
Example No matter which arbitrary value x may take on, $x^2 - a^2 = (x+a)(x-a)$ is still true.
LATIN, *arbiter*, a judge

arc any part of a **curve**. The line joining any two points. The length of an arc l is given by the formula $l = rD$, if the arc is circular (r is the radius of curvature, and D is the size of the angle in **radians** **subtended** by the arc at the centre of the circle).

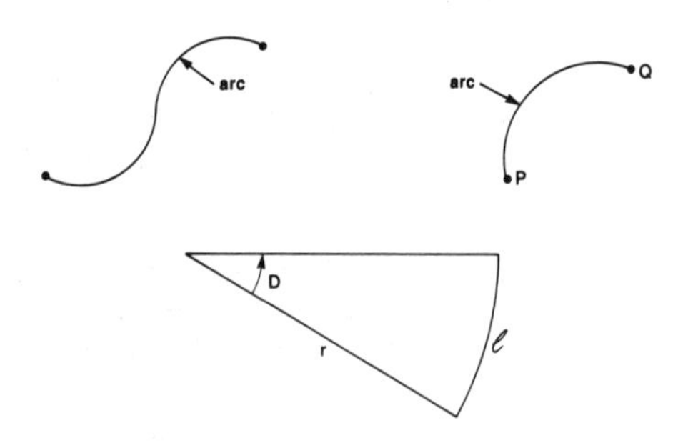

LATIN, *arcus*, a bow or arc

Archimedes (287–212 BC) said to be the greatest mathematician of antiquity. A Greek mathematician who spent a lot of his life in Alexandria. Archimedes' works are prolific—including calculation of the value of π, finding areas and volumes by the method of exhaustion, finding centres of gravity of **plane** and **solid** figures, hydrostatics and physical principles of the lever. Archimedes was also fascinated by the geometry of **conics** and successfully solved the problem of the **quadrature** (i.e. finding a square whose area is the same as the area of a given circle).

area the size of a **surface**. The surface can be flat or curved. Area is always measured in square units.
Example

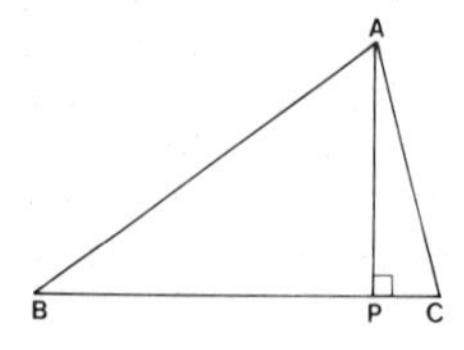

The area of triangle ABC is equal to ½ base × height (½$BC \times AP$).

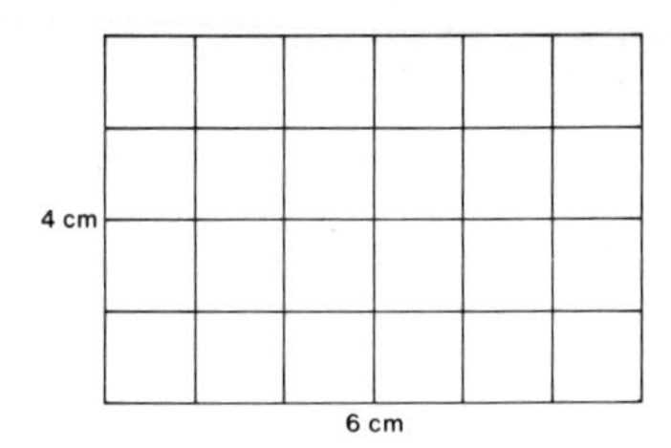

The area of a rectangle with length 6 cm and breadth 4 cm is 24 cm^2.
LATIN, *area*, an open field

Argand, Jean-Robert (1768–1822) Swiss mathematician who was responsible for invention of the geometrical interpretation of complex numbers on an **Argand diagram**. As well Argand represented any complex number in the form $r(\cos\theta + i\sin\theta)$ and showed how complex numbers can be added and multiplied geometrically on an Argand diagram.

Argand diagram A diagram on which can be represented **complex** numbers. It consists of two axes, the horizontal axis representing the **real** portion of a complex number, and the vertical axis the **imaginary** part. Complex number plane and z-plane are other names sometimes used instead of Argand diagram.
Example

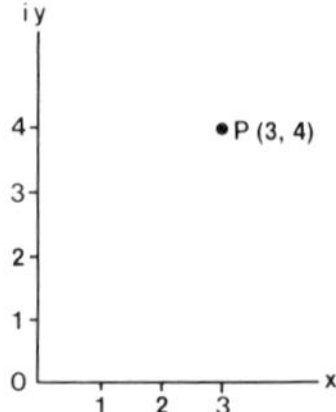

The point P represents the complex number
$z = 3 + 4i$
after J.-R. Argand.

arithmetic the science or study of numbers. It involves calculating with **real** numbers (including **fractions** and **decimals**) using **operations** such as **addition, substraction, multiplication, division**.
Example At least some arithmetic must be known before a person can handle money matters successfully.
GREEK, *arithmos*, number

arithmetic mean **average**. The number found by dividing the sum of a group of numbers by the number of **terms** or items in the group.
Example The arithmetic mean of 71, 79, 65, 82, 78 is 75

$$\frac{71 + 79 + 65 + 82 + 78}{5} \text{ is } 75$$

GREEK, *arithmos*, number; LATIN, *medius*, middle

arithmetic progression (*abbrev.* AP) a **sequence** of numbers listed one after another in which each number is larger (or smaller) than the number that comes before it by a constant amount. The general form of the progression can be written as follows: a, $a + d, a + 2d, a + 3d, a + 4d, \ldots,$ where d is the **common difference**. The sum of an AP is $\frac{n}{2}(2a + (n - 1)d)$, where a is the first term, n is the number of terms in the set, and d is the common difference. The formula for the nth term of an AP is

$$T_n = U_n = a + (n - 1)d.$$

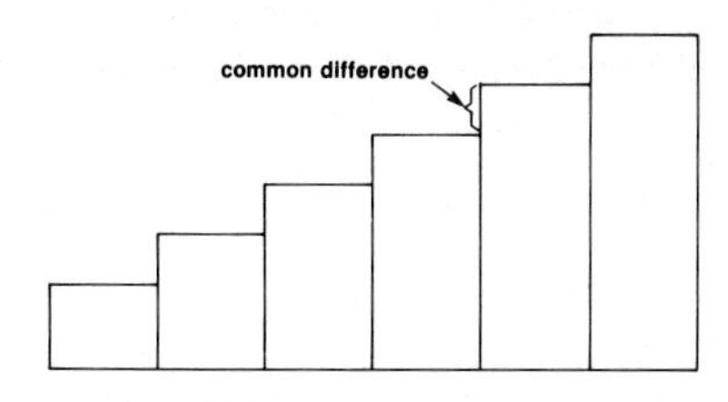

GREEK, *arithmos*, number; LATIN, *progressus*, gone forward

array a regular arrangement of elements, usually in the form of **rows** and **columns**. It is used in computer programming to store values in tables etc.

Example

$$\begin{array}{rrr} 2 & 3 & 1 \\ -2 & 0 & 7 \end{array}$$

The above is an array consisting of 2 rows and 3 columns. This is called a 2 × 3 rectangular **matrix**.

LATIN, *ad-redare*, to arrange

ascending going up.

Example We can express $(1 + x)^3$ in ascending **powers** as follows:

$$1 + 3x + 3x^2 + x^3$$

$(1 + x)^3$ can also be expressed in descending powers of x:

$$x^3 + 3x^2 + 3x + 1$$

LATIN, *ascendere*, climb, ascend

associative referring to an **operation** in which it does not matter how the quantities are grouped when combined. If * is an associative operation then $(a*b)*c = a*(b*c)$ is true.

(A + B) + C = A + (B + C) = A + B + C

Example The following example shows **addition** to be an associative operation:

$2 + (4 + 7) = (2 + 4) + 7$ (both sides equal 13).

$2 \times (4 \times 7) = (2 \times 4) \times 7$ shows **multiplication** to be associative also.

Note: $2 - (4 - 7) \neq (2 - 4) - 7$, so subtraction is not associative.

LATIN, *associare*, to associate, join to

associative law a rule applying to an **operation** such as **addition** or **multiplication** which is **associative**.

Example The associative law of multiplication is shown by $a \times (b \times c) = (a \times b) \times c$.

$2 + (4 + 7) = (2 + 4) + 7$ is an example of the associative law of addition.

LATIN, *associare*, to associate, join to; OLD ENGLISH, *lagu*, something laid down

assumption (also **postulate**) **1.** (mathematics) Something taken as being true, without it having to be proved. **2.** Something taken to be true temporarily, in order to explore the consequences.

Examples

1a. A famous assumption is the following: Through any point not on a given straight line there is only one line that can be drawn which is parallel to the given line.

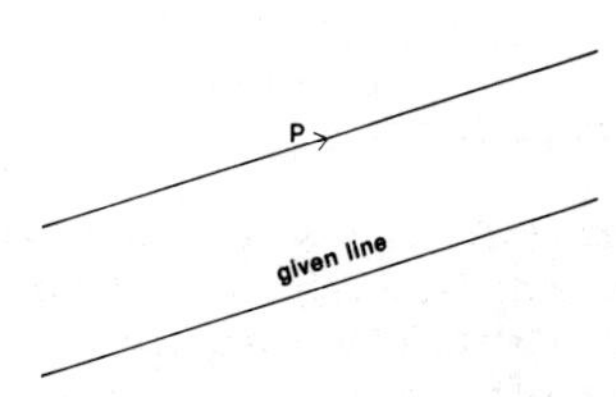

1b. Another is: The shortest distance between two points is a straight line (this works in a plane, but not when the two points are on, for example, the surface of a sphere).

2. The pattern $1 + r + r^2 + \ldots r^{n-1} = \frac{1 - r^n}{1 - r}$ appears to be true (for any value of n). For $n = 1$, it is true. We can make the assumption it is true for $n = k$, and see if it is true for $n = k + 1$, where k is any positive whole number. If it turns out the pattern is true for $n = k + 1$, then one can argue it must be true for all (positive whole) values of n. (*see* INDUCTION).

LATIN, *assumere*, to take on, assume

astroid a shape like a star with curved sides. It is constructed by tracing out the path made by a point on the circumference of a circle that rolls around the inside of a larger fixed circle.

asymptote a straight line which is approached, but never reached, by a branch of a **curve** as it stretches to infinity. It can be looked upon as a line **tangent** to the curve at infinity.

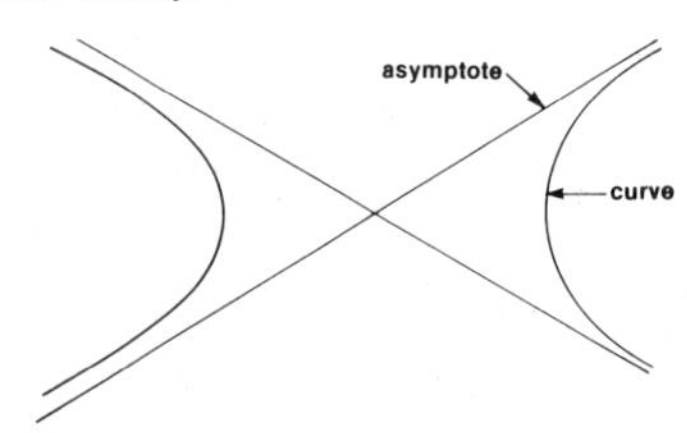

GREEK, *a*, not; *sumptotos*, falling together

average a quantity or value that represents a group. There are three important averages in mathematics: **mean, median** and **mode**. Average is commonly used instead of arithmetic mean.
Example In the group 3, 4, 5, 7, 7, 7, 7, 9 the mean is

$$\frac{3 + 4 + 5 + 7 + 7 + 7 + 9}{7} = 6$$

and the mode is 7, since 7 is the value which occurs most often in the group.
FRENCH, *avarie*, damage to ship or cargo *f.* ARABIC, *awārìya*, damaged goods, pl. of *awār*, damage at sea, loss [at one time there was an equitable distribution of the expense or loss of damaged goods amongst all interested parties; hence the present meaning].

axiom an agreed-upon **assumption**, something assumed to be true, a statement that does not require proof (*see* ASSUMPTION).
Example "Two quantities which are equal to a third are equal to each other" is an example of an axiom. (i.e. if $a = b$ and $c = d$, then $a = c$). "Adding 0 to any number does not change that number" is another axiom (this can be written as $a + 0 = a$).
GREEK, *axios*, worthy

axis (*pl.* **axes**) **1.** One of the principal lines through the centre of a figure or solid, especially a line which divides the figure or solid **symmetrically**. **2.** A fixed line along which distances are measured or to which positions are referred (*see* ANALYTICAL GEOMETRY).
Examples
1.

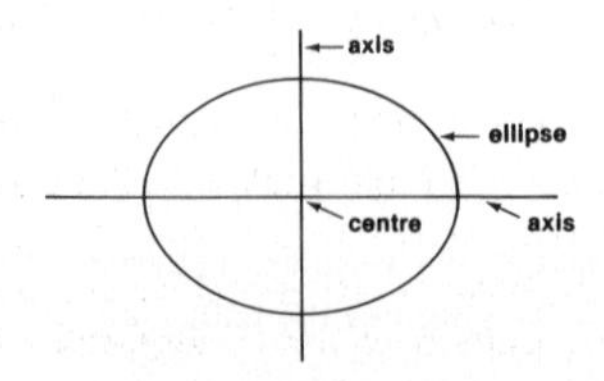

2.

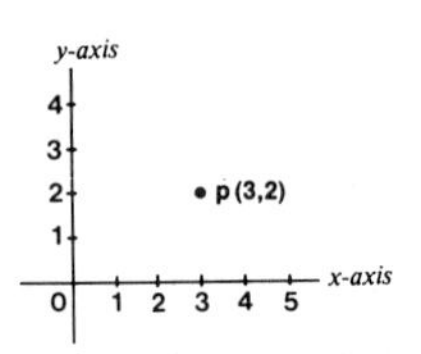

Point P is 3 units from the y-axis, and 2 units from the x-axis.
LATIN, *axis*, pivot or axis

azimuth **1.** (astronomy) The arc of the heavens extending from the zenith (point directly above the observer on the celestial sphere) to the horizon and cutting it at right angles. **2.** The angular distance from a given fixed line in the **horizontal** to the line joining the point on the horizon where the **great circle** passing through the object cuts the horizon.
Examples
1.

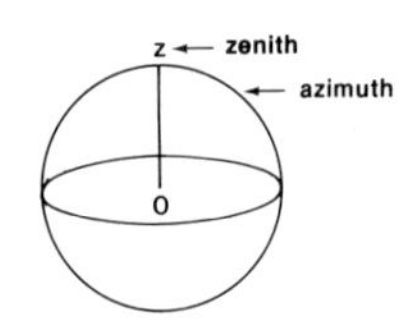

2.

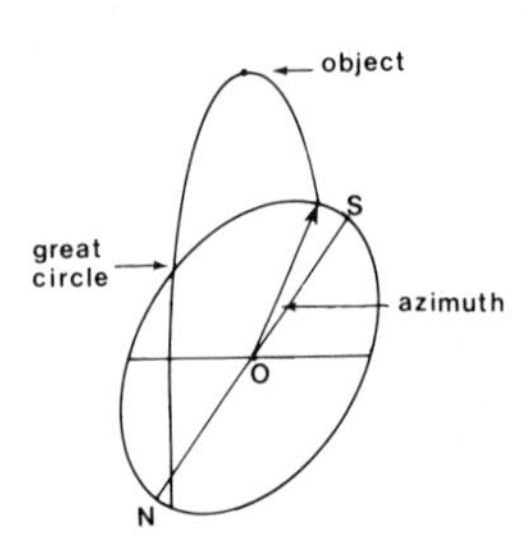

ARABIC, *al-sumut*, the way

Bb

B (b) the second letter in the English alphabet. The capital letter is often used to name a point in a geometric figure.
Example *b* is used in mathematics to represent an **unknown constant**, quantity or value generally. It can also be used to represent any variable. In the **expression** $ax + by$, both a and b are understood to be **constants** which do not yet have a known numeric value. In the **equation** $b^2 - 7 = 0$, b is regarded as an unknown quantity to be worked out (by solving the equation).

In the diagram above, *B* represents one of the **vertices** of the triangle *ABC*.
GREEK, β beta [second letter in the Greek alphabet — the Hebrews and the Phoenicians used a similar letter].

bar chart a chart or **graph** which shows information using bars or columns. The heights, or lengths, of the bars or columns correspond to the size they represent (e.g. if 1 cm = 100 units, a bar of 2 cm will represent 200 units).
Example

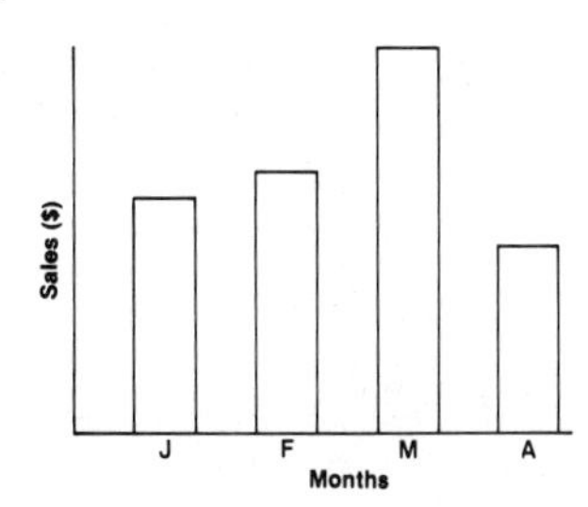

LATIN, *barra*, bar; *charta*, leaf of papyrus, paper

base 1. (geometry) The line or surface on which a figure is said to stand. **2.** (arithmetic) The number which is the size of the group used in counting. **3.** In **logarithms**, the fixed number to which is applied the **power** (index) to produce the given power.
Examples
1. In the following triangle, AB is the base.

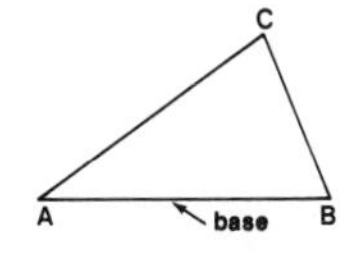

2. 10 is the base of the **decimal system**. The 10 digits are 0, 1, 2, 3, 4, 5, 6, 7, 8, 9.
2 is the base of the **binary system**. The 2 digits are 0, 1. To convert from a number in the **decimal system** to one in the binary system [e.g. $(27)_{\text{base } 10} = 2 \times 10 + 7$] split the given number into **multiples** of 2, i.e.

$$16 + 8 + 2 + 1 = 1 \times 2^4 + 1 \times 2^3 + 0 \times 2^2 + 1 + 2^1 + 1 \times 2^0$$
$$= (1\ 1\ 0\ 1\ 1)_{\text{base } 2}$$

3. In common logarithms, the base is 10. For example $2 = 10^{.3010}$ (this means the number 2 can be represented as 10 to the power .301). .301 is the logarithm (to the base 10) in this case.
(*see* NUMBER SYSTEM)
GREEK, *basis*, a stepping, tread, base

BASIC a computer language originally introduced for beginners to learn to program computers. It is one of the simpler languages, and is commonly used on smaller computers.
Example The following are some BASIC statements which accept a number, square it, and print it.

```
10 INPUT A
20 LET S = A * * 2
30 PRINT S
```

from *B*eginners *A*llpurpose *S*ymbolic *I*nstruction *C*ode

bearing the angle measured clockwise from north to the object.
Example

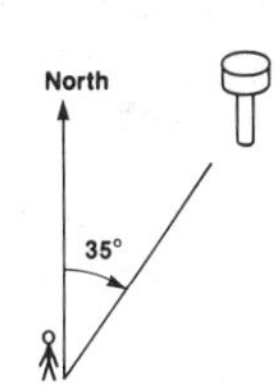

The tower is North 35° East. The bearing is sometimes written in shorthand N35°E, or true bearing T35°.
OLD ENGLISH, *beran*, to carry

Bernouilli, James (1655–1705) and **John** (1667–1748) German brothers. Both were famous mathematicians and both became professors of mathematics. The Bernouillis worked on the development of the **calculus** and corresponded with **Leibniz**. John Bernouilli produced a **theorem** for obtaining the **limit** approached by a **fraction** with a **zero denominator** and **numerator**. L'Hôpital, a pupil of John, published this theorem, now known as L'Hôpital's Rule.

Bhaskara (1114–1185) A Hindu mathematician representative of the peak of Hindu mathematical development in the period AD 200–1200. Bhaskara studied **negative** as well as **positive numbers, irrational** numbers, quite complicated **algebraic** problems and the **geometry** of the **Pythagorean theorem**.

bias **1.** An **oblique** or slanting **line**, a leaning. **2.** (statistics) A definite alteration up or down to a set of results (as opposed to chance alterations which tend to cancel each other out).
Examples
1. The cloth should be cut on the bias.

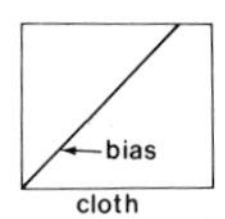

2. We threw the die 10 times but because it was weighted on the '1' side it came up '6' eight times. The bias in the results obviously indicated the die was 'loaded'.

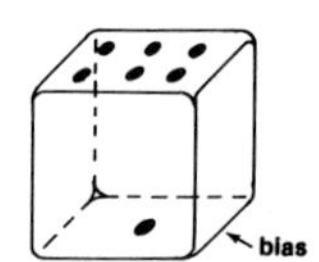

OLD FRENCH, *biais*, slant, oblique *f.* LATIN, *bifax*, looking two ways, or *biaxius*, having two axes

binary **1.** Relating to two. **2.** Dealing with counting using 0, 1 only.
Examples
1. 2 + 4 is an example of a binary **operation** where the result is a single number (6).
2. In binary arithmetic 101 means $1 + 2^2 + 0 \times 2^1 + 1 \times 2^0$.
LATIN, *bini*, two together

binary number a number represented by a **pattern** of 1 and 0 combinations.
Example 1101 is a binary number. In **decimal** arithmetic it is 13 ($1 \times 2^3 + 1 \times 2^2 + 0 \times 2 + 1$). 100011100101 is a bigger binary number.

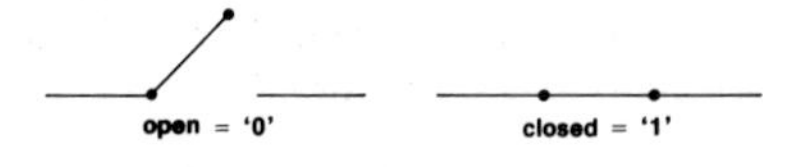

A switch can represent a binary number.
LATIN, *bini*, two together; *numerus*, number

binary operation an operation which combines two **elements** into one element.
Example Multiplication (×) is a binary operation on the numbers

$$3 \times 2 = 6, \tfrac{1}{3} \times 6 = 2, 1 \times 7 = 7.$$

LATIN, *bini*, two together; *operari*, to work

binary system a way of counting using groups of two, where the position indicates which group (whether 2, 2 × 2, 2 × 2 × 2 etc.). Binary numbers can be added, subtracted, multiplied, divided in a similar way as **decimal** numbers.
Examples

1. $101011 = 1 \times 2^5 + 0 \times 2^4 + 1 \times 2^3 + 0 \times 2^2 + 1 \times 2 + 1$
$= 1 \times 32 + 0 \times 16 + 1 \times 8 + 0 \times 4 + 1 \times 2 + 1$

2. $0.1101 = 1 \times \tfrac{1}{2} + 1 \times \tfrac{1}{4} + 0 \times \tfrac{1}{8} + 1 \times \tfrac{1}{16}$

LATIN, *bini*, two together; GREEK, *sustema*, set up

binomial **1.** Consisting of a pair of **terms** or **numbers**. **2.** A pair of terms joined by + or −.
Examples
1. $2x - y$, $a + 3b$ are binomial **expressions**, or binomials.
2. $5x + 4z$ is an example of a binomial.
LATIN, *binomius*, having two names

binomial coefficients the coefficients of the various **powers** of x in the **expansion** of the **expression** $(x + a)^n$ (*see* BINOMIAL THEOREM). The value of the **coefficient** of x^k (where k is a positive whole number) is $\frac{n!}{k!\,(n-k)!}$. This is often written $\binom{n}{k}$ for short.
Examples

1. $(x + a)^3 = x^3 + \frac{3!}{1!\,2!}x^2a + \frac{3!}{2!\,1!}xa^2 + a^3$
$= x^3 + 3x^2a + 3xa^2 + a^3$

2. $(1 - y)^4 = 1 + \frac{4!}{1!\,3!}1^3 . (-y) + \frac{4!}{2!\,2!}1^2(-y)^2 + \frac{4!}{3!\,1!}1 . (-y)^3 + (-y)^4$
$= 1 - 4y + 6y^2 - 4y^3 + y^4$

LATIN, *binomius*, having two names; LATIN, *coefficiens*, effecting together

binomial distribution a **distribution** of the **probabilities** of possible outcomes of **events** which can occur or not. The distribution is based on the **binomial theorem**, in the case where $x + a = 1$.
x represents the probability that the event will occur, and a the probability that it will not. The probability of either occurring is therefore $x + a$, which is equal to 1.
Example Suppose we toss a coin six times. Let h be the probability that heads comes up, and t the probability that tails will come up. In six tosses we could have 1 head and 5 tails (ht^5), 2 heads and 4 tails (h^2t^4), 3 heads and 3 tails (h^3t^3), and so on.The probability of various combinations turning up on 6 tosses is calculated by the formula:

$$(h + t)^6 = h^6 + 6h^5t + 8h^4t^2 + 10h^3t^3 + 8h^2t^4 + 6ht^5 + t^6$$

This means there is 6 times more likelihood of 5 heads and a tail occurring than 6 heads straight, and 10 times more likely 3 heads and 3 tails will turn up than either 6 straight heads or tails.

LATIN, *binomius*, having two names; *dis*, apart; *tribuere*, assign, grant

binomial theorem the general formula for the **expansion** of any binomial (or pair of **terms**) when raised to any **power**.

$$(x + a)^n = x^n + nax^{n-1} + \frac{n(n-1)a^2x^{n-2}}{2!} + \ldots + a^n$$

see also BINOMIAL COEFFICIENTS

Examples

$$(x + a)^4 = x^4 + 4ax^3 + \frac{4 \times 3a^2x^2}{2 \times 1} + \frac{4(3)(2)a^3x}{3 \times 2} + a^4$$
$$= x^4 + 4ax^3 + 6a^2x^2 + 4a^3x + a^4$$

LATIN, *binomius*, having two names; GREEK, *theorema*, speculation, proposition

bisector a **point**, **line** or **plane** that cuts something into two equal parts.

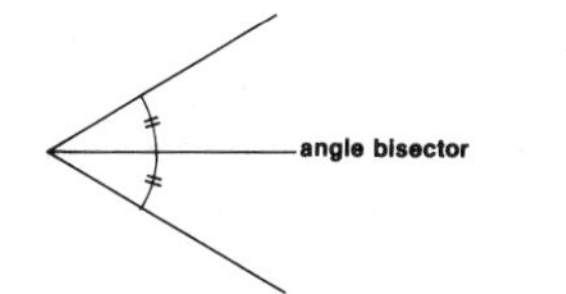

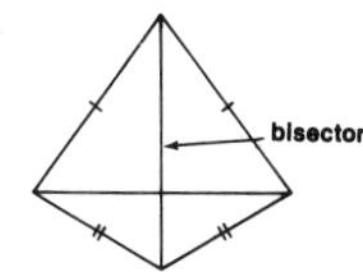

LATIN, *bis*, twice; *secere*, to cut

bit a *bi*nary digi*t*, either 0 or 1.
Example In computers a unit of information is the bit. Bit 1 usually represents a switch that is 'on', and 0 'off'.
from *bi*nary digi*t*

Boole, George (1815–1864) became professor of mathematics at Queens College in Cork (Ireland). Boole's main work was concerned with the contribution of symbolism to an **algebra** of **logic** (*see* BOOLEAN ALGEBRA).

Boolean algebra **algebra** devised by George Boole. **Operations** mainly used are AND, OR, XOR, NOT. It is used in computer logic.
Examples: 1 AND 1 = 1, 1 AND 0 = 0, 1 OR 0 = 1, 1 OR 1 = 1, 1 XOR 1 = 0, NOT 1 = 0.

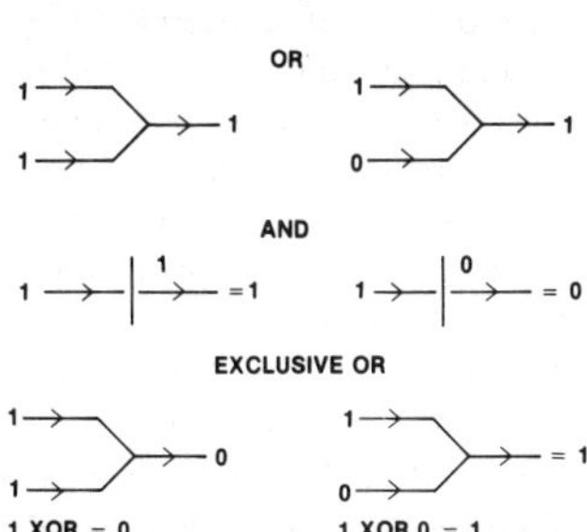

after G. Boole

boundary the line around the outside of something.
Example In a circle, the boundary is the same as the **circumference**.

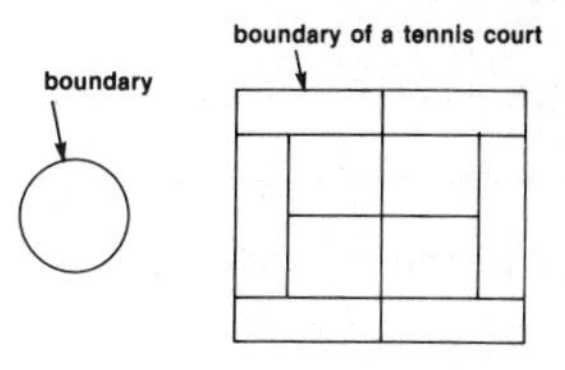

ENGLISH DIALECT, *bounder*, limit(er)

bounded limited.
Example The area was bounded to the north by a straight line, and to the south by a half circle.

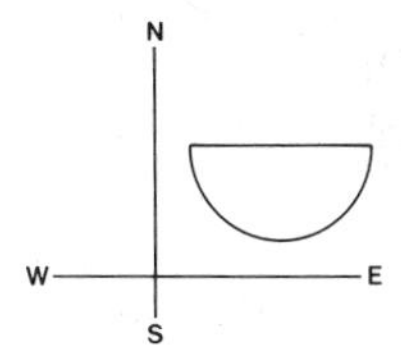

LATIN, *bodina*, boundary, limit

braces the word for { }
Example Braces are used to group things together. {...} is used to show we are dealing with a **set**. S = {1, 2, 3, 4} means the set S contains the **members** 1, 2, 3, 4.
LATIN, *bracchia*, arms

brackets symbols which tell you which things are combined first. Another name for brackets is parenthesis; symbol is [ ]. Strictly, parentheses are written ( ). Brackets also show that items or elements inside are seen as a whole.
Example [3 + 4] × 6 = 7 × 6; in other words combine 3 and 4 first before multiplying by 6.
LATIN, *bracae*, breeches (trousers).

byte a standard number of **bits** (8) which are used in many computer systems to represent a character or numeral.
Example In many computers, the sentence "The Olympic Games attract much interest." can be represented by 40 bytes, since there are 40 characters (including blanks between the words) in the sentence.
from B(inar)*Y* (digi)*T E*(ight)

Cc

C (c) the third letter in the English alphabet. Often used in mathematics as an unknown **constant** quantity. Capital C is often used to name a point in a geometric figure or in Roman numerals C = 100. In physics C can also mean Celsius (temperature measure), Calorie (heat measure), Centigrade (temperature measure).
Examples
In the **equation** $y = x + c$, c represents a constant not yet determined.

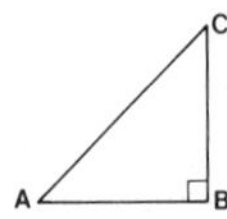

C represents the point of **intersection** of the lines AC and BC.
GREEK, γ gamma [third letter of Greek alphabet]

calculator a machine for working out **additions, subtractions, multiplications, divisions, powers** etc. It usually has a place to enter numbers (keyboard), and a place to show the result (on a display or on paper).
Examples There are many calculators sold today. Some are so small they are the size of a wrist watch. Hand held calculators are used in schools and business more and more nowadays.
LATIN, *calculus*, a small stone (used in calculating)

calculus 1. Any system of calculation. 2. A system of **analysis** dealing with the **rates** of change of a **variable** quantity.
Examples
1. The calculus of **probabilities** is a way of working out the likelihood of different **events** occurring.
2. The calculus invented by Sir Isaac Newton is also called **differential** calculus, as it deals with rates of change. If $y = x^2$, then the rate of change of y with respect to x is calculated to be $2x$ (*see* DERIVATIVE).

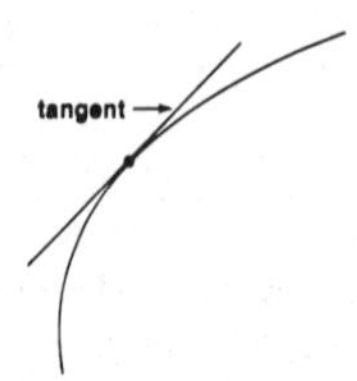

The **gradient** of the **tangent** to the curve represents the rate of change of the curve at this point.
LATIN, *calculus*, a small stone (used in calculating).

cancel the act of dividing both the top number **(numerator)** and the bottom number **(denominator)** of a **fraction** by a **common factor**. Also the act of dividing both sides of an **equation** by a common factor.
Examples
1. $\frac{15}{21} = \frac{\not{3} \times 5}{\not{3} \times 7} = \frac{5}{7}$
2. $2(x + 1) = 4$ gives $x + 1 = 2$ on dividing both sides by 2.
LATIN, *cancellare*, to make like a lattice, to strike out a writing

canonical form a standard or **normal** form of a mathematical **expression, equation** or **matrix**.
Examples
1. The canonical form of the equation of a plane is $lx + my + nz - d = 0$ where l, m, and n are the **direction cosines** of the **normal** to the plane and d is the length of the normal from the **origin** to the **plane**.
2. The canonical form of a matrix is one considered the simplest and most convenient to which square matrices of a certain class can be reduced (it has non-zero **elements** only in the principal **diagonal**).
$\begin{bmatrix} a & o & o \\ o & b & o \\ o & o & c \end{bmatrix}$, a, b, c are the non-zero elements
GREEK, *kanon*, rule; LATIN, *forma*, shape or form

cardinal 1. Of foremost importance. 2. (of a number) A number used to show quantity, not order.
Examples
1. A cardinal point is one of the four principal directions on a compass.
2. The cardinal numbers are 0, 1, 2, 3, 4, 5, 6, 7, . . . (as compared to the **ordinal** numbers 1st, 2nd, 3rd, 4th, . . .)

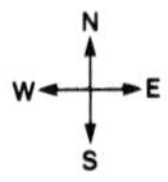

LATIN, *cardinalis*, pertaining to a hinge

Carroll diagrams diagrams used in the description of sets, as are **Venn** diagrams.
Example The shaded part of the diagram represents $A \cap B$ (the intersection of A and B).

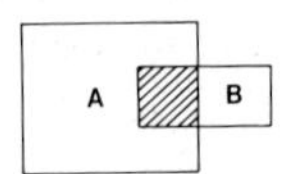

after Lewis Carroll, pen-name of Charles L. Dodgson, author whose profession was mathematics.

cartesian **1.** Of or referring to the philosophy or methods of Rene **Descartes**, a 17th century French philosopher and mathematician. **2.** Referring specifically to **analytical** or **coordinate** geometry.
Example Cartesian coordinates are the pair of numbers which locate a point with respect to 2 axes or reference frames in a plane. Three numbers are used to locate a point in general space. The cartesian **equation** of a circle where centre is at the origin and whose radius is of length r is $x^2 + y^2 = r^2$ where the axes intersect at right angles. NOTE: the 2 axes of reference frames do not necessarily have to intersect at right angles in order to locate points.

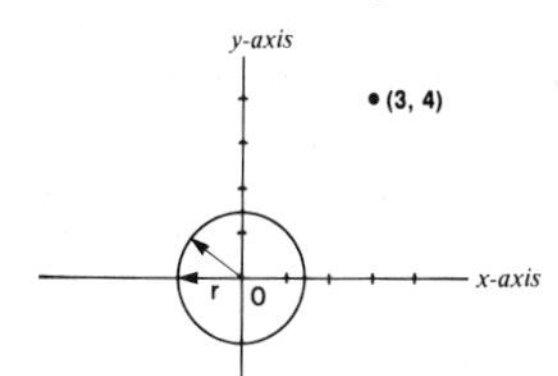

from Rene Descartes

cartesian product (also **cartesian cross-products**) **1.** The set of all possible **ordered pairs** of numbers in the form (x, y) such that x and y are numbers of the form 1, 2, 3, 4, 5, 6, 7, . . . The symbol used for cross-product is ×. **2.** The **vector** in space formed from two other vectors by the following rule: The length of the cross-product vector is the lengths of the two vectors multiplied together and also multiplied by the **sine** of the angle between the two vectors. The direction is **perpendicular** to the plane formed by the two vectors.
Examples
1. If $A = \{1,2,3,4\}$ and $B = \{1,2,3\}$, then

$A \times B$ = (1,1), (1,2), (1,3)
(2,1), (2,2), (2,3)
(3,1). (3,2), (3,3)
(4,1), (4,2), (4,3).

Note: If $\{x\}$ represented all the real numbers on a line, and $\{y\}$ represented all the real numbers on another line at right angles to the first, then $\{x,y\}$ represents all the points in the plane formed by the two lines.

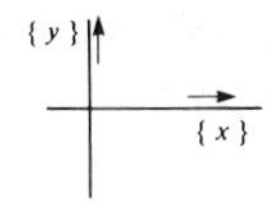

2. If a and b are vectors as shown, then $a \times b$ is the vector at right angles to the plane of a and b. The length of $a \times b$ is $|a|\ |b| \sin \theta$.

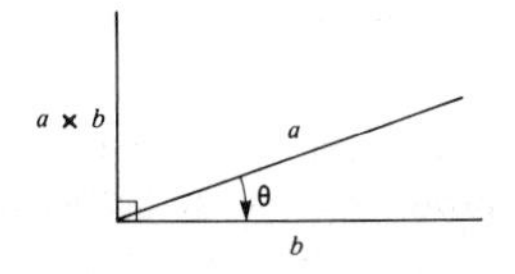

If a and b are both of length 1, and at right angles to each other, then $a \times b$ is a vector of length 1 at right angles to both a and b.
after Rene Descartes

census (statistics) a collection of data including all members of a particular **population**. It is different from a **sample**.
Example The Federal Government regularly conducts a census of the entire population with respect to age, sex, religion, employment etc.
LATIN, *census*, a registration of citizens

centimetre one hundredth of a metre. Symbol: cm.
Examples
1 cm = $\frac{1}{100}$ m
10 mm = 1 cm
LATIN, *centum*, one hundred; GREEK, *metron*, measure

central referring to the centre.
Example A central angle is an angle having radii as sides, and the centre of the circle as the **vertex**.

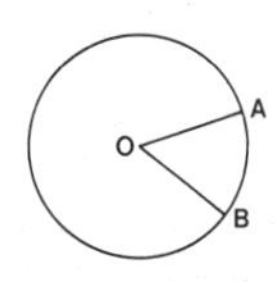

$\angle AOB$ is a central angle.
LATIN, *centrum*, centre

central tendency the idea of measuring where the greatest number of items are located in a given distribution of items. Measures of central tendency are:
(i) **mean** (the sum of all items' values divided by the number of items);
(ii) **median** (the value of the item which is located in the middle of all items); and
(iii) **mode** (the value of the item which occurs most often).
Example In a mathematics test, with scores out of 50, the following results were obtained by a class of 20 students.

Mark	40	41	42	43	44	45	46	47	48	49	50
Number of students	1	2	2	2	2	2	3	4	1	0	1

The mean (average) is 44.7. The median is 45. The mode is 47 (occurs most often).
LATIN, *centrum*, centre; *tendere*, to stretch.

centre of rotation the point which does not move when there is a **rotation**.

Example

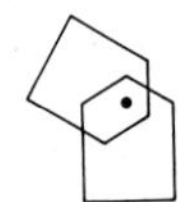

LATIN, *centrum*, centre; *rota*, a wheel

centroid the point of intersection of the **medians** of a triangle (a median is a line drawn from the **vertex** of a triangle to the midpoint of the opposite side). Note: The three medians of any triangle are **concurrent** at the centroid.
Example

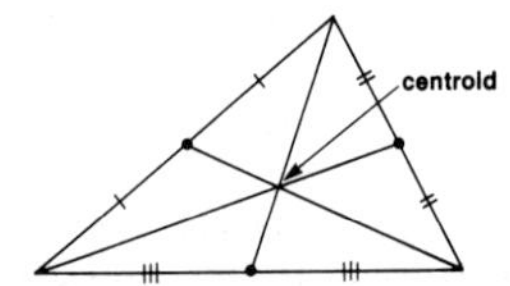

LATIN, *centrum*, centre; GREEK, *eidos*, form or shape

chain rule a rule for working out the **derivative** of a **function** of a function.

If $F(x) = f(g(x))$, then

$$\frac{dF}{dx} = \frac{df}{dg}\frac{dg}{dx} = \frac{df}{dg} \cdot g'(x)$$

Example
$F(x) = (x^2 - 2x)^3$
Let $u(x) = u = x^2 - 2x$
$\therefore F(x) = u^3$
$\therefore F'(x) = 3u^2\, du/dx$
$= 3u^2 (2x - 2)$
$= 3(x^2 - 2x)^2(2x - 2)$
LATIN, *catena*, chain; *regula*, rule

chance **1.** The happening of **events**; the way in which things fall out. **2.** (mathematics) The likelihood of an event happening; **probability**.
Example The chance of a six turning up when a die is thrown is 1/6, since there are six possible sides in all, but only 1 side containing a 6. What chance is there of rain today? It is hard to give a precise answer.
LATIN, *cadentia*, a falling

change of signs shorthand 'rules' which summarise what happens when applying 2 or more operations together.

$+ \times + = +$
$- \times + = -$
$- \times - = +$
$+ \times - = -$

Examples
$-(a - b) = -a + b$ (since $- - = +$).
$3(-a + b) = -3a + 3b$ (no sign change since $3 = +3$, and $+ - = -$, and $+ + = +$)
$-2x(x^2 - 2xy - y^2) = -2x^3 + 4x^2y + 2xy^2$
LATIN, *cambire*, barter; *signum*, a mark, token

characteristic the **whole number** part of a **logarithm** (contrast with the **mantissa**, the **decimal** or **fraction** part).
Example The logarithm of 200 = 2.3010. 2 is the characteristic.
GREEK, *kharakteristikos*, a stamp, impression

chord a straight **line** connecting two points on the circumference of a circle.
Example

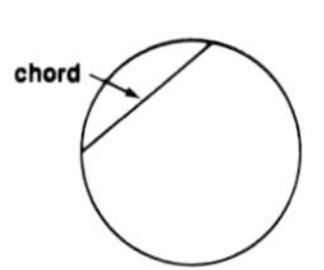

The chord divides the circle into two **segments**—the **major** and **minor** segments.
LATIN, *chorda*, a string

circle the **set** of all points in a plane situated at a fixed distance, (called the **radius**), from a given point, called the **centre** (*see also* ARC, CIRCUMFERENCE, SECTOR, SEGMENT, CHORD).
Example

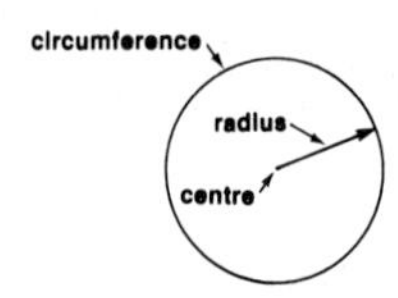

LATIN, *circulus*, a little ring

circular function another form for **trigonometric** functions. Since trigonometric functions of an angle can be related to a circle and indeed defined in terms of its radius, and the projection of the radius on the *x*-axis etc., they are often called circular functions.
Example

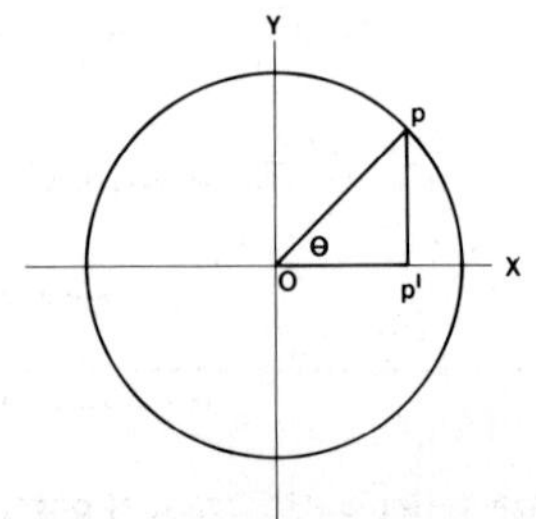

Cos θ is a circular function of the angle θ defined as the ratio of the **projection** of OP on to the x-axis (i.e. OP') to the length of OP itself. tan θ, sin θ are other circular functions.
LATIN, *circulus*, a little ring; *functio*, performance

circumcentre the centre of a circle which passes through the **vertices** of a **polygon**.
Examples

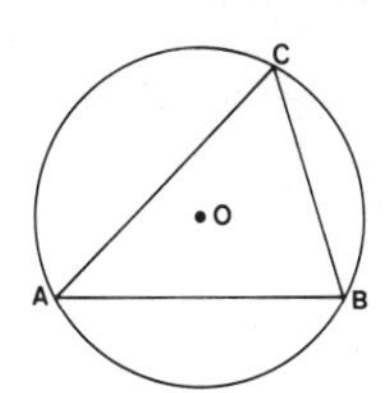

The centre of the circle which **circumscribes** ΔABC is O, and is called the circumcentre.

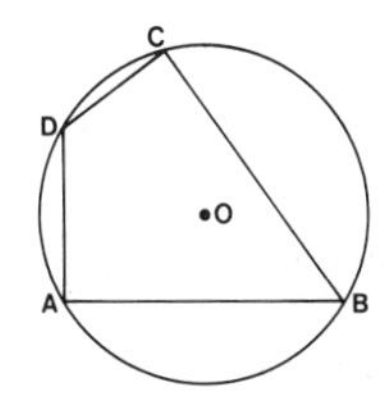

O is the circumcentre of the circle circumscribing the plane figure $ABCD$. $ABCD$ is called a **cyclic quadrilateral**.
LATIN, *circum*, around; *centrum*, centre

circumference the length around a circle. It is the special name given to the **perimeter** of a circle. It is measured as $C = 2\pi r$ (*see* PI).
Example

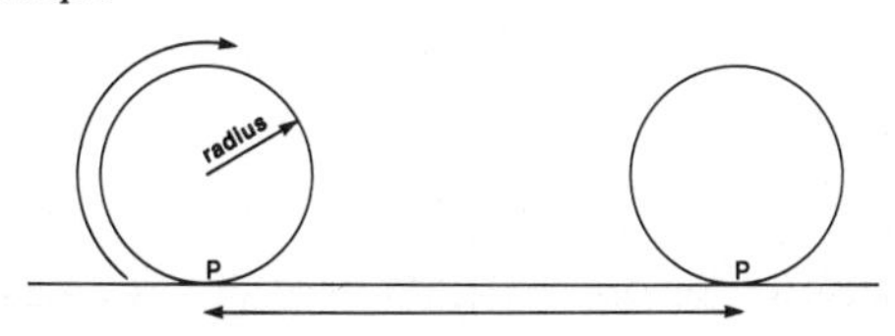

Circumference (distance rolled by a circle before coming the same way up again).
LATIN, *circum*, around; *ferre*, to carry

circumscribe to draw around (the outside of).
Example

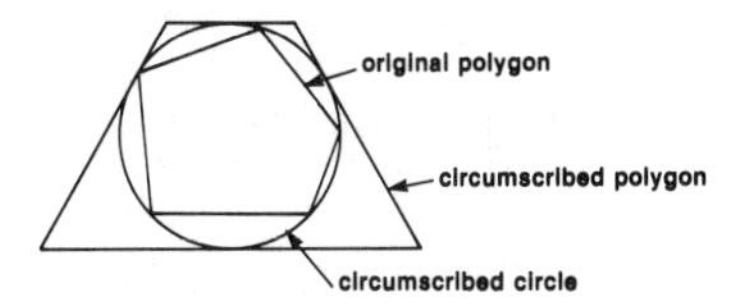

In the diagram above, the original **polygon** is circumscribed by a circle. The vertices of the polygon touch the circle. The circle, in turn, is circumscribed by another polygon all of whose sides are **tangents** to the circle.
LATIN, *circum*, around; *scribere*, to write

class **1.** A group of things or people having something in common. **2.** (statistics). One **set** of values within a number of related sets of values, or **frequency distribution**.
Examples
1. I am only interested in the class of persons who speak French.
2. In which earnings class does your father fit?

$ per week
0–100
100–200
200–300
300–400
over 400

(There are five classes in this frequency distribution).
LATIN, *classis*, one of the six divisions of the Roman people set up for taxation purposes.

class interval (statistics) one of the divisions into which a set of observations of a **random variable** has been grouped.
Example A **random variable** x has values spanning from 0 to 50. These may be grouped into class intervals 5 units wide, so that $0 < x \leqslant 5$ is the first interval, $5 < x \leqslant 10$ is the second, etc.
LATIN, *classis*, division; *intervallum*, space between ramparts

class limit the upper or lower bound of the values in the **class interval**.
Example In the class interval $0 < x < 5$, 0 and 5 are class limits.
LATIN, *classis*, division; *limes*, limit

clear (relating to a calculator) a key which when pressed cancels the operation and changes the value in the calculator counter back to zero ("clears" it). Symbol C . Note: CE is a special clear key that only clears the last entry.
Example The display on the calculator showed 273.578. On pressing "CLEAR" the display showed 0.
LATIN, *clarus*, bright, clear

clock arithmetic the rules of a special kind of arithmetic based on a clock face. A special case of **modulus** (or modular) **arithmetic**.
Example A normal clock has 12 positions for 12 hours. As soon as the hour-hand passes 12 we start from the beginning again, i.e. 7 o'clock + 6 hours = 1 o'clock, or 7 + 6 = 1. Similarly: 4 + 23 + 1 = 4; 11 + 7 = 6.

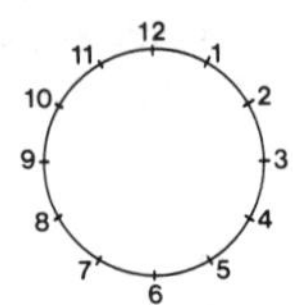

(This kind of arithmetic is called clock or modulus 12 arithmetic.)
after the working of a traditional clock

clockwise in the direction the hands of a clock move. The word derives from the traditional clock with two hands, 12 hours and a dial.
Example

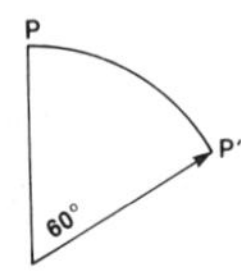

P has moved 60° clockwise to position P'.
LATIN, *clocca*, bell

closed 1. (of a curve) One whose ends are joined. 2. (of an **operation**) Relating to the situation where any two **elements** of a **set** combine to give another element of the same set. 3. See **interval**.
Example

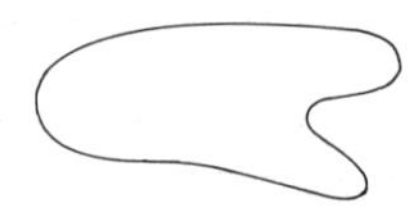

1. The diagram above is a simple closed curve.
2. Addition is a closed **operation** when dealing with **real** numbers, i.e. any 2 real numbers when added together give another real number.

$$2.4 + 5.67 = 8.07$$

LATIN, *clausus*, closed

closure the condition which describes the situation where any two **elements** of a **set** combine to give another element of the same set.
Example Closure applies to the **addition** of **positive integers**, since adding any 2 positive numbers gives another positive number: $2 + 5 = 7$. Closure does not apply to the operation of subtracting positive numbers, since one can construct a situation such as $2 - 5 = -3$, and -3 is not a positive number.
LATIN, *clausus*, closed

code 1. (computing) The instructions (written in a kind of shorthand) telling a computer what to do. 2. (general) A system of symbols which are given special meanings and used in transmitting messages.
Example
1. The following instructions are written in a code called BASIC (used to tell a computer to accept a number, add two to it, and then print out the result).

```
10 INPUT A
20 A = A + 2
30 PRINT A
```

2. There are several ways of coding messages. One is by substituting symbols for letters. A very simple code is to substitute 1 for A, 2 for B, 3 for C etc. The word CAGE is then 3175.
LATIN, *codex*, wood block, tablet, book

coefficient 1. A numeral or letter placed before an **expression** to indicate that the expression is to be multiplied by that factor. 2. In general any **factor** of a **product**.
Example
1. In $3(x + y)$, 3 is the coefficient of the expression $(x + y)$.
2. In $4ax$, 4 is the coefficient of ax, or a is the coefficient of $4x$, or x is the coefficient of $4a$.
LATIN, *coefficiens*, effecting together.

coincide 1. To occupy the same position or positions at the same time. 2. To have identical positions.
Examples
1. Two lines which are not **parallel** coincide at only one point.

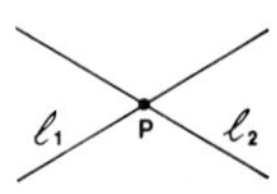

P on l_1 and l_2 coincide.

2.

C D
A B

Drawing a line CD exactly over the top of AB means AB and CD coincide.
LATIN, *co*, together; *incidere*, to happen.

cointerior angles, *see* ALLIED ANGLES.

collinear lying on the same straight line.

Example

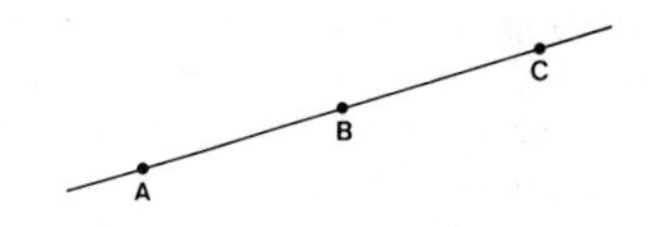

A, B, C are collinear points.
LATIN, *collineare*, to bring together

column a list of numbers (or letters) written vertically. A **vector** is usually written as a column.
Example

$$\begin{matrix} 3 \\ 1 \\ 0 \\ 9 \\ 2 \end{matrix} , \quad \begin{matrix} 2 \\ 8 \\ 7 \end{matrix}$$

LATIN, *columna*, a column

commission the sum of money paid to a sales person for selling an item or service. It is sometimes stated as a **percentage** of the selling price.
Example The commission on the first sale is 10%. On the second and subsequent sales it is 15%. This means that if the selling price is \$75.00, the first commission is 10% of \$75.00 = \$7.50, the second is 15% of \$75.00 = \$11.25.

common denominator **1.** An **integer** that is a common **multiple** of the **denominators** of two or more **fractions** grouped together. **2.** A **polynomial expression** that is a common multiple of the denominators of two or more fractions that are grouped together.
Examples
1. In

$$\frac{1}{4} + \frac{1}{6} - \frac{1}{7} = \frac{21 + 14 - 12}{84} = \frac{23}{84}$$

84 is a common denominator. In this case it is also the *L*owest *C*ommon *D*enominator. Another denominator is 168.

2.

$$\frac{1}{2\,(x + y)} - \frac{1}{x - y} = \frac{x - y - 2x - 2y}{2(x + y)(x - y)} = \frac{-(x + 3y)}{2(x + y)(x - y)}$$

$2(x + y)(x - y)$ is the lowest common denominator. ($2(x + y)^2(x - y)$ is another common denominator.)
LATIN, *communis*, common; *denominare*, to name completely

common difference the difference between any **term** and the term coming just before it in an **arithmetic progression**.
Example If 2, 5, 8, 11, 14, ... is an arithmetic progression, then, since

$$\begin{aligned} 5 - 2 &= 3 \\ 8 - 5 &= 3 \\ 11 - 8 &= 3 \\ 14 - 11 &= 3, \end{aligned}$$

3 is the common difference.
LATIN, *communis*, common; *differre*, to differ

common fraction a fraction whose top number (**numerator**) and bottom number (**denominator**) are both whole numbers.
Examples

$$\frac{1}{3}, \quad \frac{4}{7}, \quad \frac{161}{397}$$

LATIN, *communis*, common; *fractio*, a breaking

common logarithm, *see* LOGARITHM

common ratio in a **geometric progression** (GP), the common ratio is the **constant** by which each **term** is larger (or smaller) than the preceding one. Another way of saying this is as follows: In the GP a, ar, ar^2, ar^3, ar^4, ... each term is r times the previous term; r is called the common ratio.
Example In 1, ½, ¼, ⅛, 1/16, 1/32, ... ½ is the common ratio.
LATIN, *communis*, common; *ratio*, reckoning, reason

commutative (of an **operation**) relating to the situation where the order of combining two **elements** does not matter.
Examples
$3 + 4 = 4 + 3$, is an example of the commutative law for addition.
$3 \times 4 = 4 \times 3$, is an example of the commutative law for multiplication.
LATIN, *commutare*, to exchange

compass **1.** A device used to find the geographical direction. **2.** A device (usually referred to in the plural) used to draw circles.
Examples
1. The compass showed the mountain to be directly due north.

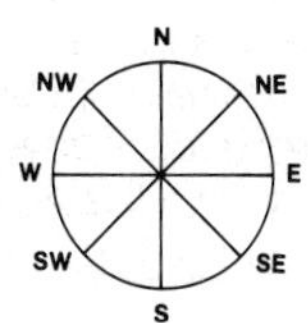

2. She should draw the circle using compasses.

LATIN, *com*, together; *passus*, pace, step

complement (of a **set**) the set of all **elements** not in the given set.
Example If S is the set of numbers 1, 3, 5, 7, 9, ... and U is the **universal set** 1, 2, 3, 4, 5, 6, 7, 8, 9, 10, ... then the complement of S is 2, 4, 6, 8, 10, ...

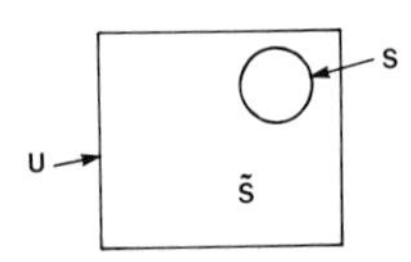

LATIN, *complere*, to fill up or complete

complementary angles two acute angles whose sum is 90°.
Example 41° and 49° are complementary angles. The complement of angle A is $(90° - A)$.

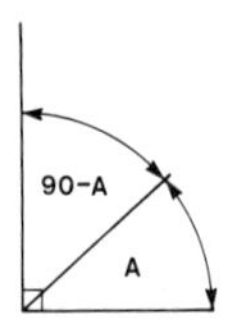

LATIN, *complere*, to fill up; *angulus*, angle or corner

completing the square the process of changing a **quadratic expression** of the form $ax^2 + bx + c$ into the form $a(x + p)^2 + q$. The latter form can be more easily used to solve equations of this kind, or determine the expression's characteristics more readily.
Example
If $E = ax^2 + bx + c$, then we can write

$$= a(x^2 + \frac{b}{a}x) + c$$

$$= a(x^2 + \frac{b}{a}x + \frac{b^2}{4a^2} - \frac{b^2}{4a^2}) + c$$

$$= a(x + \frac{b}{2a})^2 - \frac{b^2}{4a} + c$$

For example, to solve $x^2 + 3x - 6 = 0$, write

$$(x + \frac{3}{2})^2 - 6 - \frac{9}{4} = 0$$

$$(x + \frac{3}{2})^2 = 6 + \frac{9}{4}$$

$$\text{i.e. } x + \frac{3}{2} = \pm\sqrt{\frac{33}{4}}$$

$$x = -\frac{3}{2} \pm \sqrt{\frac{33}{4}}$$

Note: An alternative solution to such a quadratic equation would be to use the quadratic formula, i.e. to solve $ax^2 + bx + c = 0$, use:

$$x = \frac{-b \pm \sqrt{b^2 - 4ac}}{2a}$$

LATIN, *complere*, to fill up; *quadra*, a square

complex fraction a fraction in which either the **numerator** (top number) or **denominator** (bottom number) or both is/are a fraction or a **mixed number** (whole number + fraction).
Examples

$$\frac{3/4}{8}, \quad \frac{4}{1/2}, \quad \frac{1\frac{1}{4}}{2\frac{1}{2}}$$

Complex fractions can be simplified as follows:

$$\frac{3/4}{8} = 3/4 \div 8 = 3/4 \times 1/8 = 3/32$$

$$\frac{4}{1/2} = 4 \div 1/2 = 4 \times 2/1 = 8$$

$$\frac{1\frac{1}{4}}{2\frac{1}{2}} = 1\tfrac{1}{4} \div 2\tfrac{1}{2} = 5/4 \div 5/2 = 5/4 \times 2/5 = 1/2$$

LATIN, *complexus*, plait or fold together; *fractio*, a breaking

complex conjugate a **complex number** related to a given complex number $x + iy$ such that $x + iy$ multiplied by the complex conjugate gives $x^2 + y^2$. If $x + iy$ is the given complex number, then $x - iy$ is the complex conjugate. The symbol sometimes used to indicate the conjugate is $\overline{x + iy}$.
Examples
If $z = 4 + 3i$, then $\bar{z} = 4 - 3i$.
If $z = 3 - 2i$, then $\bar{z} = \overline{3 - 2i} = 3 + 2i$.
LATIN, *complexus*, plait or fold together; *conjugare*, yoke or join together

complex number a number of the form $a + bi$, where a and b are **real** numbers and i is the **square root** of -1. Complex numbers can be added, subtracted, multiplied or divided.
Examples

$$2 + 3i \quad , \quad -2i$$

To add $2 + 3i$ to $3 - i$ one adds the **like terms**, i.e. $(2 + 3) + (3i - i) = 5 + 2i$. To multiply $(2 + 3i)$ by $(3 - i)$ one multiplies each term by the other two and combines like terms, i.e.

$$(2 + 3i)(3 - i) = 2(3 - i) + 3i(3 - i)$$
$$= 6 - 2i + 9i - 3i^2$$
$$= 6 - 2i + 9i + 3$$
$$= 9 + 7i$$

LATIN, *complexus*, to plait or fold together; *numerus*, number

complex variable a **variable** whose values are **complex numbers** (contrast to **real** variable, whose values are real numbers).
Example If $C = F(c)$, where $c = a + ib$ (a and b are real numbers), and $i = \sqrt{-1}$, then C is a complex variable. $Z = x + iy$ (where x and y are real numbers) is also a complex variable.
LATIN, *complexus*, to plait or fold together; *varius*, changeable

component **1.** One of two or more **vectors** having a sum equal to a given (specified) vector. **2.** In a **cartesian product**, one of the **elements** of the **ordered pair**.
Examples

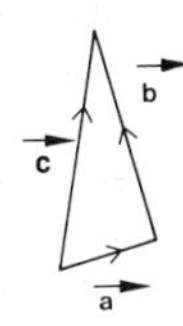

1. $\vec{a}$, $\vec{b}$ are components of $\vec{c}$.
2. In (1, 2), 1 and 2 are components.
LATIN, *componere*, to place together

composite number a whole number (**integer**) which has **factors** other than 1 and itself. (Note: **prime numbers** are numbers which have themselves and 1 as their only factors.)
Examples
1. $36 = 6 \times 6$ or 12×3 or 18×2 or 9×4
2. $21 = 7 \times 3$
LATIN, *componere*, to put together; *numerus*, number

compound expression an **expression** of the form A + (or $-$) B + (or $-$) C etc. It contains more than 2 terms.
Examples
$2x + 3y - 5$ is a compound expression.
$a + b - 2c + d$ is another.
LATIN, *componere*, to put together; *exprimere*, to press out

compound interest interest paid at given time intervals on a given sum of money where the given sum includes previous interest added to the original sum. Usually shortened to CI. $CI = P(1 + i)^n - P$ where P is the original sum, i is the interest rate as a decimal, and n is the number of time periods.
Example If the original sum is \$1000, and compound interest is paid yearly at 10% for 3 years the total compound interest is \$100 + \$110 + \$121 = \$331.

$$\text{Interest (1st year)} = 1000 \times \frac{10}{100} = \$100$$

$$\text{Interest (2nd year)} = 1100 \times \frac{10}{100} = \$110$$

$$\text{Interest (3rd year)} = 1210 \times \frac{10}{100} = \$121$$

LATIN, *componere*, to put together; *interesse*, to matter, be of concern

concavity the state of a curve as to whether it has a hollow pointing upwards or downwards. The sign + is used to represent a positive slope, and the sign − is used to represent a negative slope.
Concave upwards is like

Concave downwards is like

Example

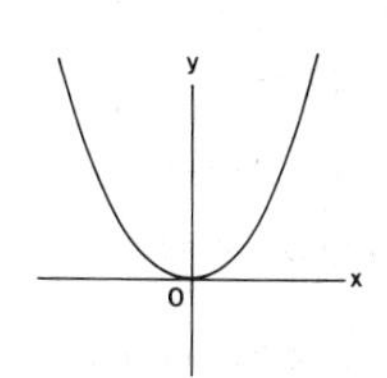

$y = x^2$ is a curve (a **parabola**) which is concave upwards

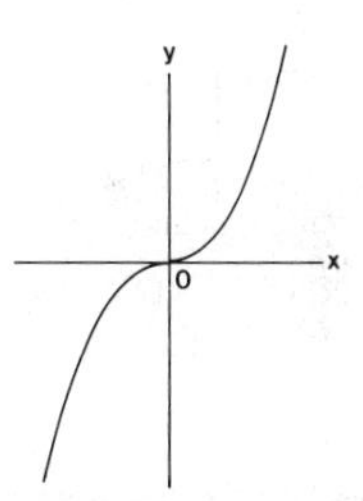

$y = x^3$ is a curve which changes concavity at $x = 0$.
LATIN, *concavus*, hollow

concentric having the same centre.
Example

C and c are concentric circles (both have the same centre O). It is also possible to have concentric **spheres**.
LATIN, *concentricus*, with the same centre

concurrent passing through a common point. *AB*, *CD*, *EF*, and *GH* are concurrent at O.

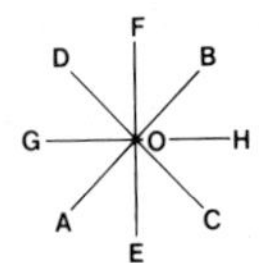

LATIN, *concurrere*, to run together

cone 1. A surface **generated** by a straight line passing through a fixed point (called the **vertex**) and moving along a **closed** curve. 2. (right circular) A solid which has a circular base and tapers to a point at the top called its **vertex**. Volume = $\frac{1}{3}$ × base area × height; Area of curved surface = π × radius × slant height.
Example
General cone

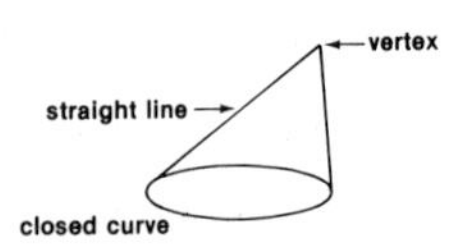

Right circular cone

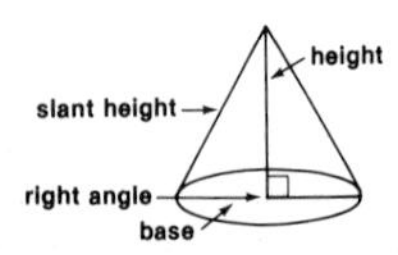

GREEK, *konos*, pine-cone etc.

confidence (statistics) the degree of certainty that something will be so.
Example I have 100% confidence in his ability to win.
LATIN, *confidentia*, trust

confidence limits (statistics) two numbers or percentages relating to a **sample distribution** of values about a **mean** or **average**. The two numbers give the likelihood that the actual mean of all the values will be inside the two limits.
Example The following table is the result of measuring the height of 10 adults from a particular town.

H (cm)	161	178	164	193	199	172	188	167	175	183

The mean or average (of this group) is 178 cm. To be 95% confident that this mean is within 2 **standard errors** from the actual mean we would calculate confidence limits as being equal to the mean + or − the standard error. Since the standard error can be defined as the **standard deviation** divided by the square root of the number of results (in this case 3.8 cm), then the 95% confidence limits are 174.2 and 181.8 cm. Other confidence limits can be similarly worked out.
LATIN, *confidentia*, trust; *limes*, a boundary

congruence the condition of things having the same size and shape. Congruence is the relationship between two or more figures where each figure has the same shape and all corresponding parts are equal.
Examples The following figures show the idea of congruence.

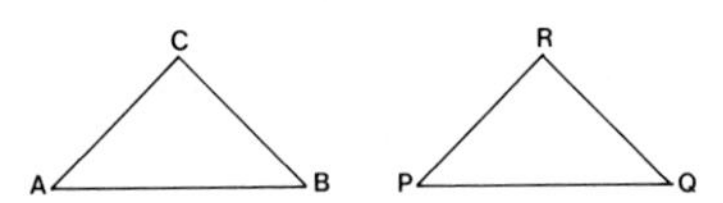

ABC and *PQR* are congruent triangles.

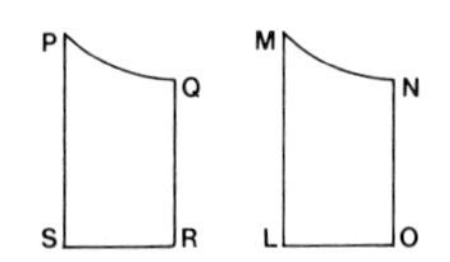

PQRS and *MNOL* are congruent figures.

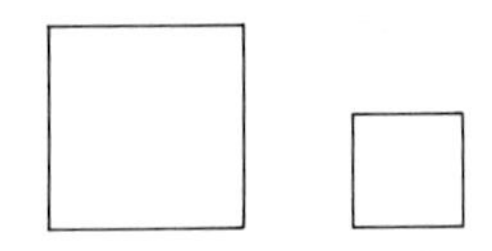

These *do not* show congruence.
LATIN, *congruentia*, agreement

congruent 1. Having the same size and shape (equal in all respects). 2. (number theory) Relating to the condition where the difference of two numbers is a multiple of a given number.
Examples
1.

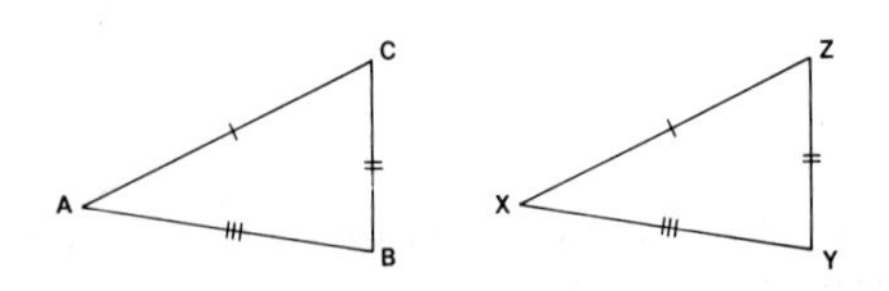

Triangles *ABC* and *XYZ* are congruent (i.e. their corresponding sizes and angles are equal, and so too are their areas and perimeters).
2. Since 12 − 6 = 6, which is divisible by 3, 12 and 6 are congruent with respect to the number 3; 18 and 15 are also congruent with respect to the number 3.
LATIN, *congruere*, agree

conic section the curve formed by a plane cutting a right circular **cone**. The curves can be of a **parabola**, **hyperbola**, **ellipse** or **circle**.
Example

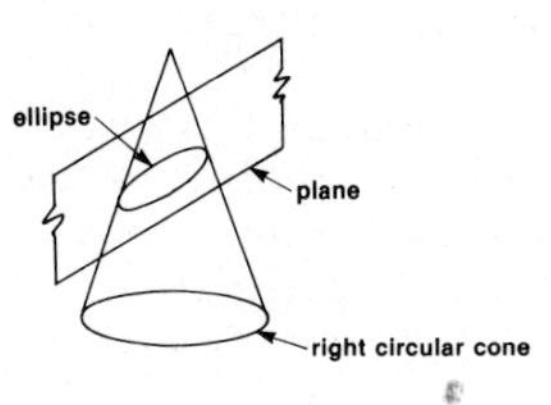

GREEK, *konos*, pine-cone etc.; LATIN, *sectio*, a cutting

constant a fixed number. A symbol whose value does not change (compare with **variable**).
Examples
c, the speed of light, is a constant in the **equation**

$$E = mc^2.$$

In the **expression** $x + 3$, 3 is a constant.
LATIN, *constare*, to stand together, remain steadfast

contact a **point** or **line** (curve) formed as a result of either a line (curve) **touching** a curve or a surface touching another surface. A **chord** of contact is the line joining the two points of contact on a **parabola** where two **tangents** from a common point meet the **parabola**.
Example

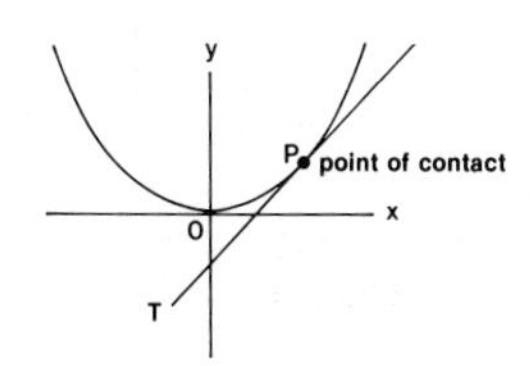

The tangent TP to the parabola $y = x^2$ touches it at the point P.

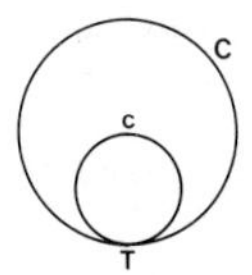

Two circles, c and C, touch at T. T is the point of contact.

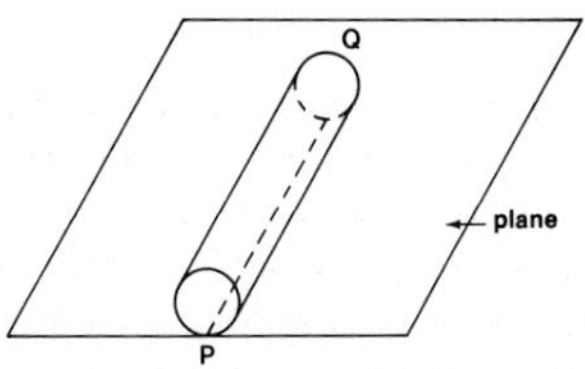

PQ is the line of contact of the cylinder on the plane.

LATIN, *contingere*, touching together

continued fraction a fraction of the form

$$a + \cfrac{b}{c + \cfrac{d}{e + \cfrac{f}{g + \dots}}}$$

with no limit. Continued fractions such as those with an observable pattern can often be reduced to a simpler form.
Example

$$1 + \cfrac{2}{1 + \cfrac{2}{1 + \cfrac{2}{\dots}}}$$

represents a continued fraction with a definite pattern. Since $1 + \frac{2}{\dots}$ repeats itself over and over we can put $x = 1 + \frac{2}{\dots}$, and then we can simplify as follows.

$$x = 1 + \frac{2}{x}$$

We can solve this equation for x as follows:

$$x^2 = x + 2$$
$$x^2 - x - 2 = 0$$
$$(x + 1)(x - 2) = 0$$
$$x = -1 \text{ or } 2$$

Since the continued fraction is positive, then $x = 2$. In other words, the value of the continued fraction is 2.
LATIN, *continere*, to hold together; *fractio*, a breaking

continuous describing a **variable** whose values are not restricted to separate numbers. In **calculus**, a function $f(x)$ is said to be continuous at $x = c$ if $f(c)$ and, $\lim_{x \to c} f(x)$ both exist, and are equal (*see* LIMIT).
Examples
Speed is a continuous quantity. As a car accelerates from start to 100 km/h, the speed must pass through *all* values between 0 and 100 km/h

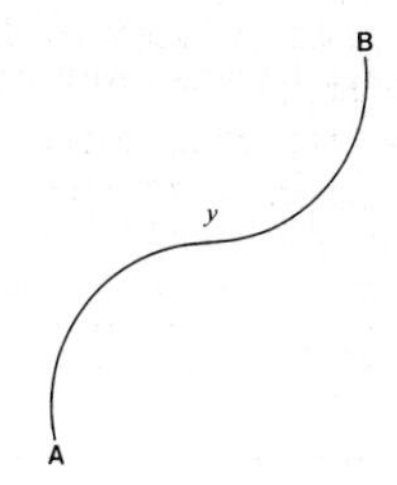

Between A and B, the curve y is said to be continuous.

Between X and Y, y is said to be *discontinuous* (i.e. there are gaps).

If $f(x) = \frac{x + x^3}{x}$, then $f(0) = \frac{0}{0}$ which is undefined (i.e. does not exist). Therefore (even though $\lim_{x \to 0} f(x) = 1$) $f(x)$ is not continuous at $x = 0$.
However, if we define $f(0) = 1$, then $f(x)$ will be continuous at $x = 0$.
LATIN, *continere*, to hold together.

convergent coming, or tending to come together, to a point or value.
Example In the graph of $y = \frac{1}{x}$, the curves and the axes are convergent at infinity.

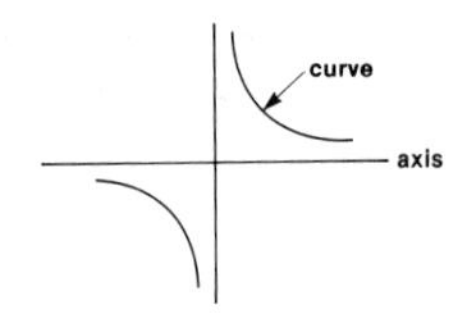

An **infinite series** of **terms** is convergent if the sum tends towards a definite value; such as the sum of

$$1 + \tfrac{1}{2} + \tfrac{1}{4} + \tfrac{1}{8} + \tfrac{1}{16} + \ldots$$

tends to the value 2.
LATIN, *convergere*, incline

conversion the act of expressing a quantity in alternative **units**. A **conversion factor** is a number which is applied to the given quantity to change it from one set of units to another.
Example The conversion of 151 mm to centimetres requires multiplying 151 mm by the conversion factor $\frac{1}{10}$.

$$151 \text{ mm} \times \frac{1}{10} = 15.1 \text{ cm}$$

LATIN, *conversio*, a turning around

convert to change the **units** with which a quantity is measured.
Examples
To convert metres into inches one needs to multiply the number of metres by 39.37. (This is only accurate to 2 decimal places.)
10 m = 10 × 39.37 inches = 393.7 inches.
To convert from Fahrenheit to Celsius one first subtracts 32, then multiplies by $\frac{5}{9}$.

$$95°\text{F} = (95 - 32) \times \frac{5}{9} °\text{C}$$
$$= 63 \times \frac{5}{9} °\text{C}$$
$$= 35°\text{C}$$

LATIN, *converto*, to turn about

convex with the **vertices** pointing outwards. A shape is convex if the line joining any two points on the shape stays inside the shape.
Example

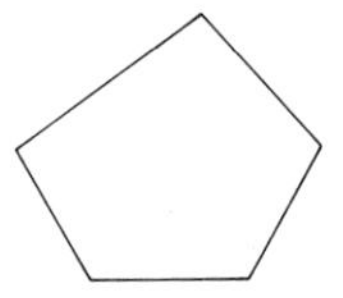

The shape above is a convex **polygon**.
LATIN, *convexus*, vaulted

coordinate any of an ordered sequence of numbers which, in relation to a given frame of reference, locates a point in space (*see* ABSCISSA, ANALYTIC GEOMETRY, ORDINATE).
Example

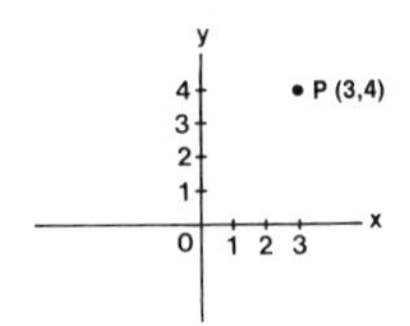

3,4 are coordinates of the point P with respect to the x–y **frame of reference**. Note: 3 refers the distance along the x-axis from 0, 4 refers the distance along the y- axis from 0. R (4,3) is a different point to P (3,4).
LATIN, *coordinare*, to set in order, regulate

coordinate geometry, *see* ANALYTICAL GEOMETRY.

correlation (in **statistics**) a relationship between two quantities such that as one increases, the other increases (or decreases). The relationship is not as precisely defined as a **function**.
Example Over a period of years the following statistics were kept of the rainfall and tonnes of wheat produced in a certain area:

Yearly Rainfall (mm)	20	23	19	26	21	28	17	29
Wheat (tonnes)	100	101	87	105	103	107	82	108

There is an apparent correlation between how much rain fell in each year and the amount of wheat harvested. Since the number of tonnes increased as the rainfall increased, the correlation is said to be positive.
LATIN, *cum*, together; *relatio*, relation

correlation coefficient a number worked out by a special formula which shows how correlated two **sets** of values are. It varies between -1 and $+1$. If the correlation coefficient is $+1$, then one set of values is directly **proportional** to the other set (i.e. if one set increases then the other set increases correspondingly). If it is -1, then the two sets are inversely proportional (i.e. as one set increases, the other decreases correspondingly). The closer the correlation coefficient is to $+1$ (or -1), the better the correlation between the two sets of values. A coefficient of 0 means there is no correlation of any kind. The symbol often used for the correlation coefficient is r.

Example In the previous example (see CORRELATION) with rainfall and wheat production, we define the following symbols:

$\overline{X}$ = Average rainfall =

$$\frac{20+23+19+26+21+28+17+29}{8}$$

$\overline{Y}$ = Average wheat produced =

$$\frac{100+101+87+105+103+107+82+108}{8}$$

σ_x = Standard deviation of rainfall =

$$\frac{\sqrt{(20-\overline{X})^2 + (23-\overline{X})^2 + \ldots + (29-\overline{X})^2}}{\sqrt{8}}$$

σ_y = Standard deviation of wheat production =

$$\frac{\sqrt{(100-\overline{Y})^2 + (101-\overline{Y})^2 + \ldots + (108-\overline{Y})^2}}{\sqrt{8}}$$

Sum of $(X-\overline{X})(Y-\overline{Y})$ =
$(20-\overline{X})(100-\overline{Y})+\ldots+(29-\overline{X})(108-\overline{Y})$

$$r = \frac{\text{Sum of } (X-\overline{X})(Y-\overline{Y})}{\sigma_x\sigma_y}$$

LATIN, *cum*, together; *relatio*, relation; *efficere*, to effect

correspondence a means of describing **relations** between one or more **points**, objects, etc.

Example

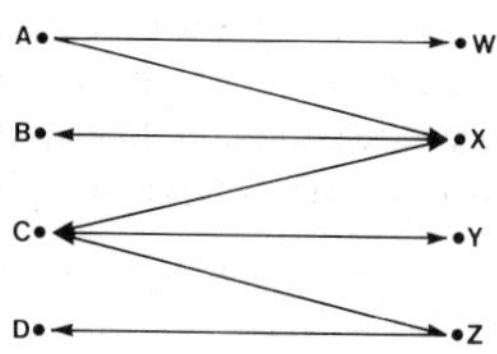

A has a relationship to W and X.
B has a relationship to X only.
C has a relationship to X, Y, and Z.
D has a relationship to Z only.
Also, W is related to A only; X is related to A, B and C etc.

LATIN, *correspondere*, pledge, answer together

corresponding angles **1.** Angles formed by a **straight line** cutting a pair of **parallel** lines located in the same relative positions. Corresponding angles are equal. **2.** Angles in the same relative position as far as two or more triangles are concerned. Note: If the lines are not parallel, then the angles can still be called corresponding, but they will not be equal.

Examples

1.

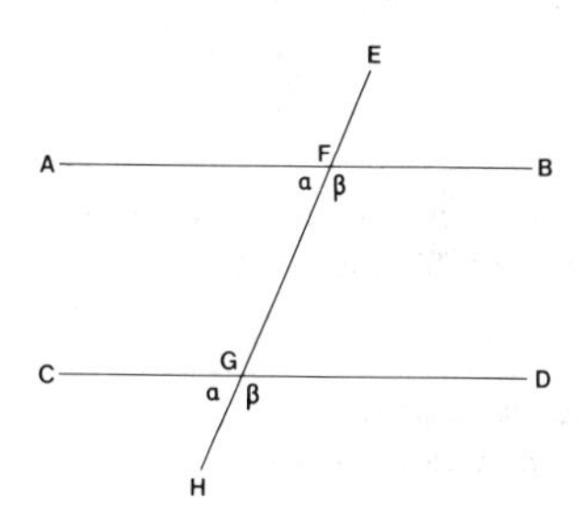

$\angle BFG = \angle DGH = \beta$ (corresponding angles)
also, $\angle AFG = \angle CGH = \alpha$ (corresponding angles)

2.

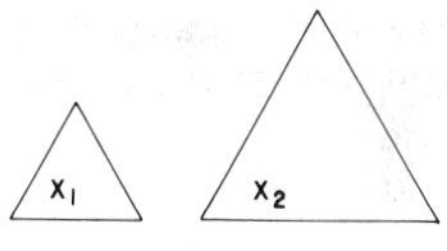

x_1, and x_2 are corresponding angles.

LATIN, *correspondere*, pledge, answer together

corresponding points points which are connected together by a **relation** such as a **reflection**, **translation**, **rotation**, etc. Points which when a **transformation** is applied to them, have an **object-image** relation between them.

Example

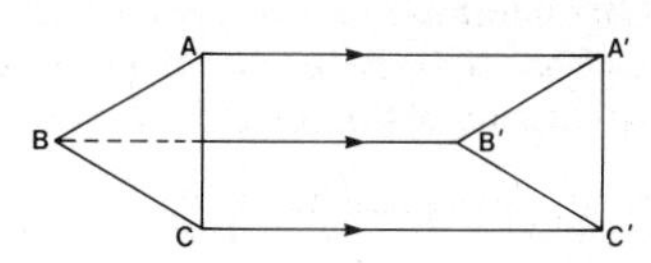

ABC is the object, and $A'B'C'$ is the image.
A and A', B and B', C and C' are corresponding points.

LATIN, *correspondere*, to respond together; *punctus*, pricked

cosecant the **ratio** of the largest side of a right angled triangle to the side opposite the angle under discussion. Abbreviated as cosec.
Example

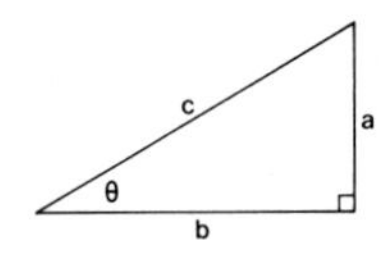

cosecant $\theta = c/a$, i.e. cosec $\theta = 1/\sin\theta$
LATIN, *cum*, together; *secare*, to cut

cosine (of an angle) the **ratio** of the side adjacent to the angle to the longest side (**hypotenuse**) in a right angled triangle. Abbreviated to cos.
Example

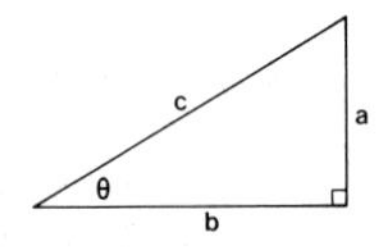

$$\cos\theta = \frac{b}{c}$$

Note: For an angle greater than 90°, the side adjacent becomes the *x*-coordinate of the point *P* as in the following diagram:

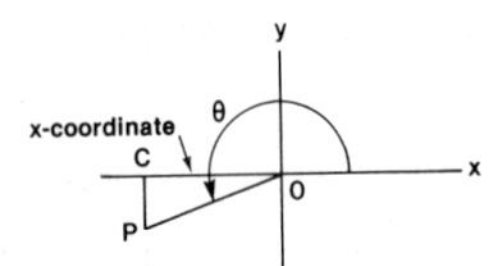

$$\cos\theta = \frac{OC}{OP}$$

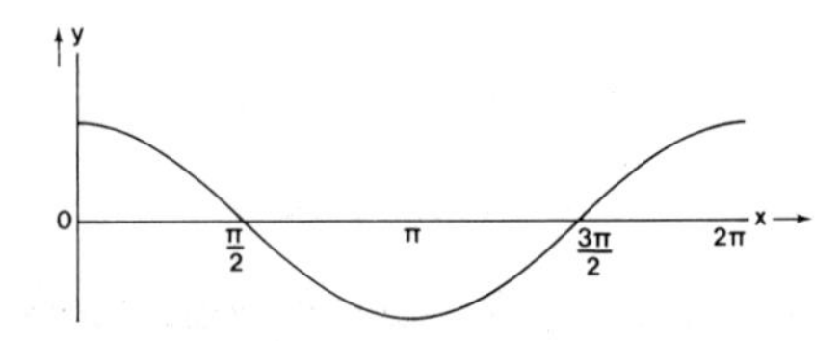

Graph of $y = \cos x$
LATIN, *cum*, together; *sinus*, a fold, bend

cosine rule a formula for relating the three lengths of the sides of a triangle to one of the angles of the triangle. The rule can be written as:

$$\cos\theta = \frac{b^2 + c^2 - a^2}{2bc}$$

or $$a^2 = b^2 + c^2 - 2bc\cos\theta$$

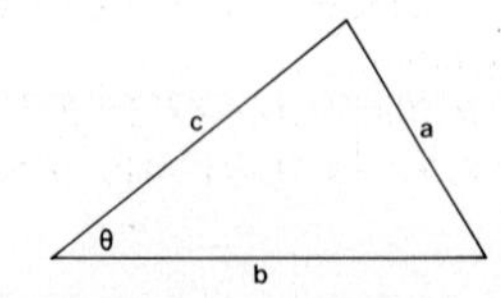

Example

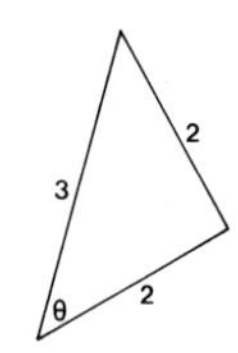

$$\cos\theta = \frac{3^2 + 2^2 - 2^2}{2 \times 3 \times 2} = \tfrac{3}{4}$$

LATIN, *cum*, together; *sinus*, a fold, bend; *regula*, a straight stick, ruler

cotangent the **ratio** of the adjacent side to the angle to the opposite side in a right angled triangle. Abbreviation cot.
Example

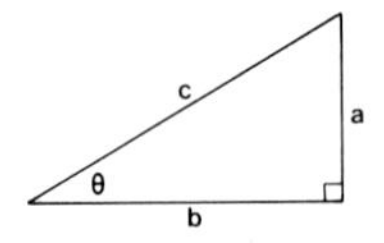

$$\cot\theta = \frac{b}{a}$$

Note: $\cot\theta = \dfrac{1}{\tan\theta}$

LATIN, *cum*, together; *tangere*, to touch

counting numbers the set of positive whole numbers.
Example 1, 2, 3, 4, 5, . . .
Note: This is a subset of the set of **integers** or whole numbers.
LATIN, *computare*, to sum up, reckon; *numerus*, number

critical point a point which makes a **function**'s **derivative** zero or undefined. In symbols, if $f(x)$ is a function of x, then if $f'(x) = 0$ or ∞ (i.e. undefined) for $x = x_1, x_2, x_3$, then x_1, x_2, x_3 are critical points.
Example

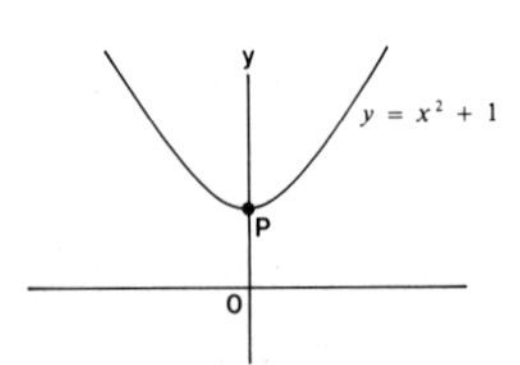

At $x = 0, f'(x) = 0$, so that $P(0,1)$ is a critical point. It is also, in this case, a **minimum turning point**.

If $y = f(x) = \sqrt[3]{(x-1)^2}$, then

$$\frac{dy}{dx} = f'(x) = \frac{2}{3}(x-1)^{-1/3} = \frac{2}{3}\frac{1}{\sqrt[3]{(x-1)}}$$

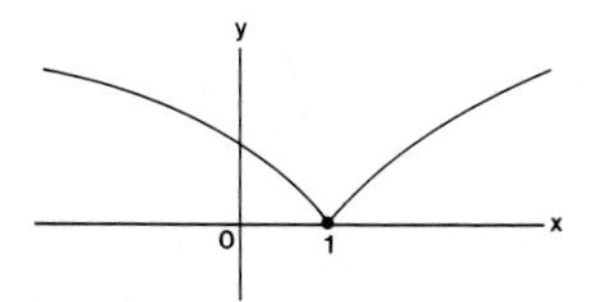

At $x = 1, f'(x) = \infty$ (i.e. undefined), therefore the point $x = 1, y = 0$ is a critical point. (It is *not* called a turning point, since the curve has an abrupt change at $x = 1$ known as a **discontinuity**.
GREEK, *kritikos*, able to discern, critical; LATIN, *punctus*, pricked

cross multiply in an **equation** containing **fractions** the **numerators** of each side are multiplied by the opposite **denominators**. It is a shorthand way of multiplying both sides of the equation by the product of the two denominators.
Example

$$\frac{x}{3} = \frac{7}{8}$$
$$\frac{x}{3} \times \frac{7}{8}$$
$$8x = 21$$
$$x = \frac{21}{8}$$

An alternative method of solution, is to multiply both sides of the fractional equation by the lowest common denominator (LCD), i.e.

$$\frac{x}{3} = \frac{7}{8}$$
$$\text{LCD} = 24$$
$$24 \times \frac{x}{3} = 24 \times \frac{7}{8}$$
$$8x = 21$$
$$x = \frac{21}{8}$$

LATIN, *crux*, cross; *multiplicare*, to multiply

cross product *see* CARTESIAN PRODUCT, VECTOR PRODUCT

cross-ratio a **ratio** of the form $\dfrac{(a-b)(c-d)}{(b-c)(d-a)}$ where a, b, c, d, are any four numbers. If the numbers represent distances of four points on a straight line from an origin point on the line, then any four points (lying on a straight line) coming from a **projection** of the original four points will have the same ratio. A common way of writing this ratio is $(ABCD)$ or $(XYZW)$ where A, B, C, D are four points on a straight line.

Example

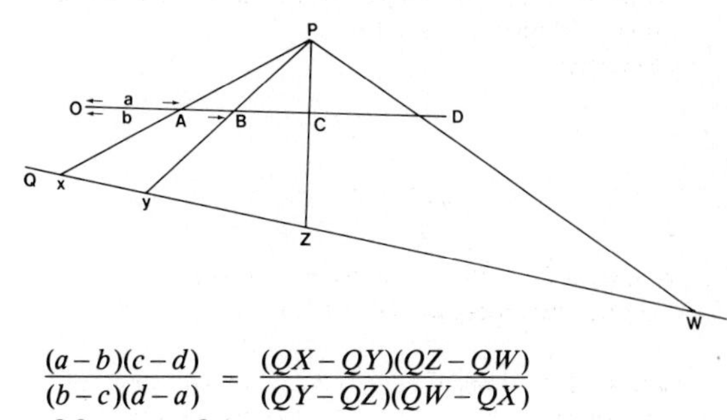

$$\frac{(a-b)(c-d)}{(b-c)(d-a)} = \frac{(QX-QY)(QZ-QW)}{(QY-QZ)(QW-QX)}$$
$$OC = c \quad OA = a$$
$$OD = d \quad OB = b$$

LATIN, *crux*, cross; *ratio*, computation

cross-section **1.** A section formed by a plane cutting through a solid. **2.** A sample selected from a group of things, people etc. which is representative of the group.
Examples
1. The cross-section of a cylinder is a circle when the plane cuts the cylinder at right angles to the cylinder's axis, and is an ellipse when the plane cuts the cylinder at an angle.

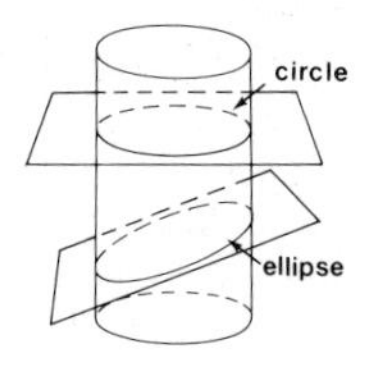

2. He studied the attitudes of a cross-section of the swinging voters.
LATIN, *crux*, cross, *sectus*, cut-off

cube **1.** (geometry) A regular solid whose faces are six squares. **2.** (arithmetic and algebra) The third **power** of a quantity.
Examples
1.

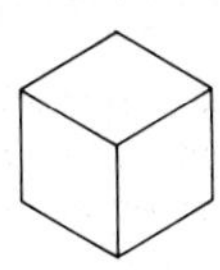

2. $2 \times 2 \times 2 = 2^3$; $a \times a \times a = a^3$
GREEK, *kubos*, a cube or die

cube root the number which, when cubed, gives the number in question.
Example The cube root of 8 is 2: i.e. $2 \times 2 \times 2 = 8$
GREEK, *kubos*, a cube or die; OLD NORSE, *rot*, branch or root

cubic centimetre a **unit** of **volume** the same as that of a cube whose edges are one centimetre.
Example The volume of the box below is
$2 \times 2 \times 5 \text{ cm}^3 = 20 \text{ cm}^3$

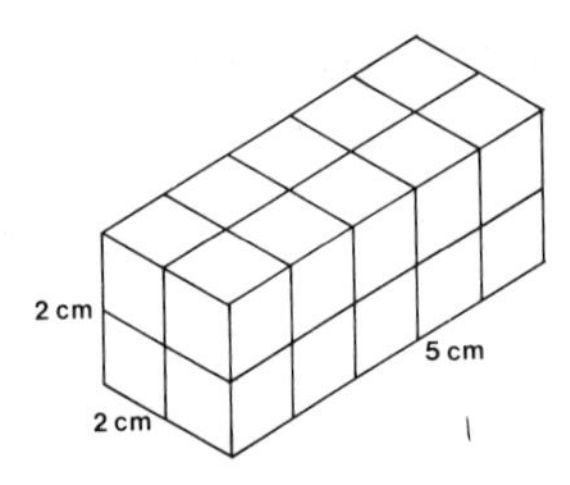

There are 20 cubes with 1 cm edges in this solid box, i.e. volume $= 20 \text{ cm}^3$.
GREEK, *kubos*, a cube or die; LATIN, *centum*, a hundred; GREEK, *metron*, measure

cubic metre a **unit** for measuring **volume**. It is the volume of a cube whose sides are one metre in length.
Example

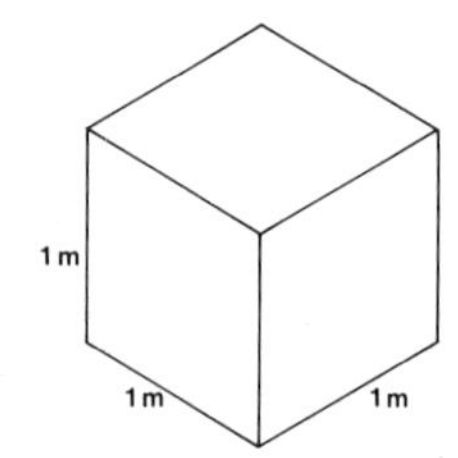

$V = 1 \text{ m}^3$
GREEK, *kubos*, a cube or die; *metron*, a measure

cubit a measure of length used in ancient times. It was the distance from the elbow to the tip of the middle finger, approximately 46 cm to 56 cm.
Example Many measurements in the Bible are quoted in cubits.
LATIN, *cubitum*, elbow

cuboid a solid which has rectangles for all of its faces.
Example

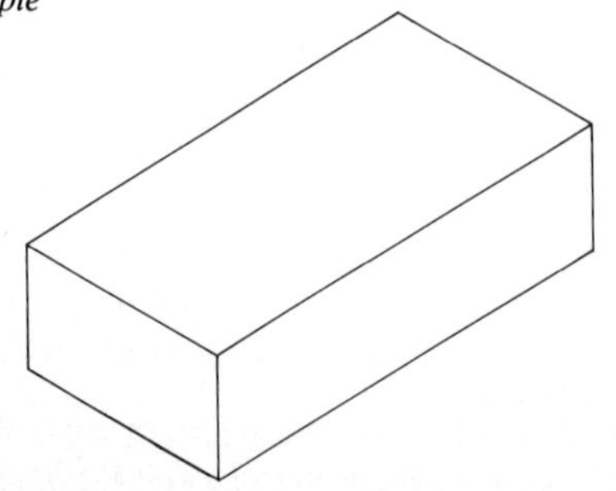

GREEK, *kubos*, a cube or die; *eidos*, form or shape

cumulative frequency the sum of the **frequencies** at or below a given value.
Example In a survey 100 students were asked which mark (to the nearest 10) they obtained in a mathematics examination.

Exam Result	*Frequency*	*Cumulative Frequency*
0–10	0	0
10–20	1	1
20–30	4	5
30–40	6	11
40–50	14	25
50–60	22	47
60–70	26	73
70–80	19	92
80–90	7	99
90–100	1	100

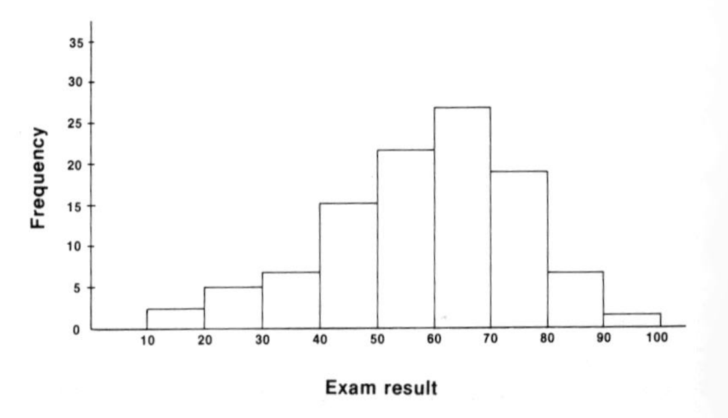

The frequency of exam results in the 50–60 group is 22. However the cumulative frequency of results 50–60 is 47, meaning there are 47 students whose mark was either 50–60 or below.
LATIN, *cumulare*, to pile up; *frequentia*, a crowd

curve 1. A **line** (or surface) which is not straight (or plane). **2.** The line drawn on a graph. **3.** The intersection of two surfaces in **three dimensions**.
Examples
1.

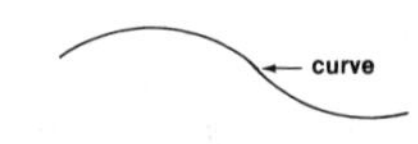

2.

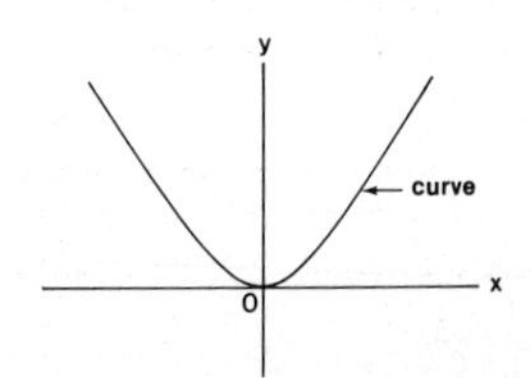

3.

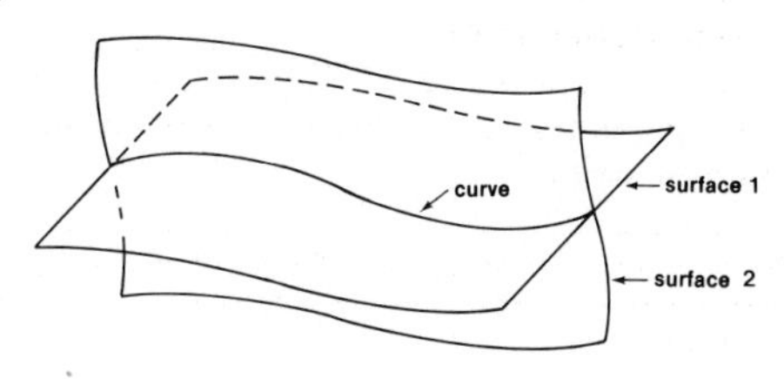

LATIN, *curvus*, bent, curved

cyclic quadrilateral a **quadrilateral** whose **vertices** lie on the circumference of a circle. It has properties such as: opposite angles are **supplementary**, **exterior angle** equals opposite **interior** angle, etc.
Example

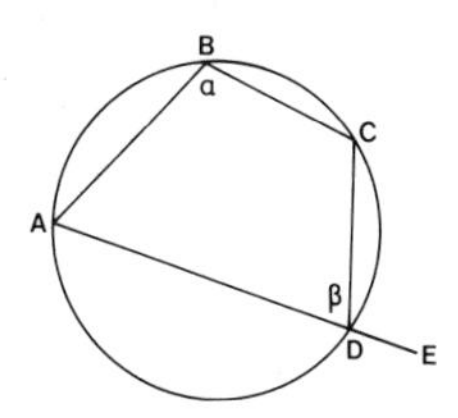

ABCD is a cyclic quadrilateral
$\alpha + \beta = 180°$
$\angle CDE = \alpha$
GREEK, *kuklos*, a circle; LATIN, *quadrus*, four; *latus*, side

cycloid (geometry) the path traced out by a point on the circumference of a circle as it rolls on a straight line.
Example

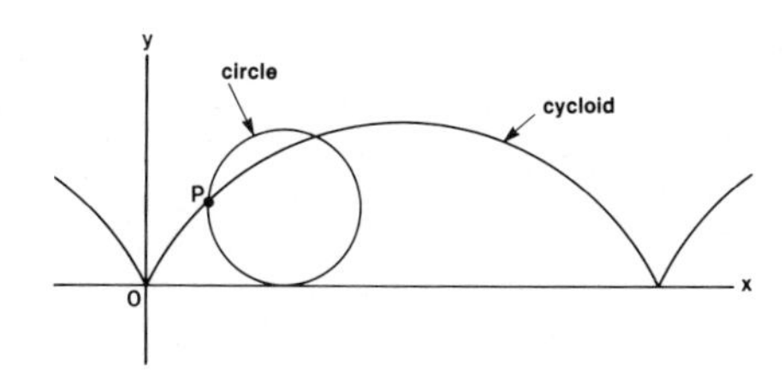

GREEK, *kuklos*, circle; *eidos*, form or shape

cylinder the surface or volume traced by one side of a **rectangle** rotating about its parallel side which serves as the axis. A **cylinder of revolution** is a right circular cylinder, as shown in the example.
Example

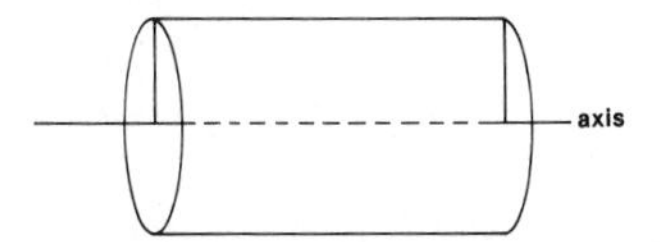

A right circular cylinder (right here means the circles are at right angles to the axis).
Volume of the cylinder = area of end (base) × height
Area of curved part of a cylinder = circumference × height
GREEK, *kulindros*, roller, cylinder

Dd

D (d) the fourth letter of the English alphabet. Used in **mathematics** to label points (**geometry**) and as part of a symbol to represent rate of change of one quantity with respect to another. Also D represents the Roman numeral for 500.
Example

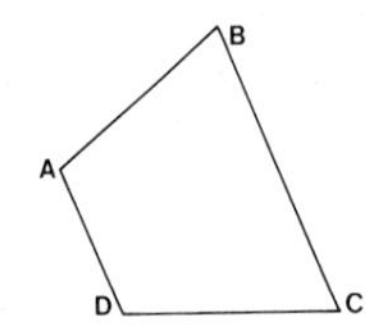

$\frac{dy}{dt}$ (the rate of change of y with respect to t) sometimes written $D_t y$ (*see* DERIVATIVE)

DXV = 515 (Roman numeral converted to **decimal**).
GREEK, Δ, δ (delta) [fourth letter of Greek alphabet]

D'Alembert, Jean Le Rond (1717–1783) French mathematician of the eighteenth century. D'Alembert studied the physics of fluids, vibrating strings and precession of the equinoxes. He was interested in the development of the **calculus** and especially in the theory of **limits**. Responsible for the study of solutions of ordinary and partial **differential equations**.

data a group of numbers collected for study. Data can be displayed by **tables**, charts, **graphs** etc.
Examples 10 students were asked how many children were in each family. The answers were:
2, 3, 1, 4, 1, 3, 2, 1, 5, 2
This is sometimes called *raw data* ("raw", because it has not yet been worked upon).

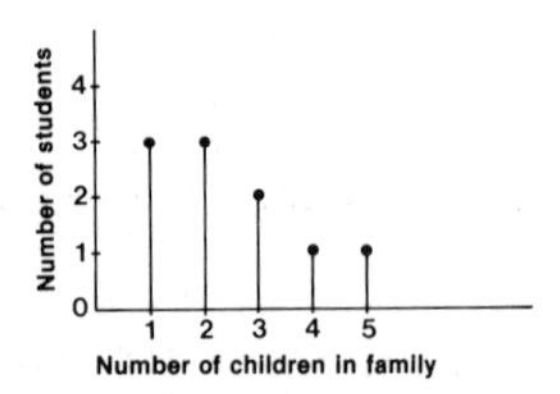

This is another way of displaying the data.
LATIN, *datum* (pl.), a thing given

datum **1.** A fact or assumed fact from which a conclusion or decision can be made. **2.** A real or assumed point or line from which measurements can be made. A reference point, line or surface.
Examples
1. People can get into trouble if they use the datum "All men are the same".
2.

Taking sea level as the datum line, how high is the balloon?
LATIN, *datum*, a thing given

de Breteuil, Emilie (1706–1749) French mathematician who is especially remembered for her translation of Isaac Newton's *Principia* into French, making Newton's study of the **calculus** available to French mathematicians.

debug to remove the "bugs" or errors which prevent a computer program (or similar) from working.
Example In debugging the program we found errors such as the same symbol being used for two different values.
AMERICAN ENGLISH, *debug*, remove electronic listening devices, applied by analogy to computers.

decagon a **polygon** with ten straight sides and ten angles.
Example

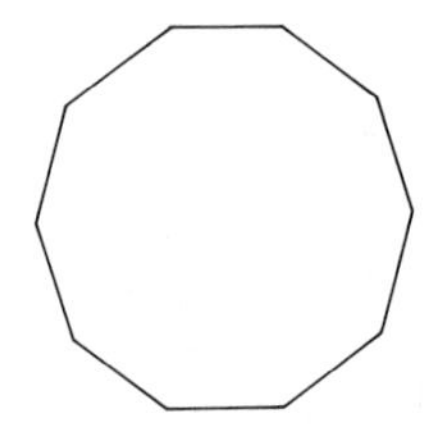

Note: A **regular** decagon has all sides equal and all angles equal.
GREEK, *dekagōnon*, one having ten angles

decimal **1.** A number written to the **base** ten. The number can include a fractional portion. **2.** Referring to the shorthand form used to write numbers in groups of ten.
Examples
1. 3.26, 57.1, 609.38 are decimals.
2. $3 \times 100 + 2 \times 10 + 4$ can be written in the decimal way as follows: 324.
LATIN, *decimus*, a tenth

decimal fraction a fraction with a **denominator** of 10, or 100, or 1000 etc.
Examples

$$\frac{2}{10}, \frac{19}{100}, \frac{251}{1000}, \frac{63}{10000}$$

Decimal fractions can also be written without denominators using the decimal point as follows:

$$\frac{2}{10} = 0.2 \quad \frac{19}{100} = 0.19 \quad \frac{251}{1000} = 0.251$$

$$\frac{63}{10000} = 0.0063$$

LATIN, *decimus*, a tenth; *fractio*, a breaking

decimal point a point used to separate the whole number from the fraction part of a number written in the shorthand form known as **decimal**. The position of each number symbol indicates the particular power of ten. The first position to the left of the point indicates units, the second indicates tens, the third hundreds, and so on. The first position to the right indicates tenths, the seconds hundredths, and so on.
Example 43.26 is a shorthand way of writing
$4 \times 10 + 3 \times 1 + 2 \times {}^1\!/_{10} + 6 \times {}^1\!/_{100}$
LATIN, *decimus*, a tenth; *punctus*, pricked

decimal system a way of counting using units, tens, hundreds, and so on. A point is used to separate the units from the tenths. The position to the left or the right of the point indicates the particular "ten group" (units, tens, hundreds, and so on, or tenths, hundredths, and so on).
Example Two hundred and seventy three can be written 273 ($= 2 \times 100 + 7 \times 10 + 3 \times 1$).
LATIN, *decimus*, a tenth; GREEK, *sustema*, organized whole

decode to convert a coded message into an understandable one.
Example Can you decode the message TFOE IFMQ? Replace each letter by the letter before it: i.e., T → S, F → E, O → N, E → D, etc. gives SEND HELP.
LATIN, *de*, from; *codex*, manuscript volume

deduce to reach (a conclusion) by reasoning.
Example If $x + 6 = 10$, then by subtracting 6 from both sides, we can deduce that $x = 4$. (*see also* DEDUCTION).
LATIN, *deducere*, to lead away; infer logically

deduction the process of reasoning or drawing conclusions from stated **assumptions** or **principles**. The act of going from a general situation or condition to a specific one.
Example From the principles of **congruent** triangles, one can deduce that a triangle with 2 sides equal must have its 2 opposite angles equal.

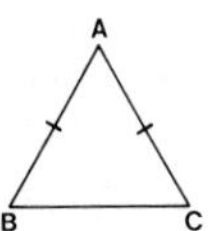

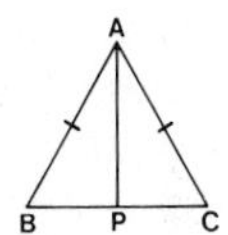

Since AP (drawn to bisect $\angle BAC$) is common to both triangles BAP and CAP and these two triangles are congruent, i.e. $\Delta\ BAP \equiv \Delta\ CAP$ (*S.A.S.*—*s*ide *a*ngle *s*ide), then $\angle B = \angle C$.
LATIN, *deducere*, to lead away, infer logically

degree **1.** In **geometry** and **trigonometry**, a 360th part of the **circumference** of the circle, or the angle obtained by dividing a circle into 360 equal parts. **2.** In **algebra**, of an **equation**: the highest **power** of any **term** contained in the equation.
Examples
1.

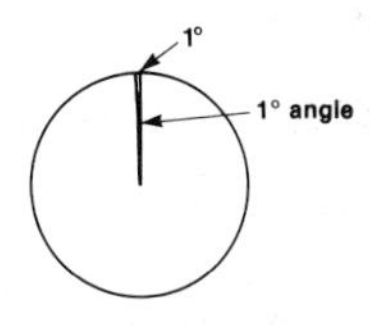

2. $3x + 5y = 7$ is an equation of degree 1, since both x and y are to the first power. $4x^2 + 3y^3 = 1$ is an equation of degree 3, since y is of the third power, which is higher than x's power.
Note: The degree of the term $5x^2y$ is 3, since power of x (2) together with power of y (1) is 3.
LATIN, *degradus*, a step down

degree(s) of freedom (statistics) any **independent variable** in a set of variables which represent statistical data. In any set of n values the number of degrees of freedom is $n - 1$.
Example In the following table of 10 observations (where each observation represents a variable) there are 9 degrees of freedom, since knowing the **average** and 9 other values means the 10th value is determined.

10	1	7	16	13	2	5	19	17	4

LATIN, *degradus*, a step down; OLD ENGLISH, *freodom*, freedom

de Moivre, Abraham (1667–1754) French mathematician, contemporary and friend of Isaac **Newton**. De Moivre's work includes developments in the study of **probability**, **calculus** and **trigonometry**. He was familiar with the equation of the normal **frequency** used in **statistics**. His most famous discovery was the formula that is now known as his theorem.

de Moivre's theorem a formula for working out the **power** of a **complex number** expressed in **polar form**. If $z = r(\cos\theta + i\sin\theta)$, then $z^n = r^n(\cos n\theta + i\sin n\theta)$. Since $e^{i\theta} = \cos\theta + i\sin\theta$, then z^n can also be written as $r^n e^{in\theta}$.
Example If $z = 2(\cos A + i\sin A)$, then
$z^3 = 8(\cos 3A + i\sin 3A)$.
after de Moivre, Abraham

denary scale (or system) a way of representing figures, numbers, quantities using a fixed number 10. (*see* DECIMAL SYSTEM).
Examples 243 in the denary scale or system means: $2 \times 100 + 4 \times 10 + 3$. Similarly 0.638 means the following: $6 \times \frac{1}{10} + 3 \times \frac{1}{100} + 8 \times \frac{1}{1000}$. The position to the right or left of the decimal point has definite (and different) meanings.
LATIN, *denarius*, consisting of ten

denominator the part of the **fraction** below the fraction line.
Example 3 is the denominator of $\frac{2}{3}$; b is the denominator of $\frac{a}{b}$.
LATIN, *denominare*, to name completely

dependent that which depends on something else.
Example In $y = 2x^2$, the value of y is dependent on the value given to x.
LATIN, *dependere*, to hang down

dependent variable a **variable** whose value depends on the value given to another variable.
Example In the equation $y = x^3$, y is a dependent variable, and x is an independent variable, since the value of y depends on the value given to x.
LATIN, *dependere*, to hang down; *varius*, changeable

depression in **mathematics**, the angle measured from the **horizontal** to the line joining viewer to object.
Example

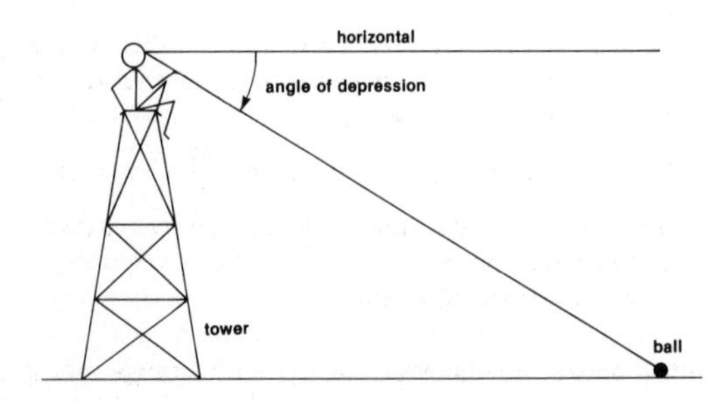

LATIN, *deprimere*, to press down

depreciation the amount of money (usually) lost due to the selling price of an item, once purchased, tending to go down (due to age, etc.). Over a number of years an item is sometimes said to depreciate at the rate of, say, 20% per annum. If r is the rate of depreciation, A the initial value, and D the depreciation, then after n years, $D = A(1 - (1 - r)^n)$. This is derived as follows: let A_1 = original value, A_2 = final value, r = rate of depreciation.
Then $A_2 = A_1(1 - r)^n$
and $D = A_1 - A_2$
$= A_1(1 - (1 - r)^n)$
Examples
1. A truck was bought in 1984 for $12 000. In 12 months it was worth $10 000. The depreciation was $2000 for one year.
2. If r = 20% and the initial value $6000, then after three years $D = 6000(1 - 0.8^3)$
$= 6000(1 - 0.512)$
$= 6000(0.488)$
$= 2928$.
LATIN, *depretium*, price going down

derivative (of a function) the rate of change of one **variable** with respect to another at some instant. If $y = f(x)$ (in other words y is a function of x) then the derivative of y with respect to x is written either $\frac{dy}{dx}$ or $f'(x)$, and is the **gradient** of the **tangent** of $y = f(x)$ at the instant in question.
Example

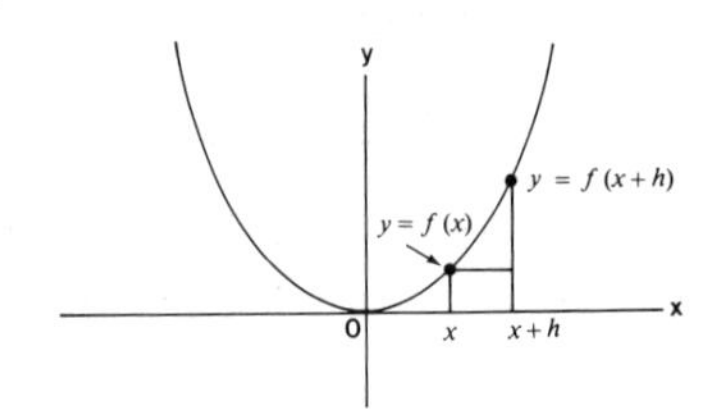

As h approaches 0, the value of
$$\frac{f(x + h) - f(x)}{h}$$
moves closer to the value of the derivative of the function $f(x)$ at the point (x,y). This becomes the value of the gradient or slope of the tangent to the curve at the point (x,y).
LATIN, *derivare*, to draw off

derive **1.** To come from or receive from a source. **2.** (mathematics) To arrive at or get by reasoning.
Examples
1. Mathematics is a word that derives from the Greek word *mathema*, meaning science.
2. We can derive the identity $a^2 - b^2 = (a+b)(a-b)$ by simply multiplying out the factors $(a+b)(a-b) = a^2 + ba - b^2 - ba = a^2 - b^2$ (since $ba = ab$).

LATIN, *derivare*, to draw off

Descartes, René (1596–1650) French mathematician with many important discoveries to his name. His most famous contribution to mathematics was his invention of **analytical geometry** which represented geometry in an algebraic manner; and the formalising of the **cartesian** number plane.

descending going down.
Example $(1 + x)^3$ can be expressed in descending powers of x: $x^3 + 3x^2 + 3x + 1$
LATIN, *descendere*, to climb down

determinant a special number or **expression** (in **algebra**) calculated from the elements of a square **matrix**.
If a matrix $M = \begin{pmatrix} x & y \\ z & w \end{pmatrix}$ then the determinant of M is $xw–yz$. Note: $|M|$ or det$M$ are other shorthand ways to represent the determinant of M.
Example
$A = \begin{pmatrix} 1 & 3 \\ 5 & 7 \end{pmatrix}$, then det$A$ = 7 − 15 = −8.
LATIN, *determinare*, to limit

deviation (statistics) **1.** The difference between one **set** of numbers and their **mean**. **2.** Any variation from a trend. (*see also* STANDARD DEVIATION).
Example Given the set

Monthly Rainfall								
mm	52	71	33	21	50	67	45	37

the **average** (mean) = 47, and the deviations are +5, +24, −14, −26, +3, +20, −2, −10. The mean deviation is the sum of the deviations divided by the number of deviations (without taking the sign into account, i.e.,

$$\text{Mean deviation} = \frac{5+24+14+26+3+20+2+10}{8} = \frac{104}{8} = 13$$

LATIN, *deviare*, to turn out of the way

diagonal **1.** A straight line drawn from one **vertex** of a **polygon** (or solid figure) to another vertex not adjacent to the first one. **2.** (of a matrix) The set of **elements** of a matrix which form a diagonal.
Examples
1.

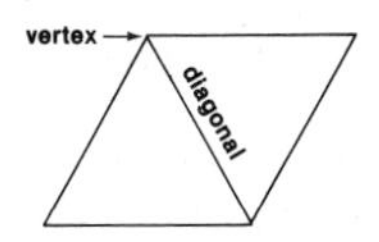

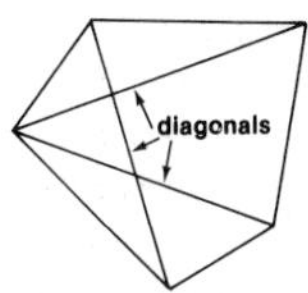

2. If $A = \begin{pmatrix} 1 & 4 & 6 \\ 7 & -8 & 3 \\ 2 & 10 & 7 \end{pmatrix}$, then $\begin{matrix} 1 \\ -8 \\ 7 \end{matrix}$ is a diagonal known as the leading diagonal).
GREEK, *diagonios*, angle to angle

diameter any line joining two points on the **circumference** of a circle and passing through the centre. All diameters of the same circle are the same length. A diameter is twice the **radius**.
Example

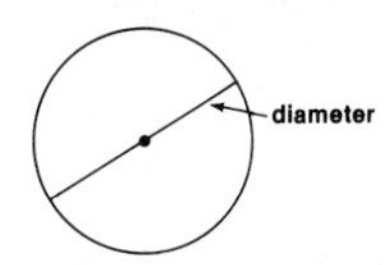

GREEK, *diametros*, measuring across

difference the amount by which one quantity is greater or less than another.
Example The difference between 10 and 7 is 3. Another way of saying this is 10 minus 7 gives a difference of 3.
LATIN, *differre*, to carry in different directions, be different

differential **1.** Of or relating to finding or working out rates of change (of **functions, variables**). **2.** A very small change in a **variable** or **function**. Symbols often used are d or ∂.
Examples
1. Differential **calculus** is the study of the rates of change of variables or functions, and was invented independently of each other by Sir Isaac **Newton** and Gottfried **Leibniz** in the 17th century.
2. $\frac{dy}{dx}$ is a shorthand way of writing the **ratio** of a very small change in the function y due to a very small change in the variable x. The symbol d is used when y is a function of x only, i.e. $y = f(x)$, which means x is the only independent variable. The symbol ∂ (e.g. $\frac{\partial y}{\partial x}$) is used to mean the ratio of a very small change in y due to a very small change in x when y is a function of more than one independent variable, i.e. $y = F(x,z,w)$ for example.
∂ is called the partial differential, and $\frac{\partial y}{\partial x}$ is the partial **derivative** of y with respect to x. (*see also* DIFFERENTIAL COEFFICIENT).
LATIN, *differens*, different

differential coefficient the rate of change of a function with respect to its **independent variable** at a particular instant or point. If $y = f(x)$, then $\frac{dy}{dx} = f'(x)$ is the symbol for the differential coeffi-

cient of $f(x)$ at the point x. Also known as the **derivative**.
Example The differential coefficient of $y = x^2$ is worked out to be $2x$, and can be represented as the gradient of the curve $y = x^2$ at the point x (*see* DERIVATIVE).

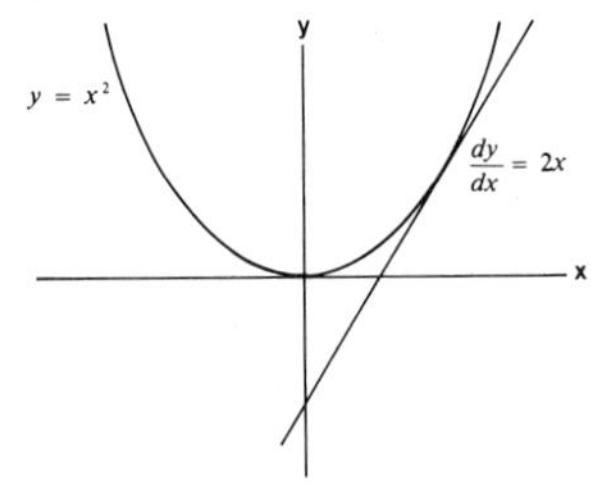

At the point $x = 3$, the differential coefficient (d.c.) of $y = x^2$ is 6.
LATIN, *differens*, different; *coefficiens*, effecting together

differential equation an equation containing **differentials** or **derivatives**.
Example $\frac{dy}{dx} = xy$, where $\frac{dy}{dx}$ stands for the derivative of the function y, with respect to x.
LATIN, *differe*, to carry in different directions, be different; *aequare*, to equal

differentiate the process of working out the **rate** of change or **derivative** of a quantity.
Example If $y = f(x)$, then we can differentiate y with respect to x (i.e. find the rate at which y changes when x is changing) by the following approach.
Let x and $x + h$ be two points close together (in a **coordinate** axis system) on the x-axis. The corresponding values of y are $f(x)$ and $f(x + h)$. The change in y due to a change in x can then be written as

$$\frac{f(x + h) - f(x)}{x + h - x} = \frac{f(x + h) - f(x)}{h}$$

As h approaches 0,

$$\frac{f(x + h) - f(x)}{h}$$

becomes the derivative of $f(x)$ at the point x, and is the rate of change of $y = f(x)$ with respect to x at that point.

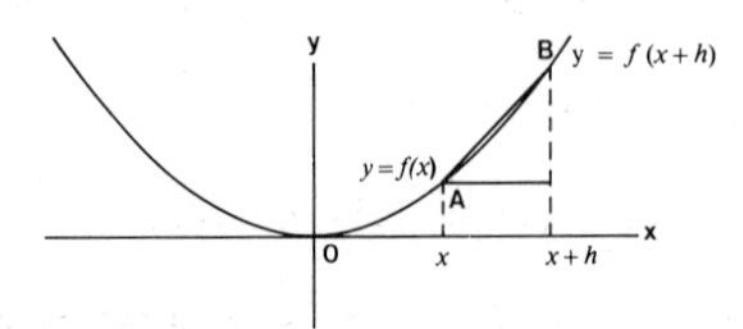

It can be seen that

$$\frac{f(x + h) - f(x)}{h}$$

is the **slope** or **gradient** of the line joining the two points A and B. As h approaches 0, this gradient becomes the gradient of the tangent at the point x.
LATIN, *differens*, different

digit a single figure used to represent a counting number (or unit).
Example 0, 1, 2, 3, 4, 5, 6, 7, 8, 9 are the digits used in the decimal system. 0, 1 are the digits used in the binary system.
LATIN, *digitus*, a finger

digital computer a machine which calculates in digits (*see also* ANALOG COMPUTER).
Example Most digital computers are designed to use the binary system of counting, i.e. only two digits, 0 and 1.
LATIN, *digitus*, a finger; *computare*, to reckon together

dihedral angle the angle between two planes. It is found by measuring the angle formed by two straight lines, one in each plane, which are perpendicular to the line formed by the intersection of the two planes.
Example

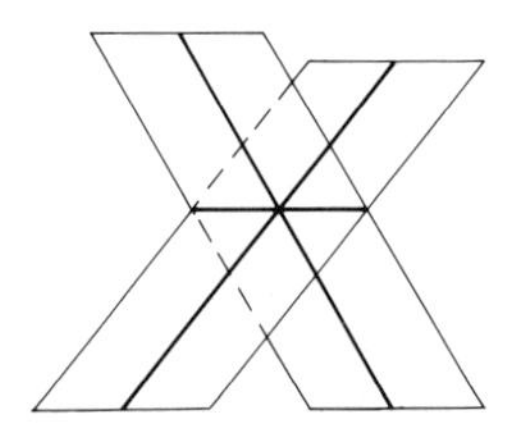

GREEK, *di*, two, twice; *hedra*, base, seat

dilation (also **dilatation**) *see* ENLARGEMENT.

dimension **1.** (general) A measure of extent in space. **2.** (mathematics) The number of coordinates required to represent the points on a line, figure or solid. A point on a line is said to be one-dimensional, a point in a plane two dimensional, a point in space three dimensional. Mathematicians generalise these ideas to talk of n-dimensional space.
Examples
1. The dimensions of a rectangle are its length and breadth.
2. A point in a plane can be located using two dimensions represented by two intersecting lines at right angles. P can be located by measuring 3½ units along the x axis, and 3 units up the y axis.

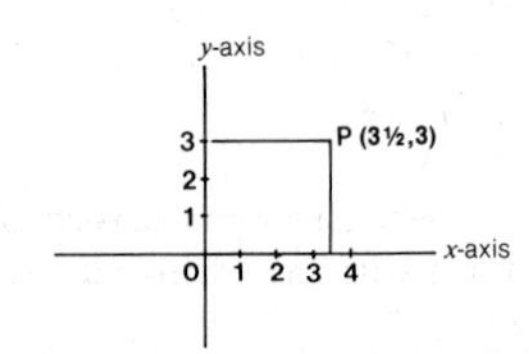

LATIN, *dimensio*, a measuring

diophantine equation an equation, containing two or more **variables**, whose solution is in terms of **integers** or **fractions**. The number of **solutions** is not limited.
Example $x - y^2 = 7$; $x = 16$, $y = 3$, etc.
after Diophantus, Greek mathematician; LATIN, *aequatio*, equation.

directed number a signed number, either positive or negative. On a number line, the numbers to the right of 0 are +, and those to the left are −.
Example

−4 −3 −2 −1 0 +1 +2 +3 +4

LATIN, *dirigere*, to steer or direct; *numerus*, number

direction angles the angles which a line makes with the positive directions of the x, y and z axes (z if in three **dimensions**). Symbols sometimes used are α, β, γ. The angles determine the line's direction with respect to three axes.
Example

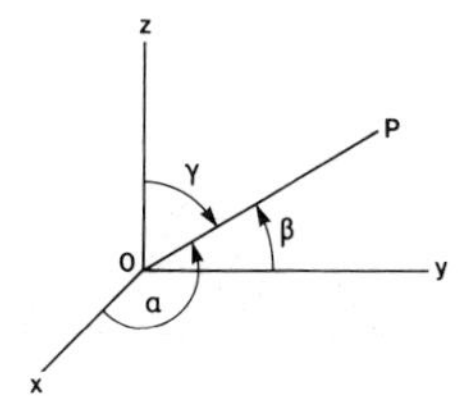

OP has the direction angles α, β, γ.
LATIN, *dirigere*, to steer; *angulus*, corner

direction cosines the **cosines** of the angles a line makes with the positive directions of the x, y and z axes. The three values determine the direction of the line in relation to the three axes (sometimes shortened to d.c.).
Example

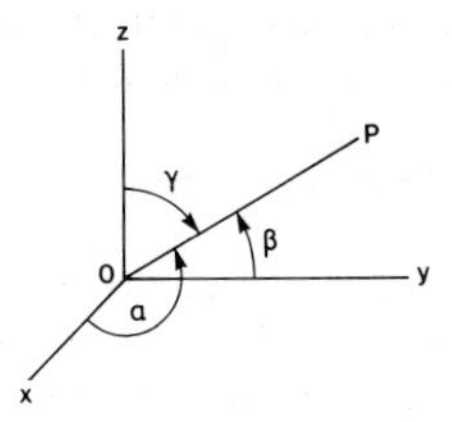

The direction cosines of OP are $p = \cos\alpha$, $q = \cos\beta$, $r = \cos\gamma$, and often written: (p, q, r).
LATIN, *dirigere*, to steer

direction ratios the ratios of the **cosines** of the three angles a line makes with the positive directions of the x, y and z axes.

Example

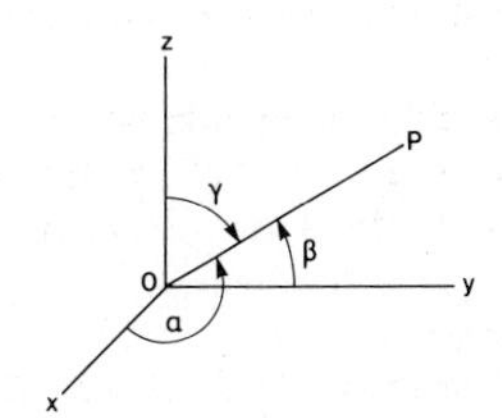

$$\text{If } \cos\alpha = p = \frac{a}{L}$$

$$\cos\beta = q = \frac{b}{L}$$

$$\cos\gamma = r = \frac{c}{L}$$

then $\cos\alpha : \cos\beta : \cos\gamma = p:q:r = a:b:c$.
$a:b:c$ are the direction ratios of the line OP, and $L = OP$.
LATIN, *dirigere*, to steer; *ratio*, reckoning, reason

directrix the fixed line which is part of the definition of a **parabola** (usually) or other **conic** sections. (A parabola is the path traced by a point moving at equal distances from a fixed line and a fixed point.)
Example

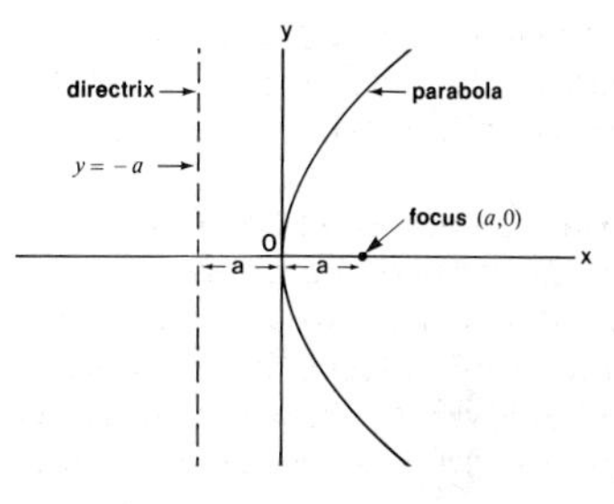

$$y^2 = 4ax$$

LATIN, *dirigere*, to direct or steer

discrete (of a **variable**) referring to separate values. A discrete variable can only take on separate values. It is the opposite to **continuous**.
Example The variable N representing the number of cars in a parking lot at any one time can only take on positive whole number values (fractions of cars are excluded). N is therefore a discrete variable. Temperature, on the other hand, is a quantity which is not discrete, but continuous.
LATIN, *discretus*, separate

discontinuous antonym of **continuous**

discount an amount to be substracted from the current or list price of an item. It is usually worked out as a percentage of the current price.
Example The list price of the tennis racquet is \$30. A discount of 30% is \$9. After taking off this discount of 30%, the discounted price was \$21.

LATIN, *dis*, apart, reverse; *computare*, to add, sum up

discriminant (of a **quadratic** equation) an **expression** of the form $b^2 - 4ac$ with reference to any equation of the form $ax^2 + bx + c = 0$. If the discriminant is positive or zero, then the equation can be solved for **real** values.
Example The discriminant of the equation $x^2 + 3x + 2 = 0$ is $3^2 - 4 \times 2 \times 1 = 9 - 8 = 1$. Therefore, the equation has real **roots**.
LATIN, *discrimen*, intervening space, distinction

disjoint having no **elements** in common.
Example The following two **sets** are disjoint;
$A = (1, 2, 3, 4, 8)$
$B = (5, 6, 7, 9)$
since they have no member in common. $A \cap B = \phi$ or { }, i.e. the **empty set**.
Disjoint sets can also be shown by diagrams;

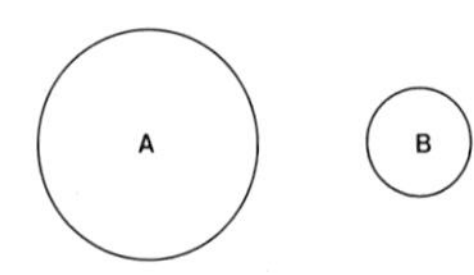

The following two sets are not disjoint, since they do have one or more members in common.

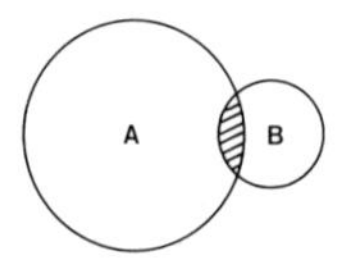

OLD FRENCH, *desjoint*, disjointed

displacement a change in position. It has distance and direction.
Example The displacement of A (with respect to T) is the distance TA and the direction is 30° below the horizontal.

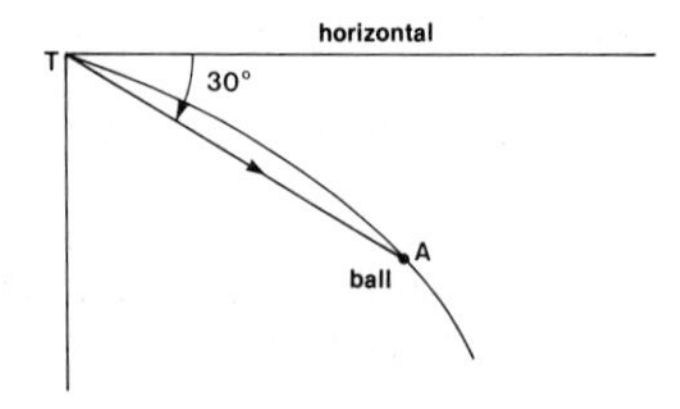

LATIN, *dis*, apart; *platea*, a broad street or place

distance formula (coordinate geometry) if two points in a plane are represented by (x_1, y_1) and (x_2, y_2) the distance S between them can be defined as:

$$\sqrt{(x_1 - x_2)^2 + (y_1 - y_2)^2}$$

The formula for the distance between two points in three dimensional space can be defined as:

$$S = \sqrt{(x_1 - x_2)^2 + (y_1 - y_2)^2 + (z_1 - z_2)^2}$$

Example

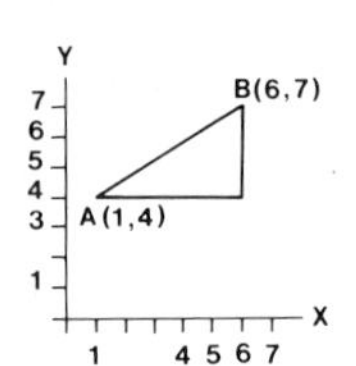

$$\text{distance } AB = \sqrt{(6 - 1)^2 + (7 - 4)^2} = \sqrt{(25 + 9)} = \sqrt{34} \text{ units}$$

If A (3,1,4) and B (4,2,1) are points in three dimensional space, then:

$$\text{distance} = \sqrt{(3 - 4)^2 + (1 - 2)^2 + (4 - 1)^2} = \sqrt{(1 + 1 + 9)} = \sqrt{11} \text{ units}$$

LATIN, *distans*, standing apart; *formula*, a small form (in the sense of a set or fixed order of words)

distribution (statistics) a **set** of numbers representing a group of observations or measurements. When grouped according to how many times a particular observation or measurement occurs, the set is called a **frequency distribution**.
Example The following set of 13 measurements is an example of a distribution:

Size of shoe	4, 7, 10, 9½, 6, 8½, 7, 8, 5½, 6, 7, 9, 8

When grouped together as below it becomes a frequency distribution.

No. of people	1	0	0	1	2	0	3	1	2	1	1	1	1
Shoe Size	4	4½	5	5½	6	6½	7	7½	8	8½	9	9½	10

LATIN, *distribuere*, to assign or grant parts of

distributive law **1.** In multiplication, the principle that allows the **multiplier** to be applied separately to each **term** in an **expression**, i.e. $a(b + c) = ab + ac$.
2. In general, the principle which states that given two binary operations * and @, if * is distributive over @, then

$$A*(B @ C) = (A*B) @ (A*C)$$

Examples
1. $4(3 + 7) = 4 \times 3 + 4 \times 7 = 12 + 28 = 40$. Note: multiplication is distributive over addition, but addition is not distributive over multiplication, as $4 + (3 \times 7) \neq (4 + 3) \times (4 + 7)$.
2. With sets, $\cap$ (intersection) is distributive over $\cup$ (union).

$$A \cap (B \cup C) = (A \cap B) \cup (A \cap C)$$

Note: $A \cap C = 0$ in this example.

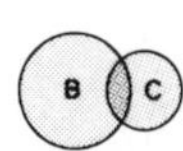

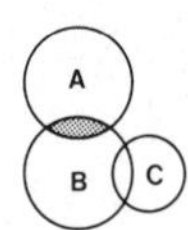

In the diagrams above, $B \cup C$, is the union of B and C.

$$A \cap (B \cup C) = (A \cap B) \cup (A \cap C)$$

In this example, $A \cap C = 0$.
LATIN, *distribuere*, to assign or grant parts of

divergent **1.** Drawing apart from a common point. **2.** Failing to approach a **limit**; not **convergent**.
Examples
1.

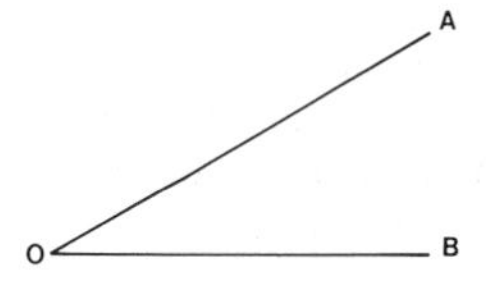

OA and *OB* are divergent.
2. The series 1 + 2 + 3 + 4 + ... is divergent since there is no limit for the sum. The series $1 + \frac{1}{2} + \frac{1}{3} + \frac{1}{4} \ldots$ can also be shown to be divergent.
LATIN, *divergere*, to turn aside

dividend **1.** The number being divided. **2.** A share of profits received by a stockholder or similar.
Examples
1. In the division

$$3\,\overline{)1983}\quad 661$$

1983 is the dividend.
2. I received a dividend of $15 from my shares in the New Blue Moon Mining Corporation N/L.
LATIN, *dividendum*, thing to be divided

division the process of finding how many times one number or quantity is contained in another (it is the **inverse** operation of **multiplication**). The sign for division is ÷.
Examples
1. The division of 12 fruit buns among 4 boys equally is 3, or 12 ÷ 4 = 3.

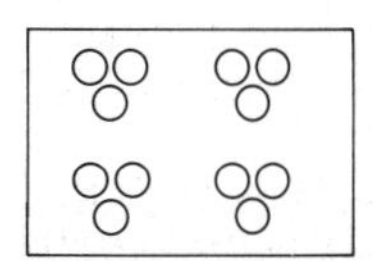

2. $6 \div \pi = 6 \times \frac{1}{\pi}$ (where $\pi = 3.1419\ldots$)

LATIN, *dividere*, to separate, divide

divisor the number you are dividing by.
Example In the division

$$3\,\overline{)1983}\quad 661$$

3 is the divisor.
LATIN, *dividere*, to separate, divide

dodecagon a **polygon** bounded by 12 straight lines.
Example

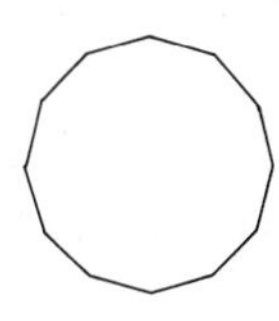

A regular dodecagon has all 12 sides equal and all interior angles equal.
GREEK, *dodeka*, twelve; *polugonos*, having many angles

dodecahedron a solid shape with 12 faces. All the faces of a regular dodecahedron are regular **pentagons**.
Example

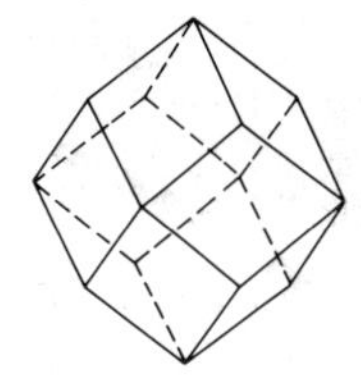

A non-regular dodecahedron
GREEK, *dodeka*, twelve; *hedron*, base, face

domain a **set** of numbers or **elements**, which a **variable** can take on.
Example If the set 1, 2, 3, 4, 5, 6, 7 is the domain of x, then $x^2 + 1$ has the values 2, 5, 10, 17, 26, 37, 50.
LATIN, *dominium*, property, ownership rights.

dot product *see* SCALAR PRODUCT.

duodecimal of or relating to twelve.The duodecimal system is a system of counting where the **base** is 12.
Example The digits of the duodecimal system are 0, 1, 2, 3, 4, 5, 6, 7, 8, 9, *t*, *e* where *t* and *e* stand for 10 and 11 in the decimal system, i. e. $2e + 3 = 25$ (base 10), and $t + e = 21$ (base 10).
LATIN, *duodecim*, twelve

Ee

E 1. The fifth letter of the English alphabet. 2. The **base** of the **natural** or Naperian system of **logarithms,** e is an **irrational** number whose value is approximately 2.71828.
Examples
If $A = e^2$, then the logarithm of A (with respect to the base e) is 2.
The logarithm of 10 with respect to the base e is 2.301... since $e^{2.301...} = 10$.
GREEK, E, ϵ (epsilon) [fifth letter of Greek alphabet]

eccentricity 1. Deviation from the normal or expected. 2. The **ratio** of the distance of any point on a **conic** section from a **focus** to its distance from the corresponding **directrix**.
Examples
1. The degree of eccentricity was high.
2.

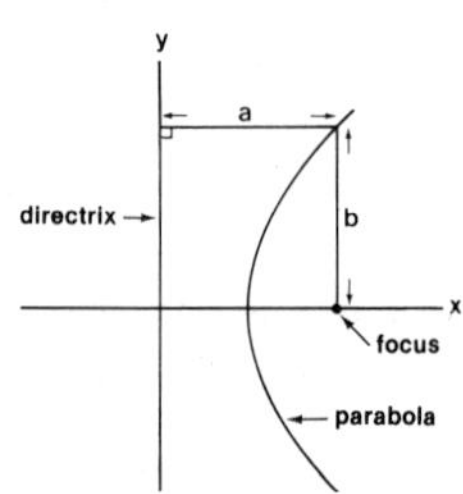

The eccentricity of the parabola is $\frac{b}{a} = 1$.

For an ellipse, $\frac{b}{a} < 1$; for an hyperbola $\frac{b}{a} > 1$.
GREEK, *ekkentros*, out of centre

element 1. A **member** of a **set**. 2. A letter or number in a **matrix**. 3. Of a cylinder or cone, the **generating** line of the surface.
Examples
1. If a set (S) consists of the last four letters of the alphabet, then the elements of S are w, y, x, z.
2. If $M = \begin{pmatrix} a & 1 \\ b & 2 \end{pmatrix}$ then a, 1, b, 2 are elements of the matrix M.
3.

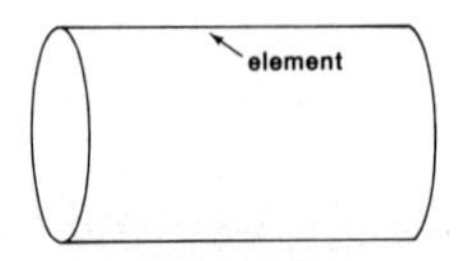

LATIN, *elementum*, a basic thing, first principle

elevation the angle measured upwards from the horizontal to an object.
Example

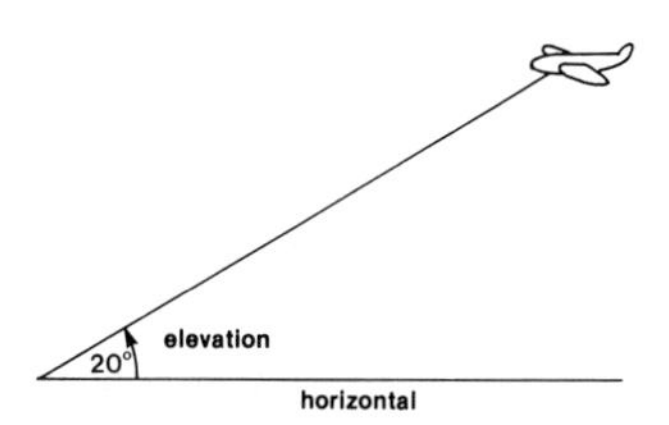

LATIN, *elevare*, to raise up, lighten up

ellipse the path of a point moving so that the sum of the distances from the moving point to two fixed points is constant. Also the curve formed by a plane intersecting a cone at an angle less than the angle between the cone's side and the base. When an ellipse is formed by a plane cutting a cone at an angle, the angle made by the plane to the cone's base is less than, or "comes short of", the angle between the cone's slant edge and its base.
Example

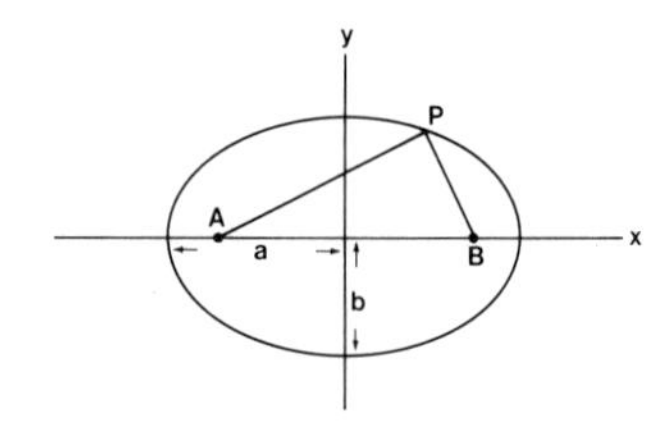

$AP + PB$ = constant. The equation of the ellipse shown is:

$$\frac{x^2}{a^2} + \frac{y^2}{b^2} = 1$$

Note: The line through A and B is called the principal axis of the ellipse.

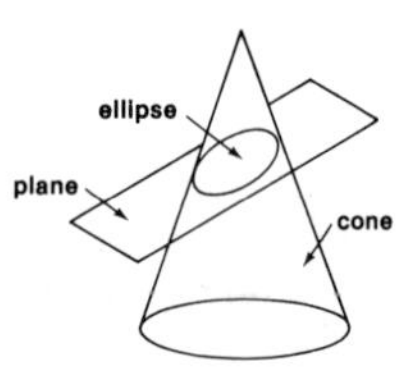

GREEK, *ellipsis*, a coming short

empirical relying upon or coming from observation or experiment. Opposite to theoretical.
Example Many mathematical rules were first discovered using empirical methods. The method of bisecting a line using a **compass** is shown as one.

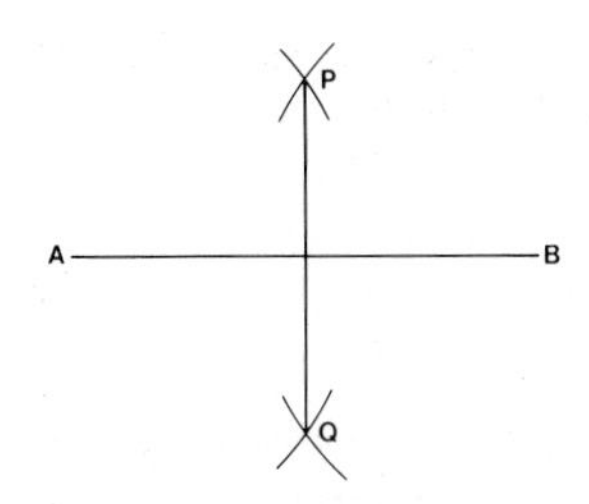

Four **arcs** (parts of a circle whose radius is bigger than half AB) are drawn using a compass to cut as shown at P and Q. The line PQ cuts AB exactly in half.
GREEK, *emperikos*, from experience

empty set a set with no **members** in it. It is represented by { }, ( ) or by the symbol ϕ. Also called the **null set.**
Example The set of all square circles is an empty set. If A is the set (2, 4, 6, 8) and B is the set (1, 3, 5, 7), then the set formed by the **intersection** of A and B is an empty set.
OLD ENGLISH, *aemtig*, at leisure, unoccupied; LATIN, *secta*, a following

end point one of the two points defining an **interval**. It may or may not be included in the interval.
Example

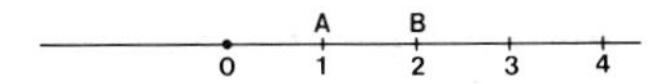

A(1) and B(2) are the end points of the interval AB consisting of all points between A and B including end point A(1) but not B(2).
OLD ENGLISH, *ende*, limit, boundary; LATIN, *punctus*, a point

enlargement a **transformation** which **maps** a shape on to a similar shape, from a fixed point P using a **scale factor** S. (Note: the scale factor is sometimes called the enlargement factor.)
Example

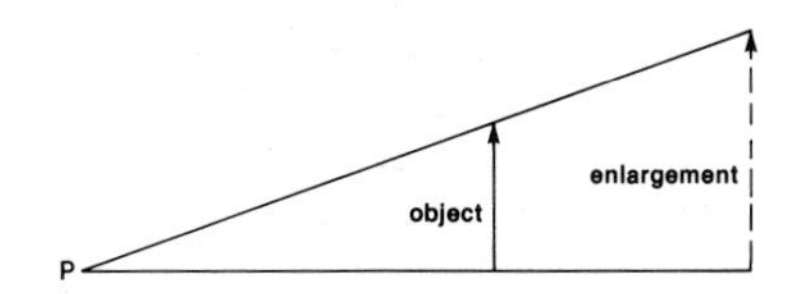

If S is greater than 1, then the transformation is an enlargement. If S is a fraction less than 1, then the transformation is a shrinking. If S is a negative, then the transformation also involves a rotation of 180° about P.
OLD FRENCH, *enlargir*, to make large

enlargement matrix a matrix of the form $\begin{pmatrix} s & 0 \\ 0 & s \end{pmatrix}$ where s is the linear **scale factor**. The area **scale** factor is s^2 (also known as the determinant of the matrix).
Example The transformation

$$\begin{pmatrix} x \\ y \end{pmatrix} \rightarrow \begin{pmatrix} s & 0 \\ 0 & s \end{pmatrix}\begin{pmatrix} x \\ y \end{pmatrix} = \begin{pmatrix} sx \\ sy \end{pmatrix}$$

represents an enlargement with centre at the origin, and scale factor s (*see* MATRIX MULTIPLICATION). i.e. OP is transformed to OQ (sx, xy).

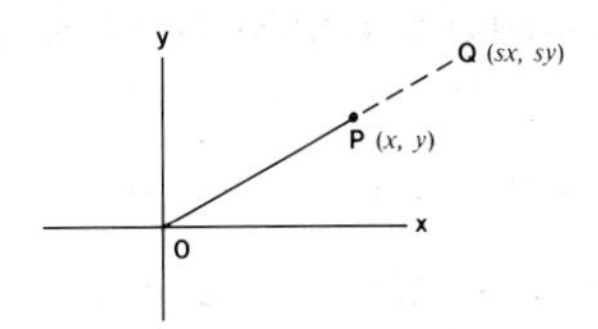

OLD FRENCH, *enlargir*, to make large; LATIN, *matrix*, womb

envelope any curve or surface which at each of its points is **tangent** to a member of a family of curves or surfaces.
Example

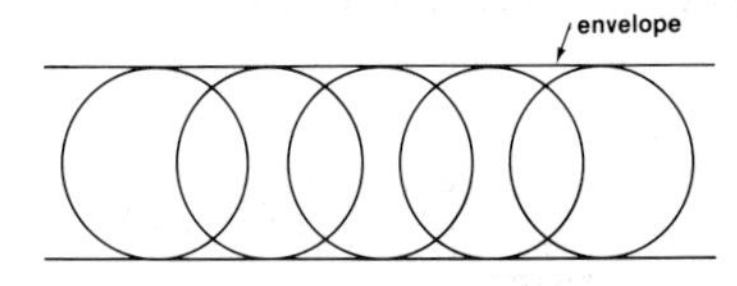

In the example the envelope of a set of circles whose centres all lie on one straight line is a pair of **parallel** lines.
OLD FRENCH, *enveloper*, to wrap up in

equation a statement which says that one **expression** is equal to another.
Examples $y = 4x + 3$, $z + 5 = 12$, $p^2 + q^2 = 3$. There are **linear** equations such as $y = 4x + 3$, **quadratic** equations such as $x^2 + 2x + 1 = 0$ or **simultaneous equations**, etc.
LATIN, *aequatio*, an equalling

equation of a line the **relation** which is satisfied by all the **coordinates** of the points of a line.
Examples

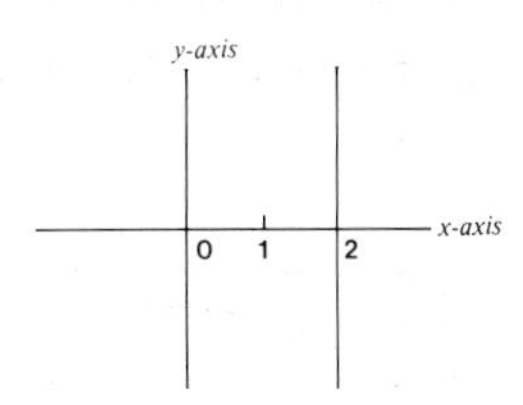

Equation of the line passing through (2,0) and parallel to the y axis is $x = 2$.

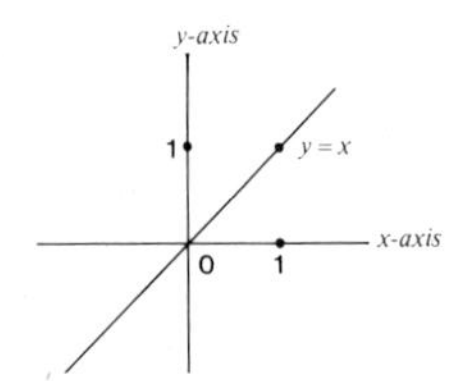

Equation of the line is $y = x$.
The general equation $y = ax + c$ of a straight line on a graph is a linear equation where c is the intercept on the y-axis and a is the slope or **gradient.**
LATIN, *aequatio*, an equalling; *linea*, a line

equi- equally, having equality.
Example Equiangular, equidistant.

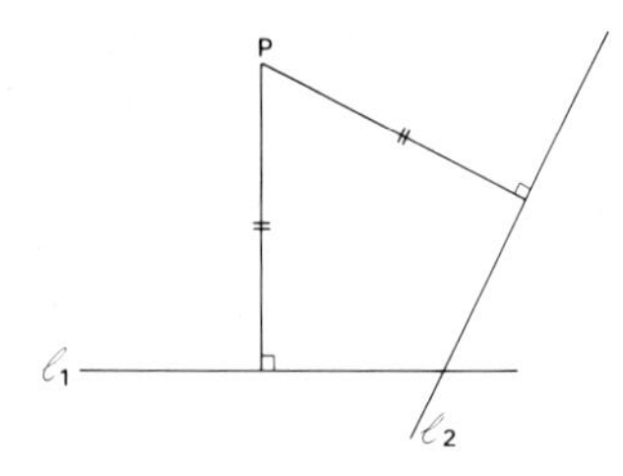

P is equidistant from lines l_1 and l_2 (i.e. the perpendiculars from P to each line are equal in length).

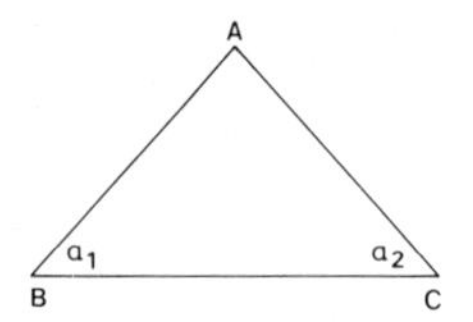

If $AB = AC$, then the angles opposite those sides are equiangular (i.e. $\alpha_1 = \alpha_2$).
LATIN, *aequus*, equal

equilateral having equal sides.
Example An equilateral triangle is a triangle all of whose sides are equal.

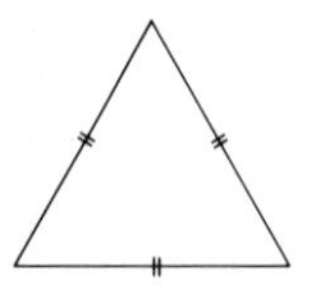

LATIN, *aequus*, equal; *latus*, side

equivalent fractions fractions which **cancel** down to the same fraction.
Example $^{15}/_{18}$ and $^{20}/_{24}$ are equivalent since both can be reduced to $^5/_6$. Similarly $^1/_3$, $^2/_6$, $^3/_9$, $^4/_{12}$, $^5/_{15}$ are equivalent.
If $\frac{x}{y}$ is any fraction, then $\frac{kx}{ky}$ is an equivalent fraction, where k is a number not equal to zero.
LATIN, *aequivalere*, to be equal in value; *fractio*, a breaking

equivalent sets two or more **sets** which each have the same number of **elements**.
Example
$A = \{a, b, c, d, e, f\}$ and $B = \{2, 4, 6, 8, 10, 12\}$ are equivalent since there are six elements in each set. Another way of seeing this is to match each element in A with a similarly positioned element in B so that every element is paired.
LATIN, *aequivalere*, to be equal in value; *secta*, a following (of people)

Eratosthenes (*c.* 230 BC) Greek mathematician who was a contemporary and friend of Archimedes. He was gifted in astronomy, geography, history, philosophy, poetry and athletics as well as mathematics. Eratosthenes is particularly remembered for his sieve, a method for finding all the **prime numbers** less than n. He also produced a mechanical solution to the "duplication of the cube" problem and a method for approximately calculating the circumference of the earth.

Euclid (*c.* 300 BC) most famous of Greek mathematicians; responsible for the writing of the *Elements*, a collection of thirteen books which contain proofs of geometrical theorems. The proofs have provided the basis for the study of **euclidean geometry** throughout the ages.

euclidean geometry the geometry of **Euclid**, which deals with **points**, **lines**, and **planes**, together with the basic assumptions: (a) through any two distant points there is a line; (b) any line can be extended indefinitely; (c) all **right angles** are equal; and (d) only one line, **parallel** to the given line, can be drawn through a point not on the given line.
Example Ordinary geometry is euclidean geometry.
after Euclid, Greek mathematician; GREEK, *geometria*, measuring land

Euler, Léonard (1707–1783) Swiss mathematician who is said to be the most prolific writer in the history of mathematics—his name appears in every branch of the subject. Euler's discoveries include the result $e^{i\theta} + 1 = 0$, Euler's formula in elementary solid geometry that $V - E + F = 2$, (where V is number of vertices, E is number of edges and F the number of faces of a given solid) and many applications of mathematics to problems in physics, astronomy and music.

Euler's formula **1.** (for **polyhedra**) vertices (V) + faces (F) = edges (E) + 2. **2.** (for **networks**) **nodes** (N) + **regions** (R) = **arcs** (A) + 2.
Examples
1.

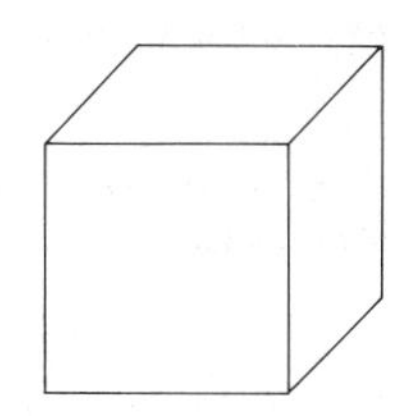

A cube has 8 vertices, 6 faces and 12 edges.
$$8 + 6 = 12 + 2$$
2.

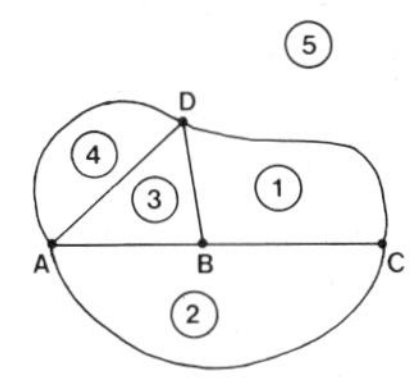

There are 4 nodes, 7 arcs and 5 regions.
$$4 + 5 = 7 + 2$$
after Euler, Swiss mathematician; LATIN, *formula*, a little form

evaluate find the value of.
Examples
If $x = 3$ and $xy = 4$, then we can evaluate $y(= \frac{4}{3})$.
If $x = 3$ and $y = 4$, then we can evaluate $2x + 3y$ (6 + 12 = 18).
LATIN, *ex*, out of; *valere*, to be worth

even function a function $f(x)$ such that $f(x) = f(-x)$.
Examples
$f(x) = x^2$ is an even function since
$(-x)^2 = x^2$
$= f(x)$.
$f(x) = x^3$ is *not* an even function since
$(-x)^3 = -x^3$
$\neq f(x)$.
Also $f(x) = x + x^2$ is not an even function since
$(-x) + (-x)^2 = -x + x^2$
$\neq f(x)$

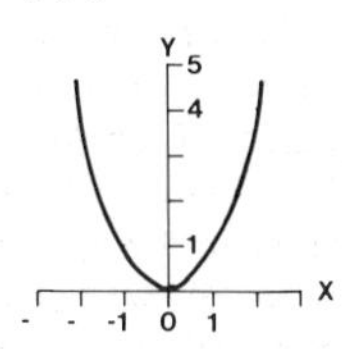

On the graph above $f(x) = x^2$ is shown: $f(x)$ is symmetrical about the line $x = 0$ (*See* SYMMETRY).

OLD ENGLISH, *efen*, even, level; LATIN, *functio*, activity, performance

even number a whole number which, when divided by 2, gives another whole number.
Example 2, 4, 6, 14, 1002 are all even numbers.
OLD ENGLISH, *efen*, even, level; LATIN, *numerus*, number

evens (probability) equal likelihood of an event taking place or not taking place.
Example The horse started at evens, but ended at 2:1 on. This means at the beginning you could outlay 1 unit to win 1 unit, but just prior to the race you could outlay 2 units to win 1.
OLD ENGLISH, *efen*, even, level

event (statistics and probability) something that happens. Also the outcome of something that occurs.
Example The event of tossing a coin has two possible results, heads or tails (assuming it does not land on its edge). The outcome of the birth event may be either a boy or a girl. It could also be said the birth of a boy (or girl) was an event.
LATIN, *evenire*, to come out, happen

expand (out) to write a compactly written **expression** or quantity in a detailed form which may contain many more **terms** or quantities.
Examples
1. We can expand $(a + b)^2$ as follows:
$(a + b)^2 = (a + b)(a + b) = a^2 + 2ab + b^2$.
Both $(a + b)(a + b)$ and $a^2 + 2ab + b^2$ are expanded forms.
2. 6! can be expanded to give $6 \times 5 \times 4 \times 3 \times 2 \times 1$.
3. $\frac{1}{1 - x}$ can be expanded to give a **series** $1 + x + x^2 + x^3 + \ldots$
LATIN, *expandere*, to spread out

expanded notation (also **extended notation**) the representation of numbers in terms of **powers** to the given base.
Examples In the **decimal** system, with 10 as the base, a number such as 347 can be written as $3 \times 10^2 + 4 \times 10 + 7$. 347 is a shorthand way of writing three hundred and forty seven. In the **octal** system (base 8), the number 1634 in expanded notation is $1 \times 8^3 + 6 \times 8^2 + 3 \times 8 + 4$.
LATIN, *expandere*, to spread out; *notare*, to mark out

expansion **1.** The result of writing a quantity or **expression** in a more spread out way, such as a sum of **terms**, or by a group of terms multiplied together. **2.** The process of obtaining this result.
Examples
1. 5 + 4 + 3 + 2 + 1 is an expansion of 15.
2. $x^3 + 3x^2 + 3x + 1$ or $(x + 1)(x + 1)(x + 1)$ are

both expansions of $(x + 1)^3$. $x^3 + 3x^2 + 3x + 1$ is obtained by the expansion of $(x + 1)^3$ i.e. multiplying $(x + 1)$ by itself three times.

3. $1 + x + x^2 + x^3 + \ldots$ is the expansion of $\frac{1}{1 - x}$

4.

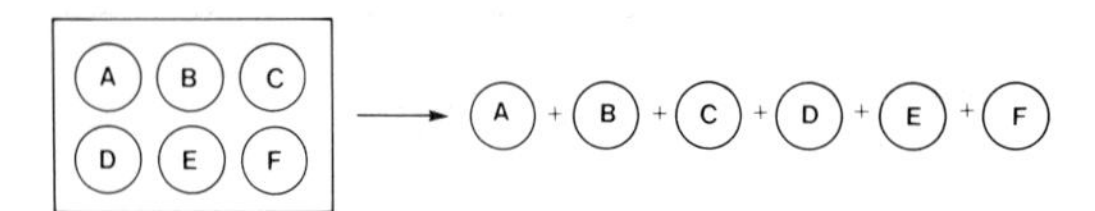

expansion

LATIN, *expandere*, to spread out

expected value the value regarded as the most likely to result from some theoretical calculation.
Example The expected value of 6 occurring in a throw of a die is ⅙ (there are 6 faces on the die, each with an equal chance of making it). From a throw of 3 coins the expected value of 3 heads turning up, 1 head and 2 tails, etc., turning up can be found from the following **expansion** of the **binomial** theorem: $(H + T)^3 = H^3 + 3H^2T + 3HT^2 + T^3$, and therefore out of 8 chances there is 1 chance of 3 heads (or 3 tails) turning up, 3 chances out of 8 that 2 heads and a tail will turn up, and 3 chances that 1 head and 2 tails will turn up.
LATIN, *exspectare*, to look out (for); *valere*, to be worth

explicit directly defined; specific.
Example The following two sets of numbers can be related by the explicit relationship $y = 2x$.

x	1	2	3	4	5	6	7	8	9
y	2	4	6	8	10	12	14	16	18

The antonym of explicit is **implicit**.
LATIN, *explicare*, to unfold

exponent the **power** to which a number is raised.
Example In $A = a^n$, the exponent is n (also known as the **logarithm** of A to the **base** a).
LATIN, *exponere*, to put forth or expose

exponential equation an equation in which the unknowns appear as exponents.
Examples
$3^x = 9$, $4^{2x+1} = 64$
Solutions may be obtained as follows: $3^x = 3^2$.
Equating indices, $x = 2$.
$4^{2x+1} = 4^3$
Equating indices gives $2x + 1 = 3$
$2x = 2$
$x = 1$
LATIN, *exponere*, to put forth; *aequatio*, an equalling

exponential function a function of the form a^x where a is a positive constant not equal to 1 and x is a variable.
Example $y = e^x$ is often called the exponential function where e (= 2.71828...) is the base of **natural logarithms**. e^x can be represented by an infinite series as follows:

$$e^x = 1 + x + \frac{x^2}{2!} + \frac{x^3}{3!} + \frac{x^4}{4!} \ldots + \frac{x^n}{n!} + \ldots$$

The graph of $y = e^x$ is as in the following diagram:

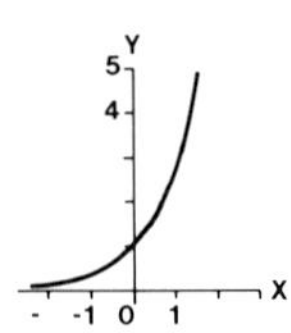

LATIN, *exponere*, to put forth; *functio*, activity, performance

expression a collection of numbers and **symbols** connected by operations.
Examples $x + 3y + 6$, $a + b + 3c - 2b$, $3 \cos x$, $x^2 + 3$. $3y$ is shorthand for 3 multiplied by y, $2b$ is also shorthand for 2 multiplied by b, x^2 is short for x multiplied by itself.
LATIN, *expressare*, to press out

exterior angle **1.** The angle between one side of a plane figure and the adjacent side extended. **2.** Any of the four angles formed outside two **parallel** lines by a third line cutting across them.
Examples
1.

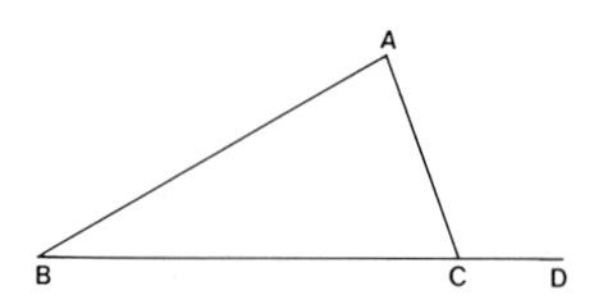

$\angle ACD$ is an exterior angle of the $\triangle ABC$ (*see* INTERIOR ANGLE).
2.

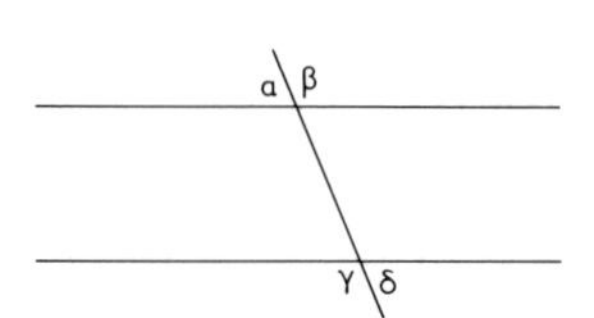

$\alpha, \beta, \gamma, \delta$ are all exterior angles.
LATIN, *exterior*, more outside; *angulus*, corner, angle

extrapolation the process of using known values of a **variable** or **function** to estimate values beyond the given values.

Example
Given the two **sets** of values for x and y as follows:

x	1	2	3	4	5	6	7	8	9
y	2	4	6	8	10	12	14	16	18

we can use extrapolation to estimate the value of y when $x = 10$. Since the known values fit the equation $y = 2x$, then we can estimate $y = 20$ when $x = 10$.

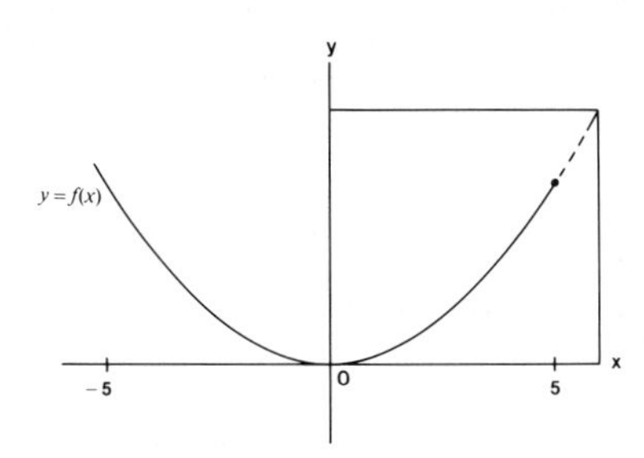

Given the curve $y = f(x)$ drawn through the points shown, to find y when $x = 6$, we can extend (or extrapolate) the curve as shown by the dotted line to estimate the value of y.
LATIN, *extra*, outside of; *polire*, to adorn, polish

extrema points on a curve where the curve changes slope from one direction just before the point to the other just after the point. **Maxima** and **minima** are names for the points.
Example

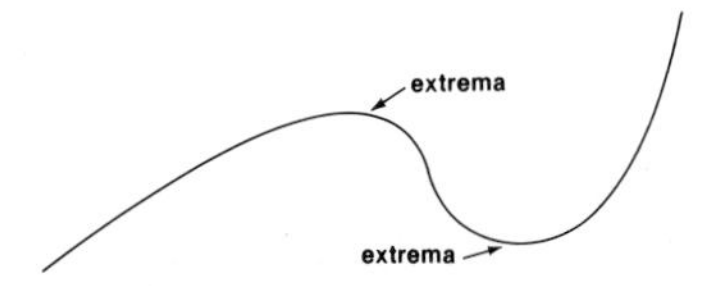

LATIN, *extremus*, outermost

extremes *see* MEANS

Ff

face the flat side of a **polyhedron**.
Example

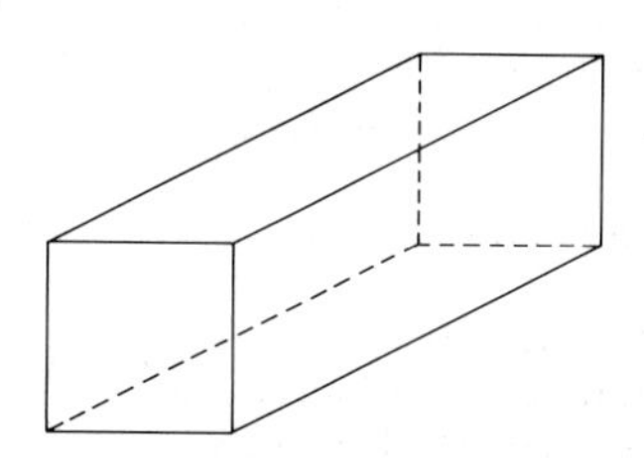

A box has 6 faces.
LATIN, *facies*, form, figure, shape

face angle an angle formed by any two adjacent edges of a **polyhedron**.
Example

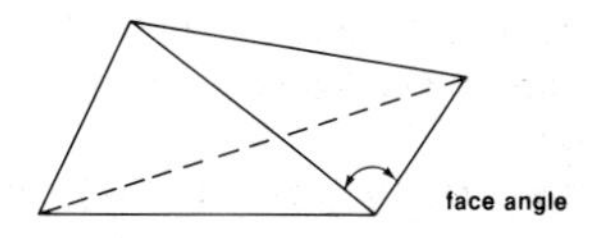

Note: The angle between two adjacent faces is called a **dihedral angle**.
LATIN, *facies*, form, figure, shape; *angulus*, angle or corner

factor **1.** Something that needs to be taken into consideration. One of the things affecting a result. **2.** One of two or more quantities which divide exactly into a given quantity. **3.** Any of the members of a product.
Examples
1. One of the important factors in playing games is the desire to win.
2a. In $2 \times 3 = 6$, 2 and 3 are factors.
b. In 12, the prime factors are 2 and 3.
3a. The product $6ab$ has the factors 6, a, b, $6a$, 2, 3, $2a$, $3a$, $2b$, $3b$, $6b$, $2ab$, etc.
b. In $xy + xz = x(y + z)$, x and $y + z$ are factors.
LATIN, *factor*, doer, maker

factorial a whole number resulting from the **product** of **integers** from 1 to the given number. The symbol or sign for factorial is ! 3! is read three factorial.
Example $5! = 1 \times 2 \times 3 \times 4 \times 5 = 120.$
LATIN, *factor*, doer, maker

factorial design (statistics) a pattern set up so that the results of a series of experiments are put into tables such that more than one **factor** may be studied within the framework of the experiments.
Example

Game	Player A	Player B	Player C	Total
X	12	15	9	36
Y	9	11	7	27
Z	14	15	13	42
Total	35	41	29	105

The table represents the results of three players playing three different games. Player B scored the highest of all players, and game Z allowed more scoring than the others.
LATIN, *factor*, doer, maker; *designare*, to mark out

factorise the separation of a given number or **expression** into two or more parts (**factors**), which, when multiplied together, give the original number or expression.
Examples
$36 = 12 \times 3$ or 6×6 or $2 \times 2 \times 3 \times 3$.
$5a + 5b = 5(a + b)$.
LATIN, *factor*, doer, maker

factor theorem if $P(x)$ is a **polynomial** and $P(a) = 0$, then $(x - a)$ is a factor of $P(x)$.
Example $P(x) = x^2 + 2x - 3$
$P(1) = 1 + 2 - 3 = 0$
therefore $x - 1$ is a factor
$P(x) = (x - 1)(x + 3)$
LATIN, *factor*, doer, maker; GREEK, *theōrēma*, speculation, proposition

factor trees a way of writing down the factors of a number showing how the factors are related to the number.
Examples

```
          20                    20
         /  \                  /  \
       4  ×  5               2  ×  10
      /       \             /        \
  2×2    ×     5           2    ×     5×2
```

LATIN, *factor*, doer, maker

Fermat, Pierre (1601–1665) French mathematician whose works enriched many branches of mathematics. Fermat's most outstanding achievement was the founding of the modern theory of numbers. Results directly related to Fermat's work include: "every odd **prime** can be expressed as the difference of two squares in one and only one way"; "there do not exist positive integers x, y, z such that $x^n + y^n = z^n$ where $n > 2$". This famous last result is known as Fermat's Last "Theorem", but has never been proved.

Fibonacci, Leonardo (1175–1250) Italian mathematician said to be the most talented mathematician of the middle ages. Fibonacci published the *Liber abaci* in 1202 which is devoted to **arithmetic** and **algebra**, and is especially important in the history of mathematics because it advances and illustrates the Hindu–Arabic system of notation. Fibonacci's other works are concerned with **geometry** and **trigonometry**. He is also remembered for his study of the **Fibonacci sequence**, 0, 1, 1, 2, 3, 5, . . . a, b, $a + b$ where each **term** is the sum of the previous pair of terms.

Fibonacci sequence (or series) a **set** of numbers each of which is the sum of the two preceding **terms**. The first two terms are 0 and 1.
$U_n = U_{n-1} + U_{n-2}$ (for $n \geqslant 2$)
Example 0, 1, 1, 2, 3, 5, 8, 13, 21, . . .
after Fibonacci, Italian mathematician; LATIN, *sequi*, to follow

field a **set** of **elements** (numbers etc.) which has the property that the result of adding or multiplying any two elements gives an element in the same set, and also that for each element e there are two elements $(-e)$, $\frac{1}{e}$ such that $e + (-e) = 0$, and also $e \times \frac{1}{e} = 1$ $(e \neq 0)$
$(-e)$ is called the addition **inverse**, and $\frac{1}{e}$ is the multiplication **inverse**.
Example The set of all real numbers (e.g. $1.2 + (-1.2) = 0$, $1.2 \times \frac{1}{1.2} = 1$). Note: The set of all positive integers is *not* a field.
OLD ENGLISH, *feld*, open field

finite limited (as opposed to **infinite**, without limit).
Example The set {1, 2, 3, 4, 5, . . . 10} is finite because it has a definite number of elements in it. The set of all **integers** is **infinite**.
LATIN, *finire*, to limit, finish

flat angle an angle of 180° (also **straight angle**).
Example

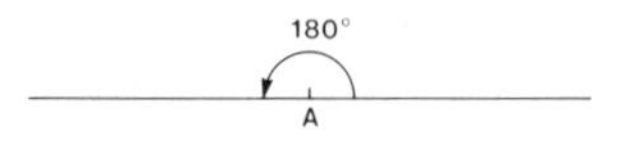

A is a flat angle.
OLD NORSE, *flatr*, flat; LATIN, *angulus*, angle, corner

flow chart (or diagram) a diagram using special shapes which show the flow or series of steps to be

taken to solve a problem. It is used extensively in computer programming.
Example What are the steps involved in putting the cat out at night? The following flow chart illustrates some of the key shapes used.

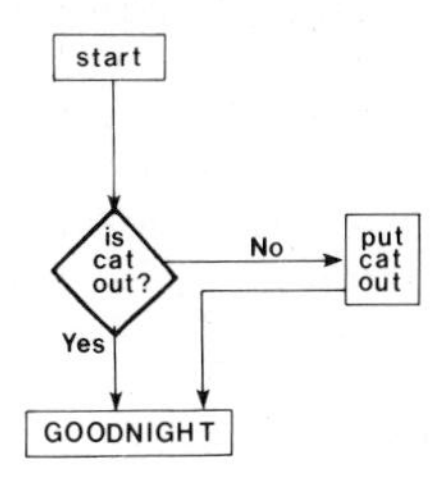

OLD ENGLISH, *flowan*, flow

focus (pl. **foci**) **1.** Either of two fixed points by which an **ellipse** is defined. They lie on the **principal axis** of the ellipse. **2.** Any of the similar fixed points for other **conic** sections such as the **parabola, hyperbola**.
Example

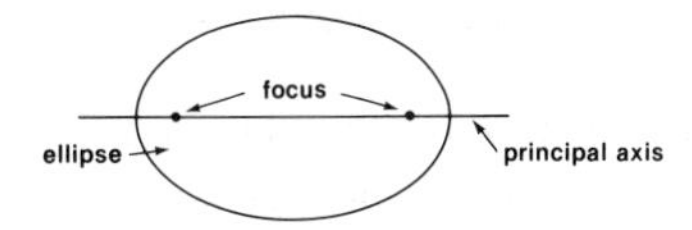

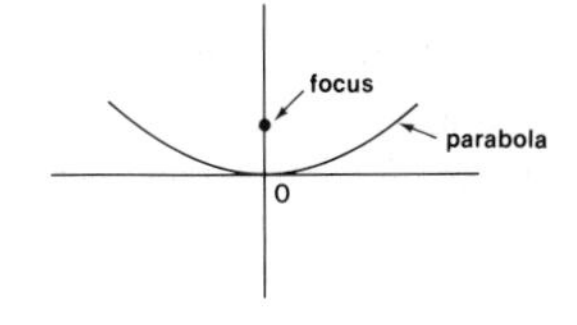

LATIN, *focus*, fireplace, hearth, centre of home

formula a general **equation** showing the connection between related quantities.
Examples
The area of a triangle (A) = ½ base × height (½b × h). The formula is $A = \frac{1}{2}bh$.

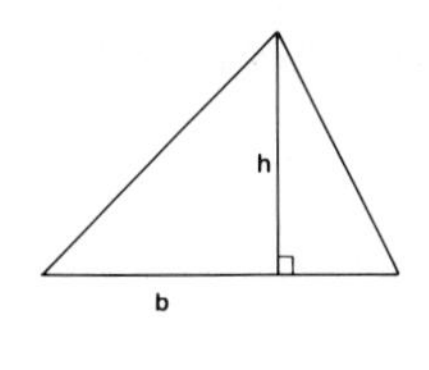

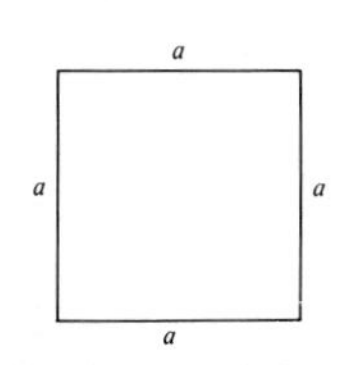

The **perimeter** of a square is four times the length of one side, i.e. $P = 4a$.
LATIN, *formula*, a little form

fraction **1.** The **ratio** of two numbers: ¾ is the fraction three quarters. **2.** In general, the result obtained by dividing one **expression** by another. $\frac{x + y}{x - y}$ is a fraction. The part above the line is called the **numerator**. The part below the line is the **denominator**.
Examples
⅔ is a **proper** fraction (the numerator is less than the denominator).
$^{15}/_{7}$ is an **improper** fraction (the numerator is greater than the denominator).
Both ⅔ and $^{15}/_{7}$ are **common** (or **vulgar**) fractions (both numerator and denominator are **integers**).
2¼ is a **mixed** fraction (it contains an integer and a fraction).
0.25 is a **decimal** fraction (represents $^{25}/_{100}$).
LATIN, *fractio*, a breaking

fraction point the dot used to separate whole numbers and fractions when using the place **notation**.
Examples
$5\,^{3}/_{10}$ = 5.3 in the **decimal** system.
2¼ = 10.01 in the **binary** system.
LATIN, *fractio*, a breaking; *punctus*, pricked

frequency the number of times something occurs.
Example 4 out of 10 students had blond hair. The frequency of blond haired students is 4.
LATIN, *frequentia*, a crowd

frequency distribution a set of values of a variable and the number of times each value occurs.
Example

Frequency	2	3	5	10	6	2	1
Height of males in village (m)	1.5	1.6	1.7	1.8	1.9	2.0	2.1

LATIN, *frequentia*, a crowd; *distribuere*, to assign or grant parts of

frequency histogram a **bar chart** which shows **frequency** and each value of a variable. The height of the bars shows the frequency of each value.
Example

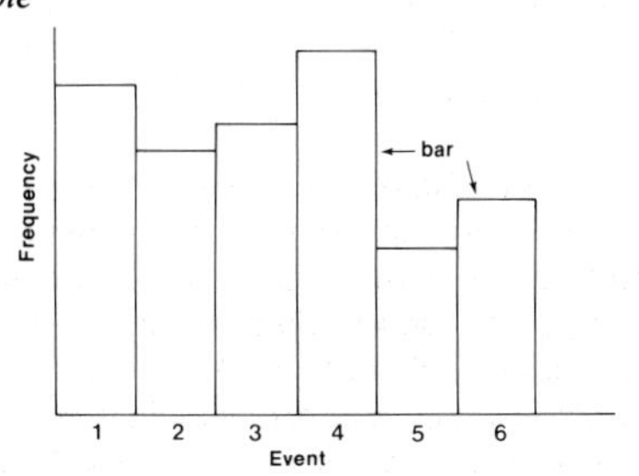

LATIN, *frequentia*, a crowd; GREEK, *histos*, mast; *gramma*, thing written

frequency polygon a **frequency histogram** drawn so that instead of bars or columns the midpoints of each bar or column is joined by straight lines making a **polygon**.
Example

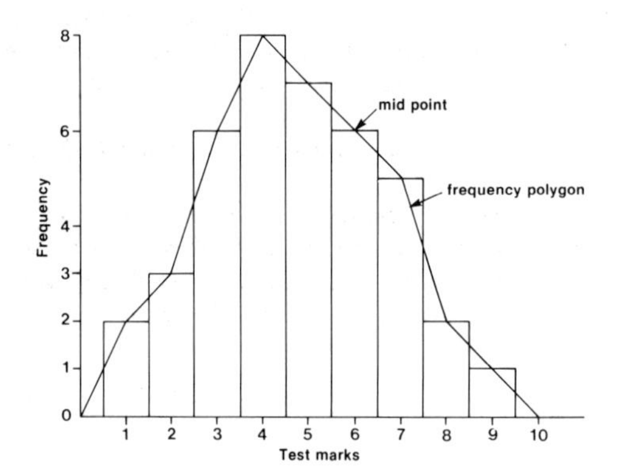

LATIN, *frequentia*, a crowd; GREEK, *polugonos*, having many sides

frustrum a part of a solid bounded by two **parallel** planes. One plane is usually the base of the solid.
Example

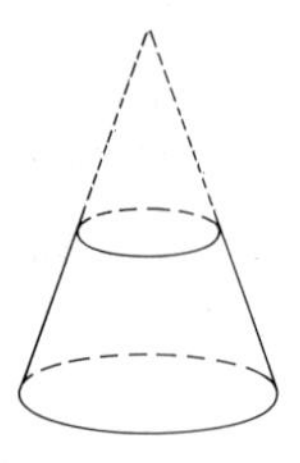

frustrum of a cone

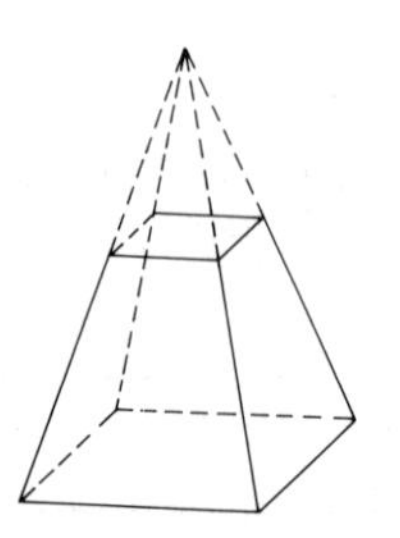

frustrum of a pyramid

LATIN, *frustrum*, piece broken off

function a **relation** which exists between two **sets** such that for each **element** in one set there is a unique element in the other set. In algebra y is said to be a function of x if for every value of x there is one and only one value of y. A function is also called a **mapping** of the elements of one set onto the elements of another.
Example

$$y = x^3 + 2, \; y = f(x) = \frac{2x^2}{x^2 + 1}$$

where $f(x)$ is the symbol for function x.
Note: The variable whose value is mapped or transformed is called the **independent** variable. The resulting or mapped value belongs to the variable called the **dependent** variable.

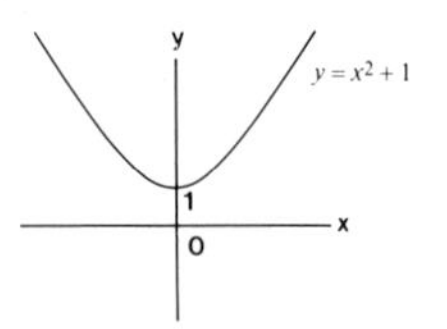

Functions can be represented as a graph as shown in the above example, $y = x^2 + 1$.
LATIN, *functio*, activity, performance

functional notation the way of writing a function of some variable in shorthand.
Examples If $y = x^3 + 2x - 1$, then y is a function of x and could be written $y = f(x) = x^3 + 2x - 1$. $f(x)$ is an example of functional notation. $\phi(x)$, $F(x)$ are other examples.
LATIN, *functio*, activity, performance; *nota*, a note

fundamental theorem of algebra a rule which states that every **polynomial** equation of the form $a_0x^n + a_1x^{n-1} + \ldots a_{n-1}x + a_n = 0$ (where $a_0, a_1 \ldots a_n$ are complex numbers) has at least one complex **root**.
Examples $3x^2 - 2x + 1 = 0$ has at least one root, which may be *real* (a subset of complex numbers) or complex. Similarly, $x^5 + 3x - 2 = 0$ has at least one complex root.
LATIN, *fundus*, bottom; GREEK, *theorema*, speculation, proposition; ARABIC, *al-jebra*, reunion of broken parts

fundamental theorem of calculus a statement which relates the process of **differentiation** and **integration**. If $\int f(x)\,dx$ exists and F is defined by $F(x) = \int f(x)\,dx$ for any x is the **closed** interval (a, b), then $F'(x) = f(x)$ for all x in the given interval.
Example If $f(x) = x^2$ and $\int_a^b F(x)\,dx$ exists, then if $F(x) = \int_a^x f(x)\,dx$ in the interval (a, b) $F'(x) = x^2$.
LATIN, *fundus*, bottom; *calculus*, a small stone; GREEK, *theorema*, speculation, proposition

Gg

games theory a mathematical approach to working out the best strategy to take to maximize one's chances of winning any game.
Example In the game noughts and crosses, there are some moves which are better than others, as follows:

x wins

draw

The games theory enables players to work out the best series of moves to increase their chances of winning.
OLD ENGLISH, *gamon*, amusement, sport; GREEK, *theorein*, to look at

Gauss, Carl Friedrich (1777–1855) regarded, along with Archimedes and Newton, as one of the three greatest mathematicians of all time.A German mathematician, Gauss' works included the discoveries that every **positive integer** is the sum of three **triangular** numbers, the proof of the **fundamental theorem of algebra**, and the development of mathematics in the fields of convergence of series, geometry of surfaces and theory of numbers.

general equation of second degree an **equation** of the form $ax^2 + 2hxy + by^2 + 2gx + 2fy + c = 0$, where a, b, c, f, g, h are constants. It represents a **conic** section.
Example $2x^2 + 3xy + y^2 + 4 = 0$ can be transformed by rotating the axes.
LATIN, *genus*, birth, race; *aequatio*, an equalling; *secondus*, second; *degradus*, a stepdown

generalise to work out a general rule or principle from observing one or more particular examples.
Example From seeing that the sum of

$$1 + 2 = 3$$
$$1 + 2 + 3 = 6$$
$$1 + 2 + 3 + 4 = 10$$

can also be worked out by taking the **average** of the first and last numbers in the group, and multiplying by the number of numbers in the group, we could generalise and say the following:

If $S = 1 + 2 + 3 + 4 + 5 + \ldots + n$, then

$$S = \frac{(1 + n)}{2} \times n$$

(Such a generalisation by itself does not constitute a proof.)
LATIN, *genus*, birth, race

generate **1.** To bring into existence, produce. **2.** (geometry) To form (a shape) by moving one or more **points, lines, curves**, or surfaces in some way.
Examples
1. Can you generate numbers in your mind without trying?
2. Anyone can generate a circle by moving a point so that it is always a fixed distance from another fixed point. Other **conic** shapes can be formed as well, as the diagram on the right below illustrates.

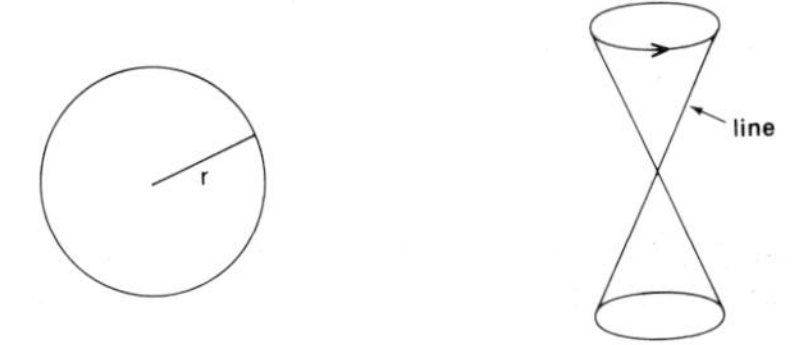

(Such a generalisation by itself does not constitute a proof.)
LATIN, *genus*, birth, race

generator (also *generative*) a point, line or surface which generates or traces out a line, surface or solid.
Examples
1. A point which moves so that it is always the same distance from a fixed point traces out a curve known as a circle.

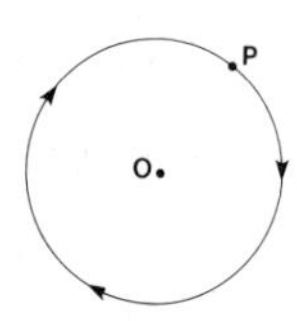

2. A rectangle rotated about one side traces out a solid cylinder. The rectangle is a generator for a cylinder.

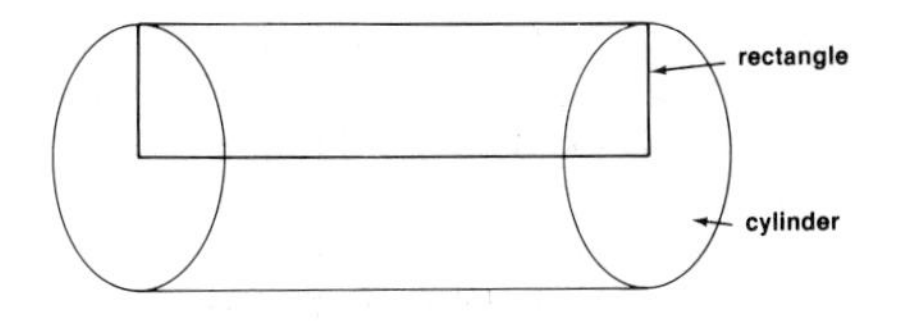

LATIN, *generare*, to generate, produce

geodesic the shortest distance between two points on a curved surface. On a sphere it is an arc of a **great circle** (a circle whose centre is the same as the sphere).
Example

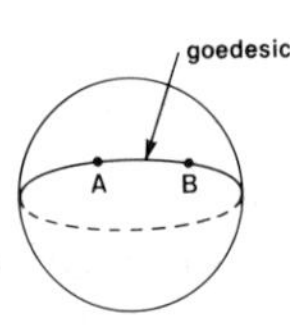

GREEK, *geo*, earth; *daisia*, to divide

geometric mean the middle **term** of three terms in a **geometric progression.** Also called the geometric average. For any two given terms, the geometric mean is the square root of the product of the two terms. In general, the geometric mean of n numbers is the nth root of the product of the n numbers.
Example The geometric mean of 3, 6, 12 is 6, since

$$\sqrt[3]{3 \times 6 \times 12} = \sqrt[3]{216} = 6$$

The geometric mean of 2, 4, 6, 8, 10 is

$$\sqrt[5]{2 \times 4 \times 6 \times 8 \times 10}$$

GREEK, *geometria*, measuring land; LATIN, *medius*, middle

geometric progression (*abbrev.* GP) a **sequence** of numbers in which each, after the first, is the product of the preceding number and a fixed number, called the **common ratio**.
Examples 1, 2, 4, 8, 16, 32, ... the common ratio is 2. In general, if a is the first term and r is the common ratio, the nth term of the progression is:

$$ar^{n-1}$$

The **sum** of the first n terms

$$S_n = a + ar + ar^2 + \ldots ar^{n-1} = \frac{a(1 - r^n)}{1 - r}$$

GREEK, *geometria*, measuring land; LATIN, *progredi*, to move forward

geometry 1. The study of the properties of **points, lines, surfaces** and solids. 2. In modern **mathematics** geometry refers to any of a number of mathematical systems in which a family of **theorems** is derived from a set of initial **axioms** or **postulates**. The axioms or derived theorems do not necessarily relate to the space of everyday experience.
Example The geometry most people have had some contact with is really the geometry developed by the Greek **Euclid**, and involves the ideas of points, line and planes. Other geometries include **projective geometry**, and **Riemannian geometry**.
Common shapes studied in ordinary geometry

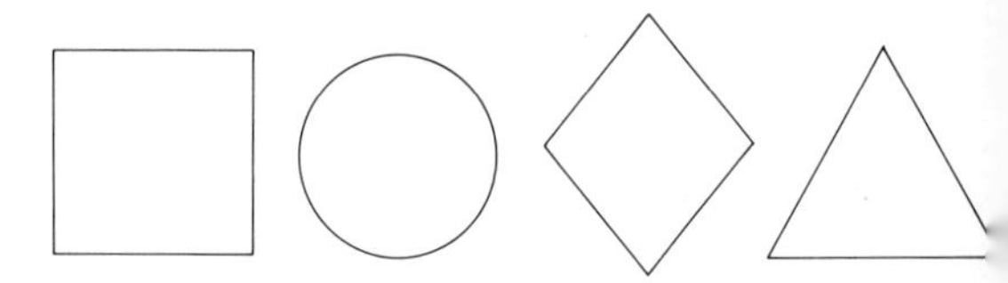

GREEK, *geometria*, measuring land

giga a prefix meaning 10^9 or 1 000 000 000 [a billion (U.S.) or 1000 million (British)].
Example 1 gigabyte is 1 000 000 000 bytes (a **byte** is a group of 8 **binary** digits).
GREEK, *gigas*, giant

Goldbach, Christian (1690–1764) a contemporary of **Euler** and the **Bernoulli** brothers, Goldbach studied many types of mathematics including convergence and divergence of series. **Goldbach's conjecture** is a proposition, which has not been proved or disproved, that every even number (except 2) is the sum of two prime numbers.
Example
6 = 3 + 3 (3 is a prime number)
10 = 7 + 3 (7 and 3 are prime numbers)
24 = 13 + 11 (13 and 11 are prime numbers)
LATIN, *conjectura*, conclusion, interpretation

golden ratio (*also* **golden section**) the ratio

$$\frac{(1 + \sqrt{5})}{2}$$

It is formed by dividing a line segment into two parts a and b such that

$$\frac{(a + b)}{a} = \frac{a}{b}$$

This proportion is found in nature and art, and is pleasing to the eye.
Example A rectangle whose sides are a and $a + b$ respectively, can be constructed from a square of length a such that

$$\frac{(a + b)}{a} = \frac{a}{b}$$

This is often called the golden rectangle since the ratio of its sides forms a golden ratio or section. The smaller rectangle *DEFC* in the diagram below is also a golden rectangle, which can be made into a square *CFGH* and another golden rectangle *DEGH*. The process can be repeated without limit.

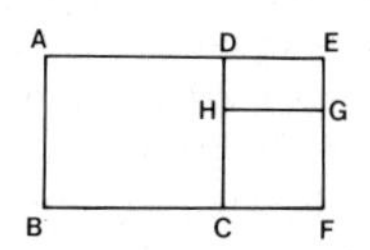

The ratio

$$\frac{1 + \sqrt{5}}{2}$$

is derived from

$$\frac{a + b}{a} = \frac{a}{b}$$

as follows:

Let $x = \frac{a}{b}$, then $\frac{a + b}{a} = 1 + \frac{b}{a}$

$$1 + \frac{1}{x} = x$$

i.e., $x + 1 = x^2$
or $x^2 - x - 1 = 0$

By **completing the square**, $x = \frac{1 \pm \sqrt{5}}{2}$

Since $1 - \sqrt{5}$ is negative and a is positive, then

$$x = \frac{(1 + \sqrt{5})}{2} \equiv \frac{a}{b}$$

OLD NORSE, *gyldin*, golden; LATIN, *ratio*, computation

goodness of fit the degree to which a set of statistical data fits an assumed theoretical **distribution**. The degree of fitness is calculated by the chi square (χ^2) test (χ^2 = sum of the squares of the difference between actual and expected values divided by expected value). Chi is used to represent the statistical variable.
Example

Height Range (cm)	Number of People: Actual	Number of People: Expected
155–155	1	1
156–160	1	1
161–165	2	1
166–170	1	1
171–175	2	2
176–180	2	2
181–185	0	1
186–190	1	1
191–195	1	1

$$= \frac{(1-1)^2}{1} + \frac{(1-1)^2}{1} + \frac{(2-1)^2}{1} + \frac{(1-1)^2}{1} + \frac{(2-2)^2}{2} + \frac{(2-2)^2}{2} + \frac{(0-1)^2}{1} + \frac{(1-1)^2}{1} + \frac{(1-1)^2}{1} = 2.$$

The smaller the value of χ^2, the better the fit. Statistical tables of χ^2 exist to allow one to determine the significance of the calculated χ^2 value.
OLD ENGLISH, *god*, good, fitting; *fitten*, to marshal troops, arrange

GP *see* **geometric progression**

gradient the number or value which tells you how steep a line or curve is at any given point. It is the ratio of the distance risen to the distance travelled **horizontally** (for a straight line). Also known as the **slope**. If the line slopes downwards, the distance risen becomes a "negative rise".
Examples

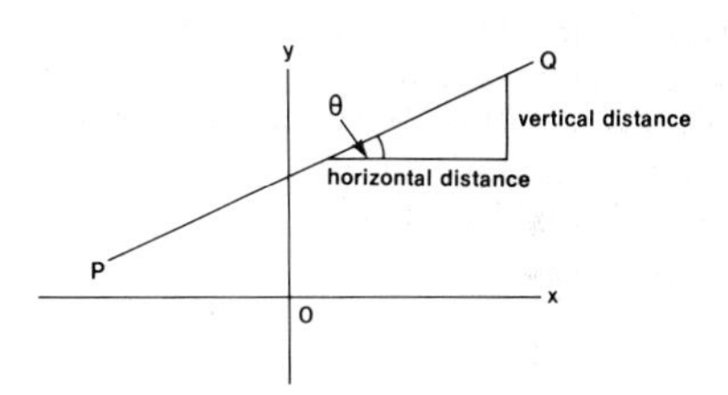

Gradient of the line $PQ = \frac{\text{vertical distance}}{\text{horizontal distance}}$ is $\tan\theta$

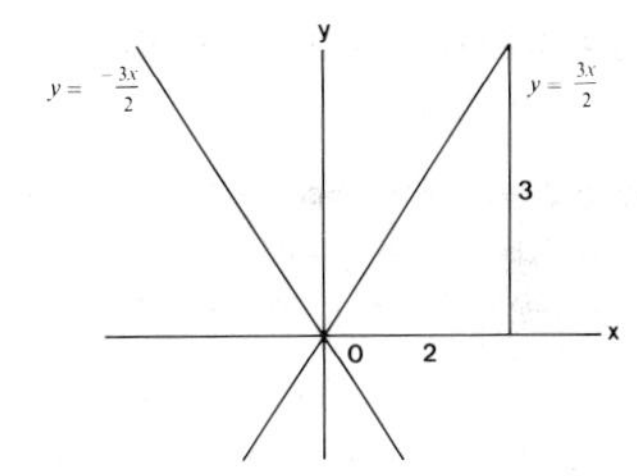

Gradient of the line $y = \frac{3}{2}x$ is $\frac{3}{2}$. The gradient of the line $y = -\frac{3}{2}x$ is $-\frac{3}{2}$.

LATIN, *gradus*, a step

gram a unit of mass. A gram weight is the weight of a mass of one gram. The symbol for gram is g.
Example An almond has a mass of about 2 grams.
LATIN, *gramma*, a small unit

graph a diagram showing how two or more quantities are connected. Data are often shown by means of **bar charts, pie charts, frequency diagrams, histograms** and **pictograms**. In **coordinate geometry** a graph is a curve drawn with reference to the two axes x and y (hence *graphic* [adj.], *graphically* [adv.]).

Examples

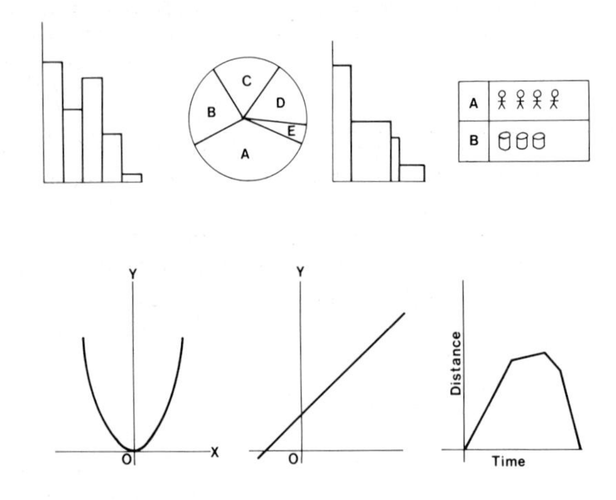

GREEK, *graphein*, to write

great circle a circle on a sphere whose centre coincides with the centre of the sphere.
Example

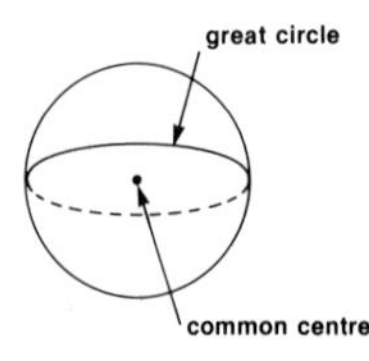

On the surface of the Earth, the equator and any **meridian** are examples of Great Circles.
OLD ENGLISH, *great*, thick, coarse; LATIN, *circus*, a ring

greater than bigger than. Symbol used is >.
Examples
$11 > 7$ (11 is greater than 7).
For what values of x is $2x + 3 > 1$? (this question suggests that there are values of x for which $2x + 3$ is not greater than 1). It can be solved as follows:
$2x + 3 - 3 > 1 - 3$ (add -3 to both sides)

$$\frac{2x}{2} > \frac{-2}{2} \text{ (divide both sides by 2)}$$

$$x > -1.$$

OLD ENGLISH, *great*, thick, coarse

greek letters letters from the Greek alphabet are often used as symbols, representing values not yet known, in mathematical expressions. The alphabet is set out in the table below, with typical uses for the letters described on the right.

Letters	Name	Typical uses
Α α	alpha	used to represent an unknown angle or other variable (or constant).
Β β	beta	also used to represent an angle (unknown) or other unknown value (variable or constant).
Γ γ	gamma	often used to represent an unknown angle or constant (or variable).
Δ δ	delta	Δ used to represent a triangle, or a small difference (see signs and symbols at the beginning of the dictionary). δ also used to represent a small difference, as in δx meaning a small change in the value of x.
Ε ϵ	epsilon	often used to mean "is a member of".
Ζ ζ	zeta	often represent unknown angles, or constants, or variables.
Η η	eta	
Θ θ	theta	
Ι Ι	iota	often represent an unknown constant, or variable (in physics, usually represent the wavelength). μ is the symbol for "micro", meaning 10^{-6}.
Κ $\varkappa$	kappa	
Λ λ	lambda	
Μ μ	mu	
Ν ν	nu	
Ξ ξ	xi	seldom used at high school.
Ο o	omicron	
Π π	pi	represents the constant ratio of the circumference of a circle to its diameter.
Ρ ϱ	rho	used in physics for quantities such as density.
Σ σ	sigma	Σ represents the sum; σ standard deviation.
Τ τ	tau	
Υ υ	upsilon	
Φ ϕ	phi	represents an unknown angle.
Χ χ	chi	used in χ test.
Ψ ψ	psi	represents an unknown angle.
Ω ω	omega	usually represents angular velocity.

group **1.** A collection of **elements**. **2.** In higher mathematics, a special word for a **set** of elements which, for a certain operation, satisfy the **closure**, **associative**, **identity** and **inverse** laws, but not necessarily the **commutative** law.
Examples
1. A group of 5 men.
2. The set of all numbers of the form a/b where a and b are integers form a group for the operation called multiplication, i.e. $\{1/1, 1/2, 1/3, \ldots 2/2, 2/3, 2/4, 2/5, \ldots\}$
ITALIAN, *gruppo*, a group

Hh

half **1.** One of two equal parts of a whole. **2.** Forming a half.
Examples
1. Half of a circle is a semicircle.

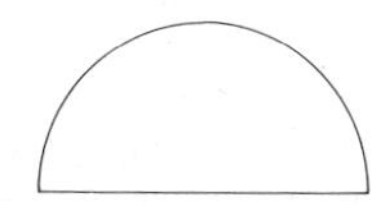

2. The shopkeeper sold me a half kilogram of flour.
OLD ENGLISH, *half*, half

halving the interval a method used to find the approximate value of a **root** of an equation in an interval.
Example
If $f(x) = x^3 - 10$
$f(2) = -2$
$f(3) = 17$
i.e. $f(x) = 0$ between x having the values 2 and 3
Take $x = \frac{2 + 3}{2} = \frac{5}{2}$
$f\left(\frac{5}{2}\right) = \frac{125}{8} - 10 = 5.6$
Since $f(2)$ is negative and $f(2\frac{1}{2})$ is positive the root must lie between 2 and $2\frac{1}{2}$.

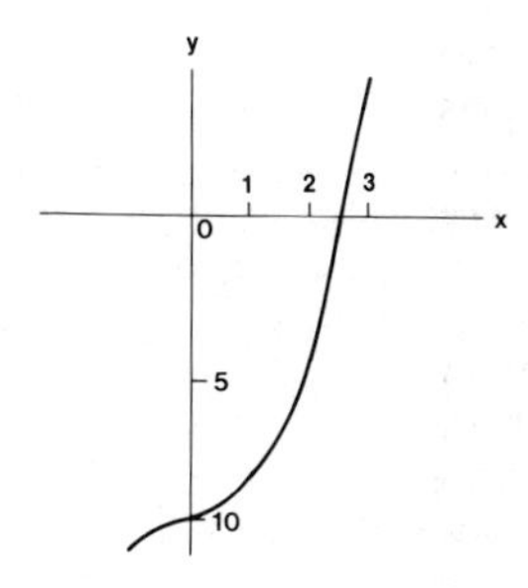

Halving the new interval, $x = \frac{2 + 2\frac{1}{2}}{2} = 2\frac{1}{4}$.
$f(2\frac{1}{4}) = (2\frac{1}{4})^3 - 10 = 1.39$.
Continuing in this way we can find a better approximation to the root of $x^3 - 10 = 0$, i.e. $\sqrt[3]{10}$.
OLD ENGLISH, *healf*, half; LATIN, *intervallum*, space between ramparts

harmonic mean the middle term of three numbers in a **harmonic progression**. The harmonic mean of two numbers is the **reciprocal** of one half of the sum of the reciprocals of the numbers.
Examples
The harmonic mean of 3 and 6 is $1 \div \frac{1}{2}(\frac{1}{3} + \frac{1}{6})$ $= 1 \div \frac{1}{4} = 4$
The harmonic mean of $\frac{1}{3}$, $\frac{1}{6}$, $\frac{1}{9}$ is $\frac{1}{6}$.
GREEK, *harmonia*, harmony; LATIN, *medius*, middle

harmonic progression a **sequence** of numbers whose **reciprocals** are in an **arithmetic progression**.
Examples $1, \frac{1}{3}, \frac{1}{5}, \frac{1}{7}, \frac{1}{9}, \ldots$
GREEK, *harmonia*, harmony; LATIN, *progredi*, to go forward

HCF *see* HIGHEST COMMON FACTOR

hecto- prefix meaning one hundred.
Example 1 hectolitre = 100 litres (note: centi- means one hundredth).
GREEK, *hekaton*, one hundred

helix a curve lying on the surface of a **cylinder** or **cone** which cuts the cylinder or cone at a constant angle.
Example

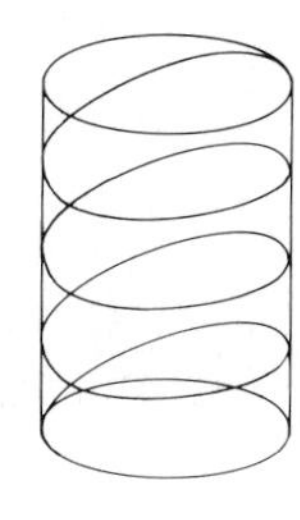

GREEK, *helix*, a spiral

hemi- prefix meaning half.
Example

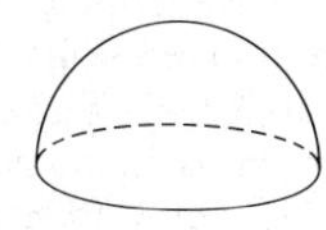

Hemisphere (half a sphere).

GREEK, *hemi*, half

hence from this (as a result).
Examples
Prove that $x^2 + 2x + 1 = (x + 1)^2$
Hence solve $x^2 + 2x = -1$.

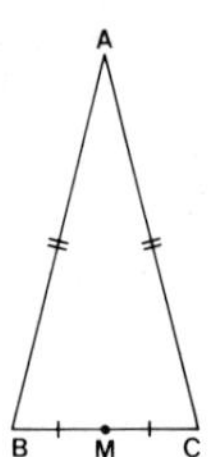

In the ΔABC prove that AM (where M is the mid point of BC) divided ΔABC into two **congruent** Δs, and hence show that AM is also perpendicular to BC.
OLD ENGLISH, *heonane*, from here, away

heptagon a **polygon** with seven sides and seven angles.
Example

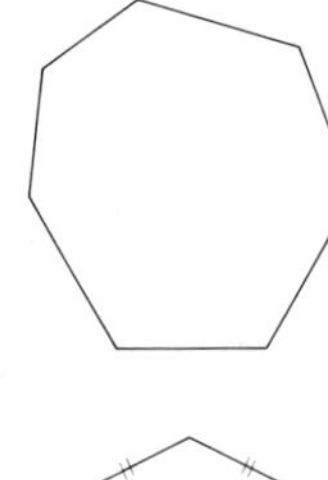

This is a **regular** heptagon.
GREEK, *hepta*, seven

Hero (*c.* AD 75) a Greek mathematician who added to the rigorous **geometry** of **Euclid** and **Archimedes** and also studied the solution of algebraic **equations**. He is particularly remembered for the formula for the area of a triangle (Hero's formula). This formula uses only the lengths of the three sides. It is:

$$A = \sqrt{s(s - a)(s - b)(s - c)}$$

where A is the area of the triangle, s is $\dfrac{a + b + c}{2}$, and a, b and c are the lengths of the three sides.
Example For a triangle with sides $a = 25$, $b = 24$, $c = 7$, $s = 28$ and:

$$\begin{aligned} A &= \sqrt{28(28 - 25)(28 - 24)(28 - 7)} \\ &= \sqrt{28 \times 3 \times 4 \times 21} \\ &= \sqrt{7^2 \times 3^2 \times 4^2} \\ &= 7 \times 3 \times 4 \\ &= 84 \end{aligned}$$

This can be verified using the other formula for the area of a triangle, area is equal to half the base by the height. $A = \frac{1}{2}(24 \times 7)$
$= 84.$

hexagon a **polygon** with six sides. A **regular** hexagon has all angles equal to 120°.
Example

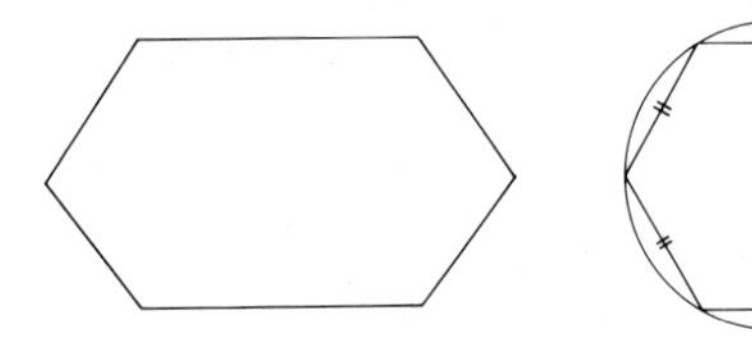

A regular hexagon inscribed in a circle.
GREEK, *hexagonos*, six angled

highest common factor the highest number (or expression) which will go into two or more other numbers (or expressions) [abbreviated as HCF].
Examples
3 is the HCF of 15 and 18, since $15 = 3 \times 5$ and $18 = 3 \times 3 \times 2$.
$(x + y)$ is the HCF of the expressions $(x + y)^2$, $(x + y)(x - y)$.
OLD ENGLISH, *hech*, high; LATIN, *communis*, common; *factor*, doer

hint a suggestion or pointer as to how to go about working out a problem.
Example
Solve $x^2 + 2x - 3 = 0$
(Hint: Factorise $x^2 + 2x - 3$)
This hint suggests that by factorising $x^2 + 2x - 3$, i.e. $x^2 + 2x - 3 = (x - 1)(x + 3)$, then by looking at $(x - 1)(x + 3) = 0$ we can say $x - 1 = 0$ or $x + 3 = 0$, so that the solution to $x^2 + 2x - 3 = 0$ is x is either 1 or -3.
OLD ENGLISH, *hentan*, to seize

histogram a bar graph used to show a **frequency distribution**. The area of the bars shows the frequency. If the widths of the bars are all the same, the heights of the bars also show the frequency distribution.
Example

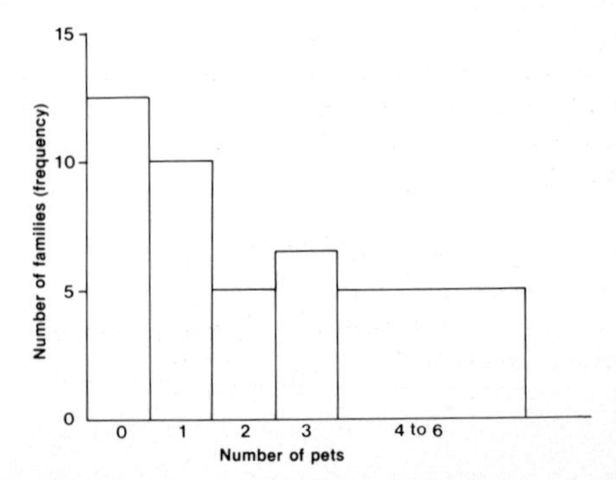

GREEK, *histos*, mast; *gramma*, thing written

Hooke's law a law (or rule) which states: "The tension in a light elastic string (or spring) is directly **proportional** to the extension beyond the natural length of the string (spring)". In symbols, $T = kx$ (T represents tension, x the displacement, and k is a constant). When using λ (modulus of elasticity), then $T = \lambda x/l$ where l is the natural length of the string (spring). λ is the force which produces an extension equal to the natural length of the string (spring).
Example Hooke's law is used in analysing the motion of a particle at the end of a spring or equivalent which is displaced from its **mean** position (*see* SIMPLE HARMONIC MOTION).
after Robert Hooke, English physicist

horizontal a line **parallel** to the Earth's skyline. The antonym of **vertical**.
Examples

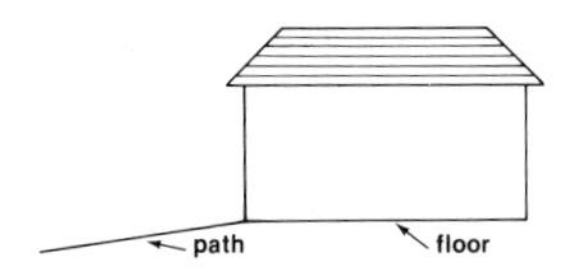

The floor of the garage was horizontal, but the path leading up to it was not. Any line drawn on the surface of water which is still is horizontal.
A horizontal **displacement** is a length measured horizontally.
GREEK, *horizein*, to divide, separate

Hypatia (AD 370–415) Greek mathematician who taught mathematics at the University of Alexandria. Hypatia studied the **geometry** of **conic sections** and solutions of **linear** and **quadratic equations**. She also was interested in the physics of flotation.

hyperbola **1.** A curve (in two parts) formed by a **plane** which cuts a right circular **cone** in such a way that the angle formed by the plane and the cone's base is greater than the angle formed by the plane and the cone's side. **2.** In **analytical geometry**, the path traced out by a point such that the difference between the distances from the point to two given fixed points is constant.
Example
1.

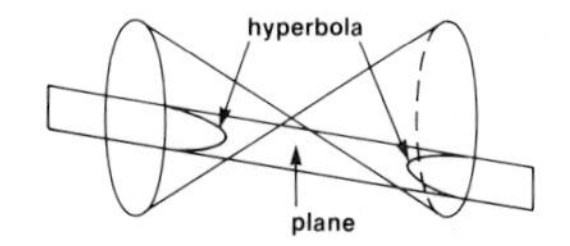

2.

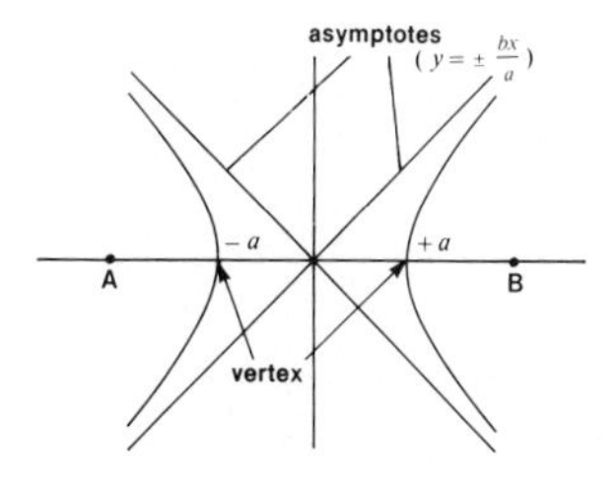

A and B are the two fixed points (called **foci**). The equation of a hyperbola is

$$\frac{x^2}{a^2} - \frac{y^2}{b^2} = 1$$

GREEK, *huperballein*, to throw beyond, exceed

hypotenuse the side opposite the right angle in a right angled triangle. It is also the longest side in a right angled triangle.
Example

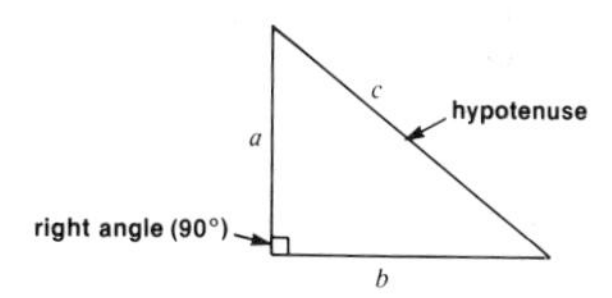

Note: The square of the hypotenuse of a right angled triangle is equal to the sum of the squares of the other two sides (**Pythagoras's theorem**).
GREEK, *hupotenusa*, the line or side subtending the right angle

hypothesis a statement that is made to explain a set of facts and to form the basis for further investigation. It is understood that the statement is subject to proof or checking.
Example The **null hypothesis** in statistics says that, between any two samples, there is really no significant difference, and that any difference which does occur is due only to errors of taking samples at random. For example, taking two samples of the heights of 200 people at a time from 2000 people, we should find the **mean (average)** height to be almost the same, and the spread of deviations to be consistent. If the means of the two samples are sufficiently different, we should then suspect the samples to come from different groups or populations.
GREEK, *hupothesis*, foundation

Ii

I (i) the ninth letter of the English alphabet. It is used to represent the Roman one (I) or the square root of -1 (i).
Examples
I + V = VI in Roman numerals.
$3 + 2i$ is a **complex** number which is made up of 3 and $2\sqrt{-1}$.
GREEK, ι (*iota*) [ninth letter in the Greek alphabet]

icosahedron a solid with twenty faces. A regular icosahedron has faces in the shape of an equilateral triangle.
Example

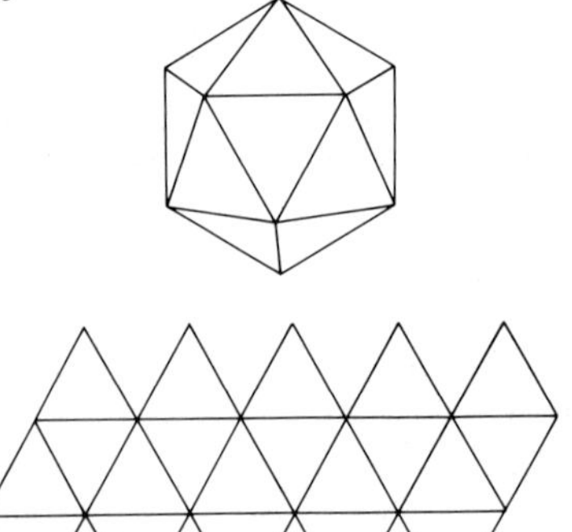

The above shape can be folded into an icosohedron.
GREEK, *eikosaedron*, twenty-sided

identical having the same shape and size.
Example

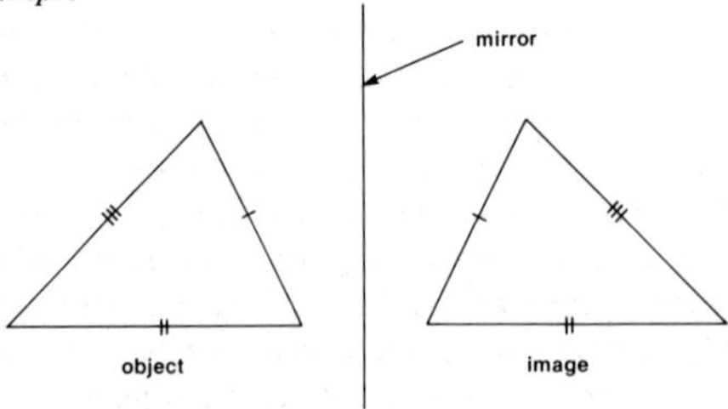

The two triangles are identical.
LATIN, *identitas,* identity

identity **1.** A statement that two or more **terms** or **expressions** are equal for all values (hence *identical*, adj.). **2.** In a mathematical system, an element which leaves unchanged any other element with which it is combined.

Examples
1. $(a + b)^2 = a^2 + 2ab + b^2$
$x \cdot x = x^2$ [• means multiply here]
2. 0 is the identity element for the addition operation
i.e. $a + 0 = a$
$1 + 0 = 1$
1 is the identify element for multiplication,
i.e. $a \cdot 1 = a$
$1 \cdot 1 = 1$
LATIN, *idem*, same

identity laws statements involving the **identity** element and any other element in the set which are true for all values of the other element.
Examples
1. $a \times I = a$ is true for all values of a (I is the identity element when the operation is multiplication).
2. $a + 0 = a$ is true for all values of a (0 is the identity element when the operation is addition).
LATIN, *idem*, same; OLD ENGLISH, *lagu*, something laid upon

identity matrix a square matrix whose **diagonal** values from top left to bottom right are all "ones", and the remaining values are 0. When multiplied by another matrix of the same **degree** it leaves the second matrix unchanged.
Examples

$$I = \begin{pmatrix} 1 & 0 \\ 0 & 1 \end{pmatrix} \text{for a } 2 \times 2 \text{ matrix}$$

$$I = \begin{pmatrix} 1 & 0 & 0 \\ 0 & 1 & 0 \\ 0 & 0 & 1 \end{pmatrix} \text{for a } 3 \times 3 \text{ matrix}$$

$$I = \begin{pmatrix} 1 & 0 & . & . & . & 0 \\ 0 & 1 & & & & . \\ . & & . & & & . \\ . & & & & 1 & 0 \\ 0 & & & & 0 & 1 \end{pmatrix} \text{for an } n \times n \text{ matrix}$$

LATIN, *idem*, same; *matrix*, womb

image a **point**, number or unique **element** of a space that **corresponds** to some other point, number or unique element, when the object is **transformed** (the image element and object element are linked by some rule).
Example

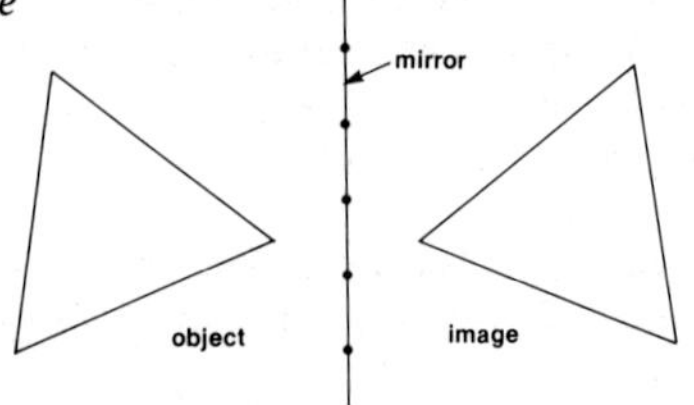

{5, 4, 3, 6} is the image set of {2, 1, 0, 3} under the transformation $x \rightarrow x + 3$.
LATIN, *imago*, an image

imaginary number a number expressed as the square root of a negative number. The symbol i represents $\sqrt{-1}$.
Examples $\sqrt{-3}$, $\sqrt{-1.2}$
Note: $\sqrt{-3}$ can be expressed $\sqrt{3}i$
LATIN, *imago*, an image; *numerus*, number

improper fraction a fraction whose **numerator** is equal to or greater than its **denominator**.
Example $\frac{4}{3}$, $\frac{7}{6}$, $\frac{16}{11}$, $\frac{9}{9}$
Note: An improper fraction can be changed into **mixed numbers** (a whole number and a proper fraction).
LATIN, *improprius*, not one's own; *fractio*, a breaking

implicit expressed indirectly (antonym of **explicit**).
Example In the relation $y^3 + xy^2 + x^3 = 1$, which cannot directly be expressed in the form $y = f(x)$, y is said to be an implicit function of x.
LATIN, *implicare*, involved, entangled

in-centre the centre of a circle that touches every side of a polygon.
Examples
1.

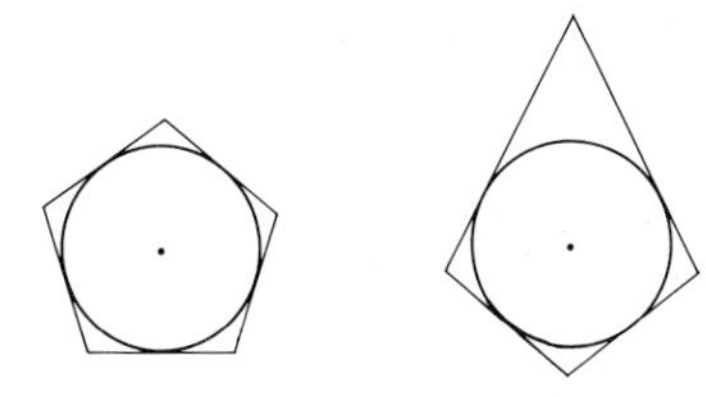

2. In a triangle the in-centre is the point where all the three bisectors of the angles meet. In ΔABC h is the in-centre.

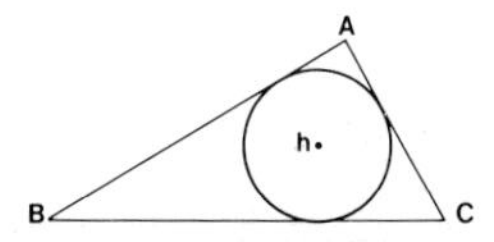

LATIN, *in*, in; *centrum*, centre

increment the amount by which a varying quantity increases from one point to another.
Example If $y = x^2$, then the increment of y when x changes from 1 to 1.1 is $(1.1)^2 - 1^2 = 0.21$.
LATIN, *increscere*, to increase

independent **1.** Not dependent on another or other variables. **2.** Not dependent on another or other events.
Examples
1. In the function $y = x^3$, x is the **independent variable** (and y is the dependent variable).
2. In two throws of a die the outcome of each throw is **independent** of each other.

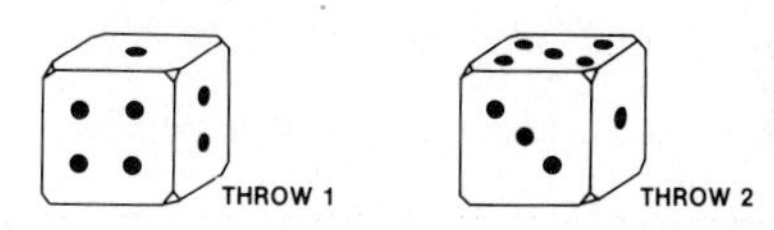

LATIN, *in*, not; *dependere*, to depend, hang from

indeterminate unable to be worked out to give an answer.
Example $\frac{0}{0}$ is indeterminate, as is $\frac{\infty}{\infty}$.
LATIN, *in*, not; *determinare*, to determine, limit

index (*pl.* **indices, indexes**) the number which indicates the **power** to which a given number or **term** is to be raised.
Example $8 = 2^3$; 3 is the index.
LATIN, *index*, forefinger, indicator

index laws (also **laws of indices**) the rules by which indices or **powers** may be combined. The main ones are:

$$a^m \times a^n = a^{m+n}$$
$$a^m \div a^n = a^{m-n}$$
$$(a^m)^n = a^{mn}$$
$$\sqrt[n]{a} = a^{\frac{1}{n}}$$

Examples

$$x^2 \times x^3 = x^5$$
$$y^5 \div y^2 = y^3$$
$$(y^{1/2})^3 = y^{3/2}$$
$$\sqrt[3]{8} = 8^{1/3} = (2^3)^{1/3} = 2^1 = 2$$

LATIN, *index*, forefinger, indicator; OLD ENGLISH, *lagu*, something laid down

index number a number indicating a relative position compared with an agreed upon base. It is usually used to compare variations in prices, wages, production etc., from one point in time to another.
Example The consumer price index is a number worked out by combining the average price of milk, bread, petrol etc. and comparing this to the value obtained at some point in time usually taken as 100.
LATIN, *index*, forefinger, indicator; *numerus*, number

induction **1.** (logic) A process of drawing general conclusions from looking at a few specific cases. **2.** (mathematics) A method of proving a general rule by proving it is true for the first case and, assuming it true for a case selected at random, proving it true for the case immediately following.

Example It is suggested that the sum of n positive whole numbers $1 + 2 + 3 + 4 + \ldots + n$ is $\frac{1}{2}n(n + 1)$. In the first case (when $n = 1$), the sum $= \frac{1}{2}(1 + 1) = 1$, which is true. Let us assume it is true for the case $n = k$, i.e.
$S = \frac{1}{2}k(k + 1)$, where k is a whole number.
The sum of

$$\begin{aligned}(k + 1) \text{ numbers} &= 1 + 2 + \ldots k + (k + 1)\\ &= \tfrac{1}{2}k(k + 1) + (k + 1)\\ &= (k + 1)\left(\frac{k + 2}{2}\right)\\ &= \frac{(k + 1)}{2}(k + 2)\end{aligned}$$

which is the same as $\frac{n}{2}(n + 1)$ when

$$n = (k + 1)$$

Since the result is true for $n = 1$, it is thus true for $n = 2, 3, 4, \ldots$ Therefore, by induction the rule is true for all n.
LATIN, *inducere*, to lead in

inequality a statement which says one quantity is not equal to another.
Example $x \neq y$, $x > y$, or $x < y$ are all inequalities ($>$ means greater than, $<$ means less than, $\neq$ means not equal to).

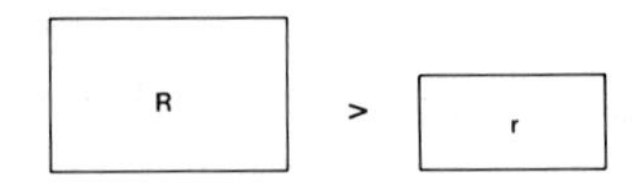

LATIN, *inaequus*, inequal

infinite, antonym of **finite**

infinite set 1. A **set** with an unlimited number of members in it. 2. A set the members of which can be put into a **one to one correspondence** with part of the set.
Example The set of whole numbers can be put into a one to one correspondence to the set of odd numbers.

$$\begin{array}{ccccccccccc}1, & 2, & 3, & 4, & 5, & 6, & 7, & 8, & 9, & \ldots\\ \updownarrow & \updownarrow & \updownarrow & \updownarrow & \updownarrow & \updownarrow & \updownarrow & \updownarrow & \updownarrow & \\ 1, & 3, & 5, & 7, & 9, & 11, & 13, & 15, & 17, & \ldots\end{array}$$

In other words, for each positive whole number we can find one odd number to correspond with it. Since the set of odd numbers is part of the set of whole numbers, it satisfies the definition of an infinite set.
LATIN, *infinitus*, without limit; *secta*, a following

infinitesimal 1. Immeasurably small. 2. An immeasurably small amount. 3. A **variable** quantity which can have values very close to zero. Symbols used are: δ, Δ, as in δx, Δx (a very small quantity of x).
Examples
1. Infinitesimal **calculus** is a branch of mathematics which deals with the rates of change of things, using very small quantities, and also the integration of things using very small quantities. Also known as **differential** and **integral calculus.**
2a. Using infinitesimals we can work out the approximate solution to **polynomial** equations such as $a_1x^n + a_2x^{n-1} + \ldots a^n = 0$ (*see* NEWTON'S METHOD).
b.

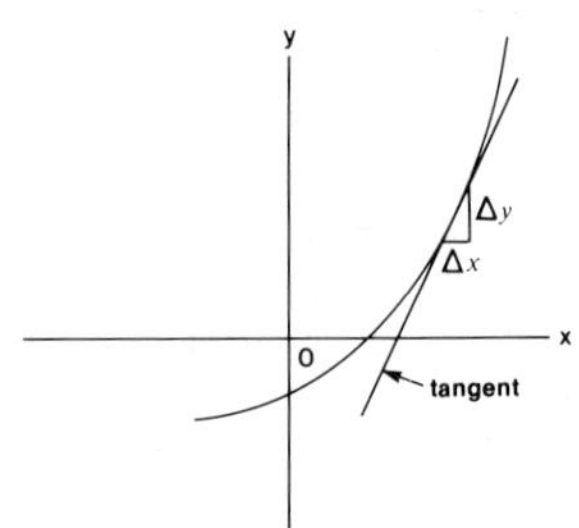

As $\Delta x \to 0$, $\frac{\Delta y}{\Delta x}$ approaches the gradient or slope of the tangent to the curve at the point x.
3. $\frac{1}{x}$ could be an infinitesimal as $x \to \infty$.
LATIN, *infinitus*, without limit

infinity 1. Space, time or quantity without limits. 2. An indefinitely large number. 3. A value greater than any finite value. Symbol is ∞.
Examples
1. The universe is said to extend to infinity.
2. For any finite value N, $\infty > N$.
3. As $x \to 0$, $\frac{1}{x} \to \infty$.

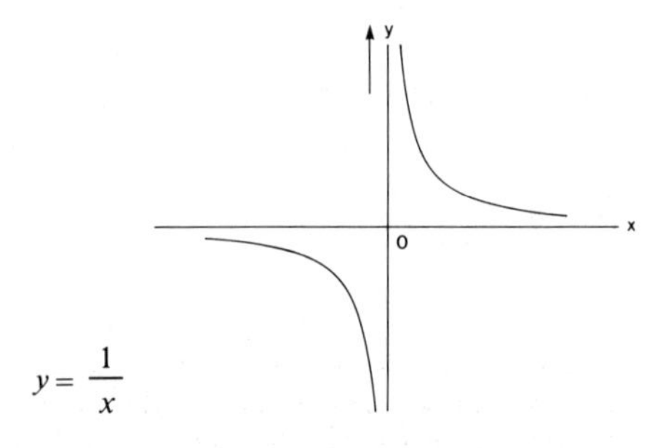

LATIN, *infinitus*, without limit

inflexion, *see* POINT OF INFLECTION

inflexional tangent the tangent to a curve at a **point of inflection** on the curve.
Examples
1. If $y = x^3 + 1$, then $\frac{dy}{dx} = 3x^2$, and $\frac{d^2y}{dx^2} = 6x$.

Both $\frac{dy}{dx}$ and $\frac{d^2y}{dx^2} = 0$ when $x = 0$, and $\frac{dy}{dx}$ is positive for all x, so $x = 0$, $y = 1$ is a horizontal point of inflection.

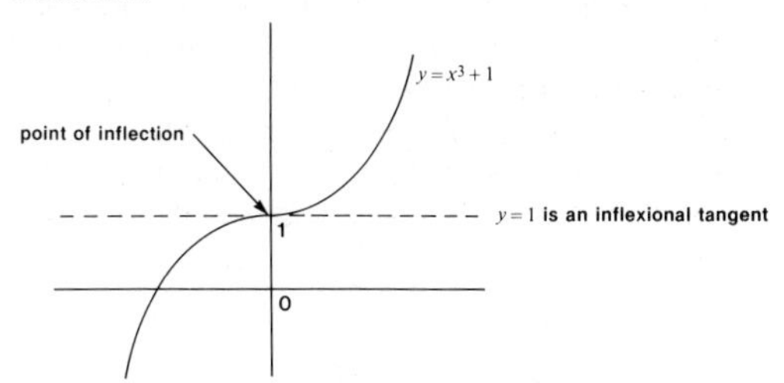

2. If $y = \tan x$, then $\frac{dy}{dx} = \sec^2 x$, and and $\frac{d^2y}{dx^2} = 2\tan x.\sec^2 x$. Since $\frac{d^2y}{dx^2}$ approaches 0 as x approaches 0, and $\frac{d^2y}{dx^2}$ is negative for $x < 1$, and positive for $x > 1$, therefore $x = 0$ is also a point of inflection (but *not* a horizontal one). The inflexional tangent is $y = x$.

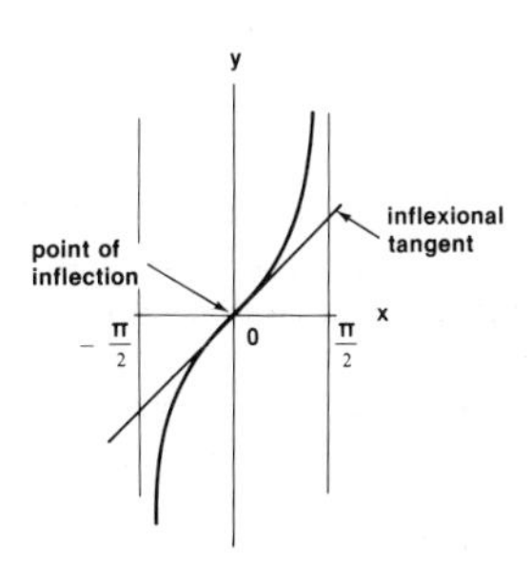

LATIN, *inflectere*, to bend; *tangere*, to touch

integer a whole number.
Example $-3, -1, 0, 1, 4, 7, \ldots$
LATIN, *integer*, whole, complete

integral the **function** obtained by the reverse of **differentiating** a function. If the **operation** is done between limits, the result is called the **definite integral**. If not, the result is called the **indefinite integral**. It is the area between the graph of the function and the x axis. Symbol for integral is $\int(\ )dx$.

Example The integral of x is $\frac{x^2}{2} + c$ (indefinite), i.e. $\int x\,dx = \frac{x^2}{2} + c$ in shorthand form (note: c is any constant), since $\frac{d}{dx}(\frac{x^2}{2} + c) = x$. The integral of x between 0 and 1 is ½.

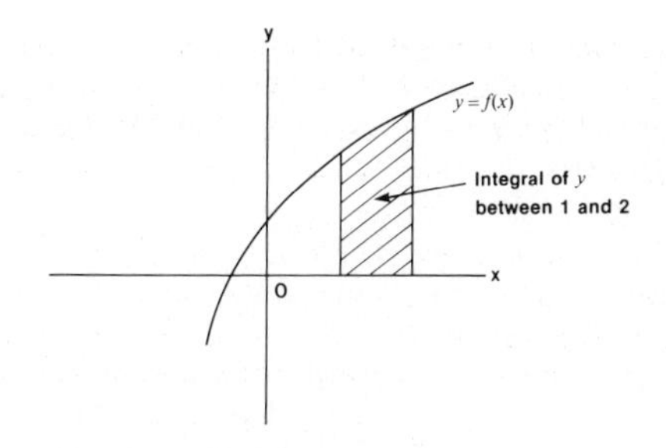

LATIN, *integer*, whole, complete

integration the **operation** of finding the value of the **definite** or **indefinite integral** of a function. The symbol for integration is $\int$. It is derived from the Greek letter Σ (sigma), used for summing or adding up.
Example If $f(x) = x^2$, then $\int f(x)dx = \frac{x^3}{3} + c$, where c is a constant. The dx is the symbol for integrating with respect to x.

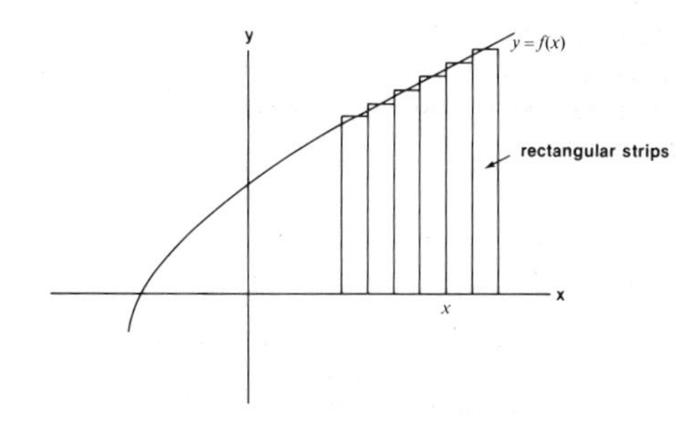

Integration can be viewed as the process of adding up very small rectangular strips between limits. As the width of the strips decreases, then the adding or summing up of them comes closer to the actual area under the curve. The symbol dx comes from the combination of δ and x, together meaning a very small width. $f(x)dx$ can be thought of as the area of a very thin rectangle at the point x up to the curve at that point.
LATIN, *integrare*, to make complete

intercept in **coordinate** or **analytical geometry** the point of intersection of a line or curve with one of the **coordinate axes**.
Example

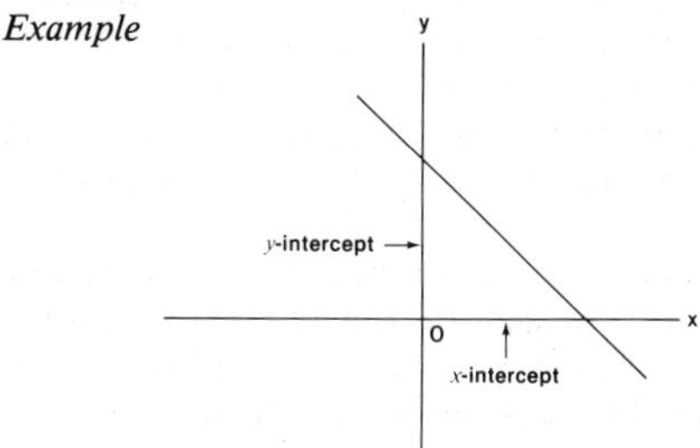

In the case of a straight line cutting the two axes as shown, the **equation** of the line can be written in terms of the intercepts a (x intercept) and b (y intercept) as follows:

$$\frac{x}{a} + \frac{y}{b} = 1$$

LATIN, *intercipere*, to intercept, seize in transit

interior angle **1.** Any angle formed within a figure by two adjacent sides of the figure. **2.** Any of four angles formed between two parallel lines cut by another line.
Examples
1a.

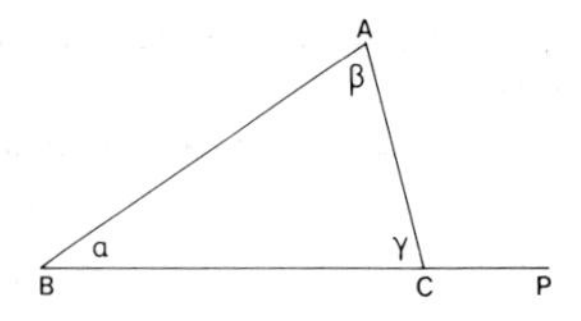

In the $\triangle ABC$, α, β and γ are all interior angles ($\angle ACP$ is an exterior angle). A property of all triangles is that the exterior angle of a triangle is equal to the sum of the two interior opposite angles, i.e. $\angle ACP = \alpha + \beta$

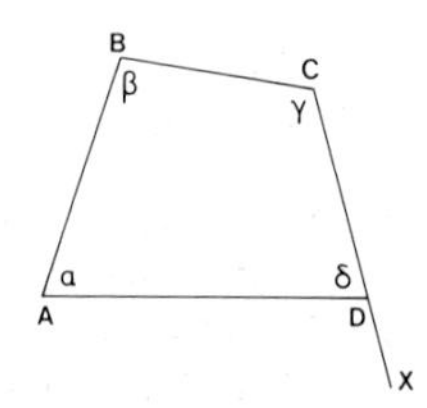

b. α, β, γ and δ are all interior angles of the **quadrilateral** $ABCD$. A property of a **cyclic quadrilateral** is that the exterior angle at any point is equal to the interior opposite angle, i.e. $\angle ADX = \beta$.
2.

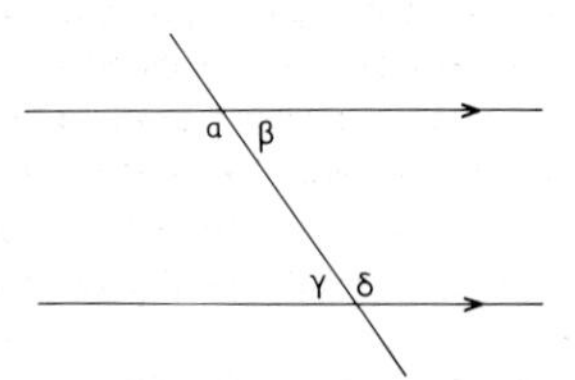

α, β, γ, and δ are interior angles.
LATIN, *interior*, more within; *angulus*, corner, angle

international system of units, *see* SI UNITS

interpolation the act of working out a value between two values in a set of values by a particular method.
Example

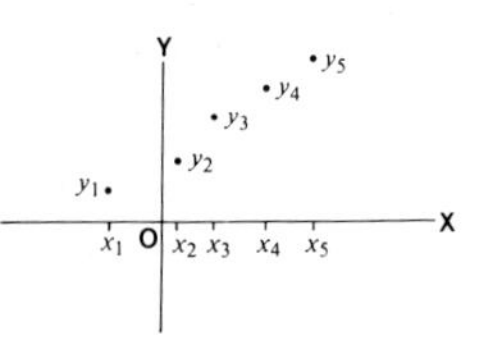

y_1, y_2, y_3, y_4, y_5 correspond to x_1, x_2, x_3, x_4, x_5
We use interpolation to estimate the value y which corresponds to x_a. One method is to use proportion as follows:

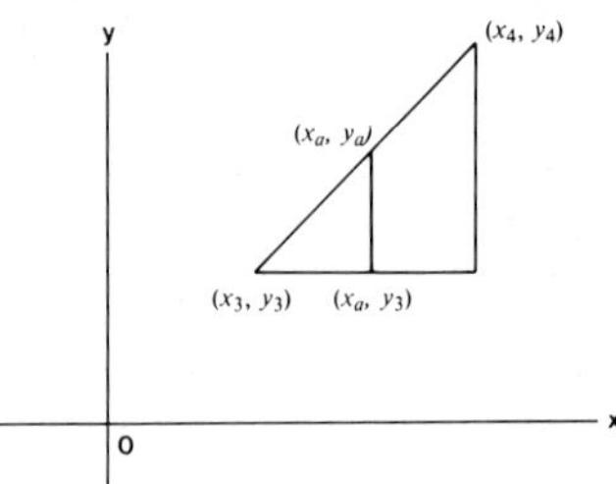

$$\frac{y_a - y_3}{y_4 - y_3} = \frac{x_a - x_3}{x_4 - x_3}$$

LATIN, *inter*, between; *polire*, to adorn, polish

interpret to explain the meaning of.
Examples Graphs often need someone to interpret them. What you can interpret from the graph below?

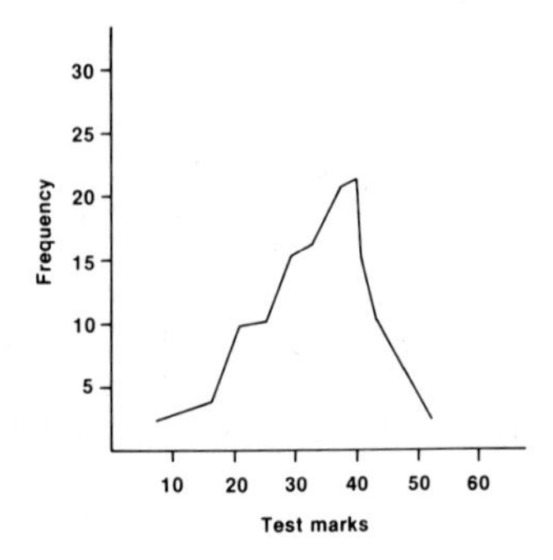

$x^2 + y^2 = 1$ can be interpreted as the equation of circle centre (0,0) and radius 1.
LATIN, *interpres*, interpreter, explainer

interquartile range the range of a set of data between the first and third **quartiles**.
Example In the set of data there are 16 individual results: 1, 7, 9, 15, 16, 21, 22, 23, 23, 23, 24, 24, 24, 25, 25, 25.

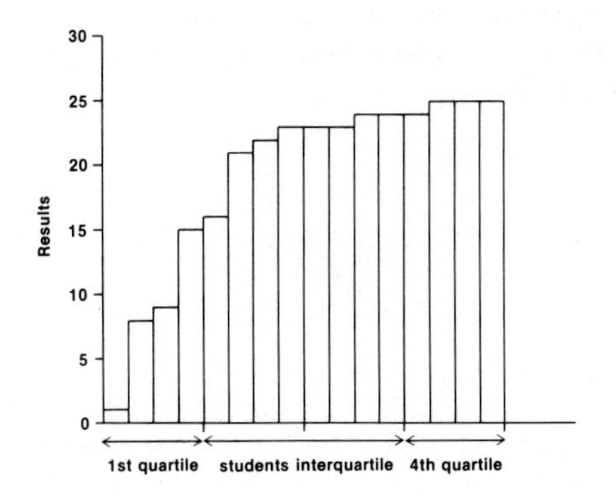

The first quartile (the first quarter of the data) is 15 or less. The third quartile (three quarters of the data) is 24 or less. The interquartile range is 15 to 24.
LATIN, *inter*, between; *quartus*, fourth

intersection 1. The action of lines and/or curves crossing or **touching**. 2. The **set** the members of which are in the sets overlapping each other.
Examples
1.

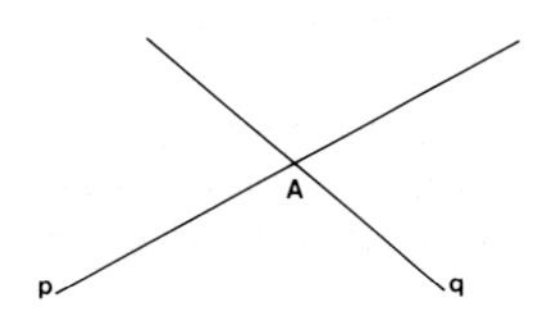

A is the intersection of lines p and q.
2. If $P = \{1, 2, 4, 8, 10\}$ and $Q = \{2, 5, 7, 8, 9, 10\}$ then $P \cap Q$ (the intersection of P and Q is $\{2, 8, 10\}$).
LATIN, *intersecere*, to mutually cut

interval 1. A space or distance between two points, objects or units. 2. The **set** of all points between two points on a line. The set may or may not include the two points at each end.
Examples

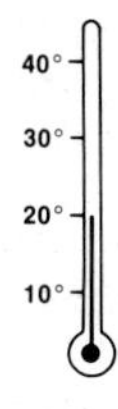

1. The interval between 10° and 40° is 30°.

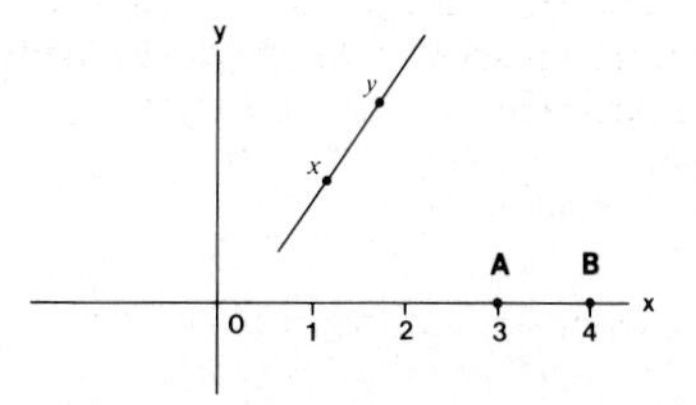

2. AB is an interval (all points between 3 and 4 inclusive). XY is another interval. When the two end points are included, an interval is said to be closed.
LATIN, *intervallum*, space between ramparts

invariant something that remains the same after a **transformation**.
Example The **area**, **perimeter**, and **angles** of a triangle remain the same after the triangle has been rotated.

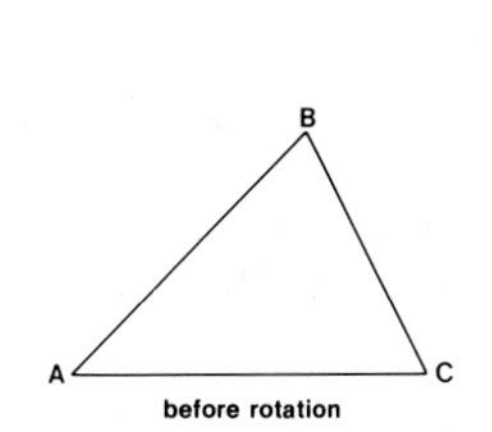

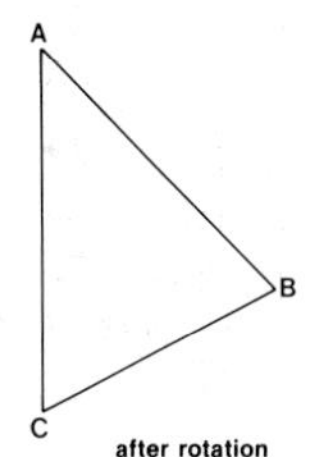

LATIN, *invarius*, not changing

inverse 1. The inverse of an element 'a' in a **set** is that **element** which when combined with 'a' gives the identity element. The combining must be a **binary operation**. 2. Referring to a reversal.
Examples
1. The inverse of 5 is -5 (when the operation is addition) since $5 + (-5) = 0$ (0 is the identity element for the addition operation). The inverse of 5 as far as multiplication is concerned is $\frac{1}{5}$, since $5 \times \frac{1}{5} = 1$, which is the identity element for the operation.
2. An inverse **mapping** is a mapping which maps each point of an image **set** back on to the corresponding point on the object set. If $x = \sin\theta$, then $\sin^{-1}x = \theta$ is the inverse function, meaning the angle whose sine is x. The -1 refers not to a power, but to the inverse operation.
LATIN, *invertere*, turn upside down

irrational number a **real** number which cannot be expressed as a ratio a/b, where a and b are whole numbers, and $b \neq 0$.
Examples $\sqrt{3}$, $\sqrt[3]{55}$, π
LATIN, *irrationalis*, not reasonable; *numerus*, number

isometry a **transformation** in which all the lengths remain the same.
Example A translation is an isometry as is a **reflection** and **rotation**.

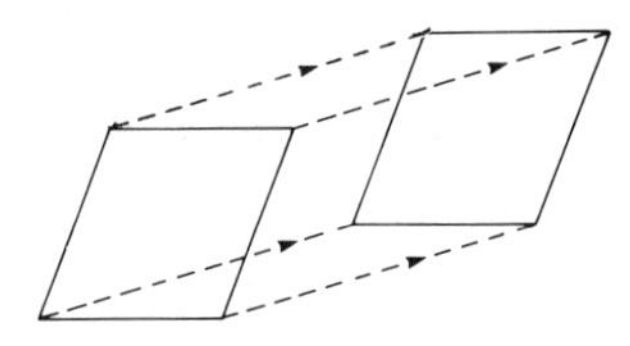

LATIN, *isometria*, of equal measure

isosceles (of a triangle) having two equal sides.
Example

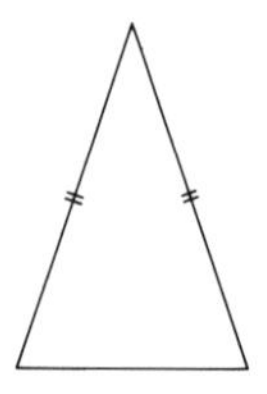

An isosceles triangle also has 2 angles equal.
GREEK, *isoskeles*, having equal legs

isomorphic relating to two or more **sets** the **elements** of one of which can be put in a **one-to-one correspondence** with elements of another, and there is a corresponding operation in each.
Example The arabic system 1, 2, 3, 4, ... and the roman system, I, II, III, IV, ... are isomorphic, also the arabic (**decimal**) system and the binary system 1, 10, 11, 100, 101, 110, 111, 1000, ... Sums and products in the arabic system correspond with sums and products in the roman and binary systems.

1,	2,	3,	4,	5,	6,	7,	...
↕	↕	↕	↕	↕	↕	↕	
1	10	11	100	101	110	111	...

GREEK, *iso*, same; *morphe*, shape or form

iteration the action or process of performing again (hence **iterative**, adj.).
Examples Iteration is a way of working out answers to many problems by first finding an approximate answer, then using this to find a more accurate answer, then using the more accurate answer to find an even more accurate answer, and so on.

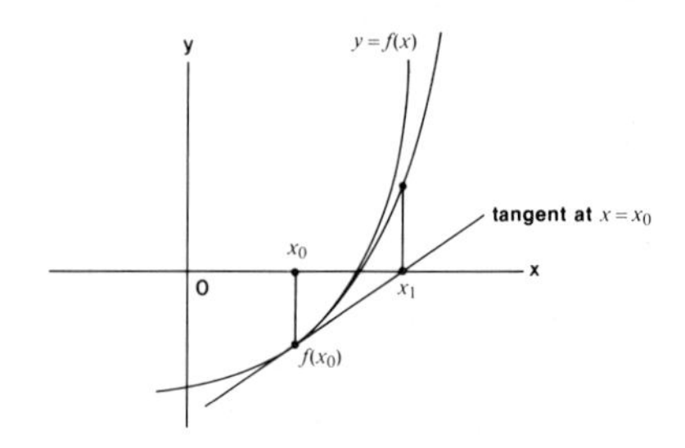

To find x so that $f(x) = 0$ we can guess a value x_0 which makes $f(x_0) \doteqdot 0$. By using a method (known as **Newton's method**) we can find x_1 which is a better approximation than x_0 (by drawing a tangent at x_0 and finding where it cuts the x-axis). By repeating the operation of drawing **tangents** we can find x_2, x_3, etc., each of which comes closer to the actual solution of $f(x) = 0$. Since computers can perform calculations over and over again very rapidly iteration is often used to solve problems which can be put into this form. A solution thus derived is said to be an iterative solution.
LATIN, *iterare*, to repeat, do again

Jj

join union.
Example The join of **sets** P and Q (written $P \cup Q$) is a set of things which are in P or Q or both.

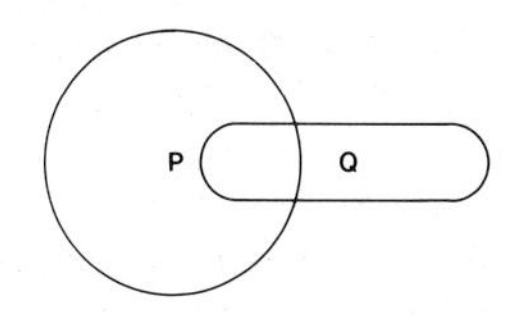

LATIN, *jungere*, to join

joule a unit of energy. It is amount of work done when a force of 1 **newton** moves through a distance of 1 **metre**.
Example To heat a kettle requires many joules of electrical energy.
after James Joule, English physicist

Kk

kilo one thousand.
Example kilogram (1000 grams); kilometre (1000 metres).
GREEK, *khilioi*, a thousand

kite a four sided figure with two pairs of sides equal. The equal sides are adjacent.
Example

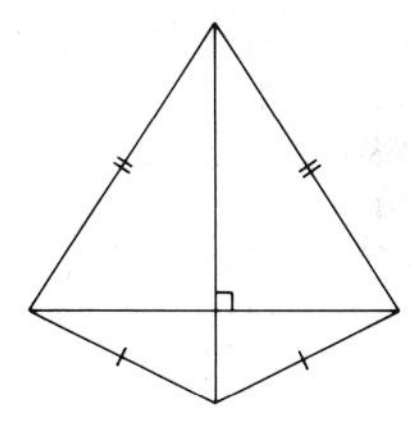

Note: The diagonals intersect at right angles. There is only one axis of **symmetry** in a kite.
OLD ENGLISH, *cyta*, kite (bird of prey)

Klein, Felix (1849–1925) German mathematician who was especially interested in the study of non-**Euclidean geometry**. Klein showed that non-Euclidean and Euclidean geometries were really special cases of **projective geometry**.

königsberg bridge problem a famous problem in **topology**, posed by the Swiss mathematician **Euler**, involving the seven bridges of the city of Königsberg. Euler proved that it was impossible for a person to start anywhere and cross each bridge only once.
Example

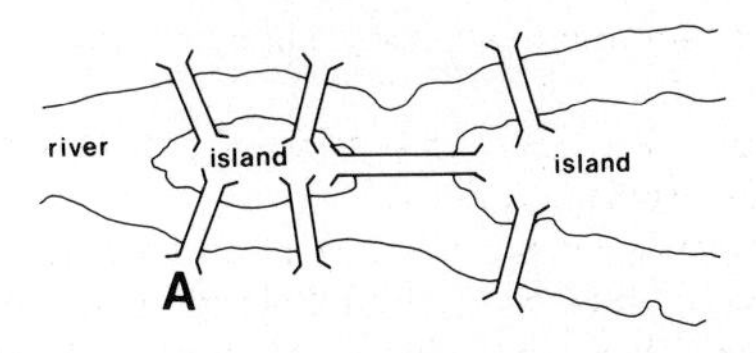

If one starts at A, one can cross six bridges only once, but to cross the last requires crossing at least one twice.

Ll

label **1.** (computing) A letter or group of letters, symbols which are used to identify a value or an **address** in a computer. **2.** (graphs) A symbol, word, or phrase used to describe what quantity is being plotted on an axis or sector of a graph.
Examples
1. In the following group of computer instructions,

```
READ A
GOTO OUT
OUT: PRINT A
```

there is a label called OUT which represents the address of the PRINT instruction.
2.

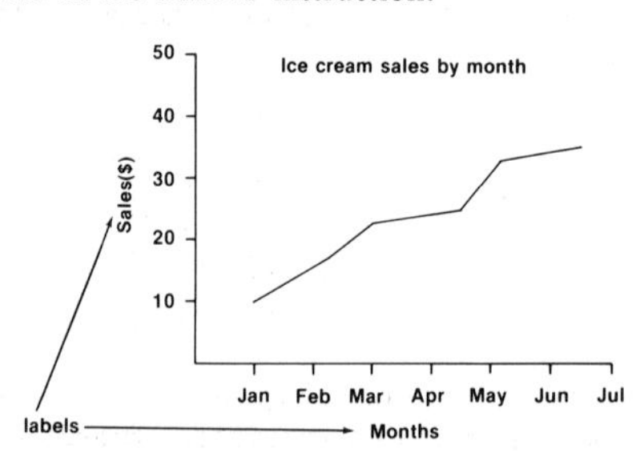

OLD FRENCH, *label*, strip, ribbon

latin square a square made up of n rows and n columns containing n^2 locations. Each location contains one of n specified numbers so that every number appears once and only once in each row and each column.
Example For $n = 4$:

0	1	2	3
1	2	3	0
2	3	0	1
3	0	1	2

The above square is a latin square for $n = 4$. Patterns represented by latin squares are useful in designing experiments to minimize the number of tests required to evaluate the effects of various factors. In the case above, 0, 1, 2, 3 could represent 4 different wines, the rows 4 different wine tasters, and the 4 columns 4 glasses each made of different material. The latin square allows one to set up the minimum number of tastings to work out how the wine tasters and glasses affect results.
LATIN, *latinus*, latin; *quadra*, square

latitude one of the two angles used in navigation. It is the number of degrees in an **arc** of **meridian** north or south of the equator. Lines of latitude are lines parallel to the equator.
Example

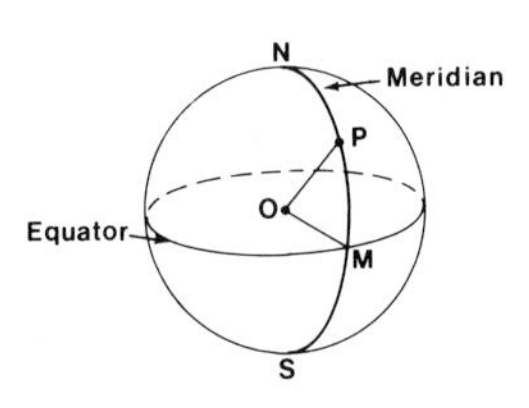

$\angle POM$ is the latitude of P.
LATIN, *latitudo*, breadth

lattice points points formed by one group of **parallel** lines cutting another group of parallel lines. The two groups of lines are usually at right angles to each other.
Example

A	B	C	D
E	F	G	H
I	J	K	L
M	N	O	P

$A, B, C, D, \ldots P$ are lattice points. These points can be described by the two numbers which refer to the distances of the point from two axes as shown.

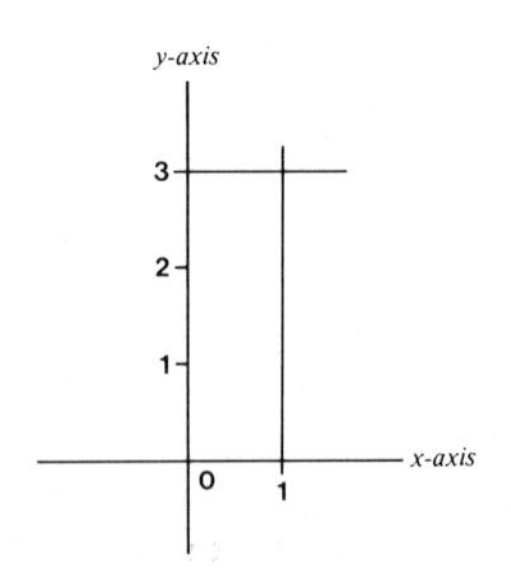

Point A is 1 unit from the y axis, and 3 units from the x axis. This is written in shorthand as (1,3). The point A is often described by the pair (1,3).
OLD FRENCH, *lattis*, lath, strip; LATIN, *punctus*, pricked

latus rectum a straight line drawn through the **focus** of a **parabola** (or other **conic** section) at right angles to the principal axis.

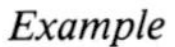
Example

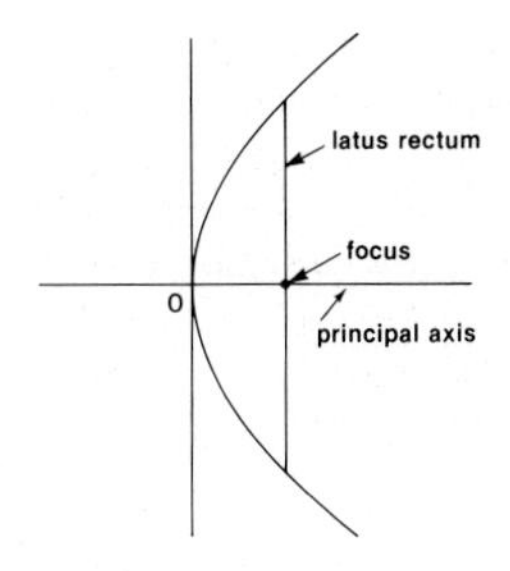

For the parabola $y^2 = 4ax$, length of latus rectum $= 4a$.
LATIN, *latus*, side; *rectum*, straight

law of errors (also **Gaussian law of errors**) an assumption that chance errors in measurement tend to be distributed about a mean according to the "**normal**" pattern. The diagram below illustrates the normal pattern or curve.

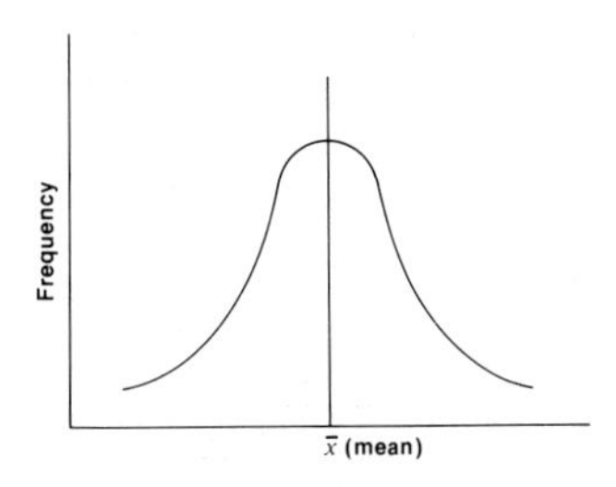

Example The following measurements of the speed of light were taken over successive days.

no. of times measured	2	5	20	4	1
measured speed (km/s × 10^5)	2.96	2.97	2.98	2.99	3.0

The mean is apparently 2.98×10^5 km/sec, and the other results are spread about the mean in a way that approximates the **normal** curve.
OLD ENGLISH, *lagu*, something laid down; LATIN, *errare*, to wander. **Gauss**, a famous eighteenth century mathematician, was one of the first to work on this subject.

law of large numbers (statistics) an assumption that the larger the number of observations, the more closely they will tend to be representative of the total population from which they were drawn.
Example Out of a town of 2000 people, the heights of 20, then 200, were measured. The results were:

20	No.	0	1	6	7	5	1
	H (cm)	131–140	141–150	151–160	161–170	171–180	181–190
200	No.	3	17	34	69	50	27
	H (cm)	131–140	141–150	151–160	161–170	171–180	181–190

The law of large numbers says that the second group is more likely to be representative of the actual population.
OLD ENGLISH, *lagu*, something laid down; LATIN, *largus*, copious; *numerus*, number

laws of indices, *see* INDEX LAWS

LCM *see* LOWEST COMMON MULTIPLE

leading diagonal the line of numbers or letters in a table or **matrix** which go from the top left corner to the bottom right corner.
Example

$$\begin{bmatrix} 1 & 2 & 3 & 4 & 5 \\ 3 & 4 & 5 & 6 & 7 \\ 5 & 6 & 7 & 8 & 9 \\ 7 & 8 & 9 & 10 & 11 \\ 9 & 10 & 11 & 12 & 13 \end{bmatrix}$$

leading diagonal

OLD ENGLISH, *laedan*, to lead; GREEK, *diagonios*, from angle to angle

least **1.** Smallest or lowest. **2.** *(At) least*—no less than.
Examples
1. The least **common denominator** of the fractions $\frac{5}{12}$ and $\frac{2}{9}$ is 36 (since 12 and 9 both go into 36, but not both into any smaller number).
2. Find at least two sets of positive numbers x and y so that $3x + 2y = 12$.
(Let $x = 1$, then $3 + 2y = 12$, $y = 4\frac{1}{2}$ which is positive)
(Let $x = 2$, then $6 + 2y = 12$, $y = 3$ which is positive)
OLD ENGLISH, *laest*, least

least squares method (statistics) a method used to work out a straight line which best fits a set of statistical data. The method was first developed by a French mathematician, Legendre, in the nineteenth century.
Example

Sales	95	207	231	320	364	610
Month	J	F	M	A	M	J

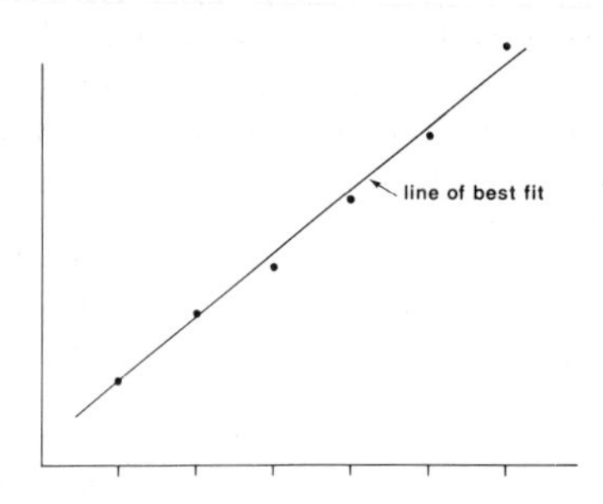

The **graph** shows the tables of values plotted (sales against months). The least squares method allows one to work out the line of best fit as drawn. This is useful when working out trends.
OLD ENGLISH, *laest*, least; LATIN, *quadra*, a square; GREEK, *methodos*, a going after, pursuit

Leibniz, Gottfried (1646–1716) German mathematician of great distinction jointly responsible with, though independent of, Isaac **Newton** for the invention of calculus. Leibniz invented the current notations dx, dy and $\frac{dy}{dx}$

less than smaller than (symbol is $<$).
Examples $7 < 11$, $-2 < 1$.
For what values of x is $3x + 1 < 7$?
This can be solved as follows:

$$3x + 1 - 1 < 7 - 1 \text{ (add } -1 \text{ to each side)}$$
$$\frac{3x}{3} < \frac{6}{3} \text{ (divide both sides by 3)}$$
$$x < 2$$

OLD ENGLISH, *laes*, less, smaller

like terms terms which contain the same combination of letter symbols or powers of symbols. The combination is usually made up of letters multiplied or divided in groups.
Examples
$3x^2y$, $2x^2y$ are like terms, since both contain x to the second power, and y to the first power.
$\frac{2a}{b^3}$, $\frac{10a}{b^3}$ are like terms.
$6x^2y$, $7x^2y^2$ are unlike terms, since the first term contains y to the first power, and the second term contains y to the second power.
OLD ENGLISH, *gelic*, similar; LATIN, *terminus*, end, limit

limit **1.** (of a sequence) A number which is approached ever more closely, but never is reached by the successive **terms** of a **convergent** infinite sequence. **2.** A value which is approached ever more closely by a function of a given **variable** as that variable approaches a given number (symbol is lim).
Examples
1. The limit of the unending sequence 1, $\frac{1}{2}$, $\frac{1}{3}$, $\frac{1}{4}$, ... is 0.
2a. The limit of x^2 as x approaches 2 is 4. This is sometimes written $\lim_{x \to 2} f(x) = 4$, where $f(x) = x^2$ in this case.
b. A not so obvious example is $\lim_{x \to 0} \frac{1 + \frac{1}{x}}{x + \frac{1}{x}} = 1$.
LATIN, *limes*, boundary, frontier

limiting sum the sum of an infinite number of terms in a **geometric series** of **progression**. Also known as the sum to infinity. The limiting sum exists only if the common ratio is less than 1.
Examples $\frac{1}{3} + \frac{1}{9} + \frac{1}{27} + \frac{1}{81} + \ldots$ has a limiting sum of $\frac{1}{2}$ (limiting sum is symbolised as lim S) where:

$$\lim S = \frac{a}{1 - r},$$
$$S = a + ar + ar^2 + \ldots \text{ (see GEOMETRIC SERIES)}$$
$$\text{i.e. } \lim S = \frac{\frac{1}{3}}{1 - \frac{1}{3}} = \frac{\frac{1}{3}}{\frac{2}{3}} = \frac{1}{2}$$

LATIN, *limes*, boundary, frontier; *summus*, highest, topmost

line (straight) the extension of the shortest length between two points on a plane. A line extends infinitely in both directions. A line has length but no breadth or width (hence **linear** adj., *see* LINEAR EQUATION).
Example The straight line between points P and Q is called a line segment.

LATIN, *linea*, thread, line

line of symmetry a line dividing a shape in half in such a way that one half is the mirror image of the other.
Example

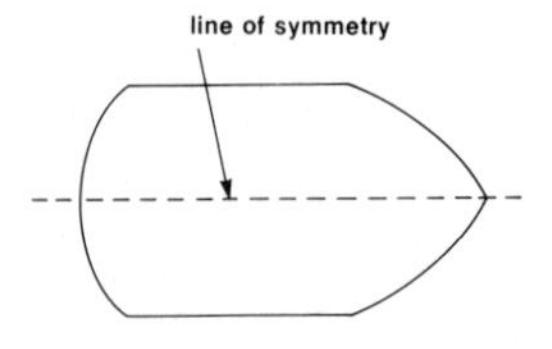

LATIN, *linea*, line, thread; GREEK, *summetros*, of like measure

linear equation the **equation** of a straight **line**. It contains one or more **variables** which are raised to the first power only.
Example $y = mx + c$ is the general equation of a straight line. Note: m represents the slope of the line, and c is the distance from 0 where the line cuts the y-axis (the y-intercept).

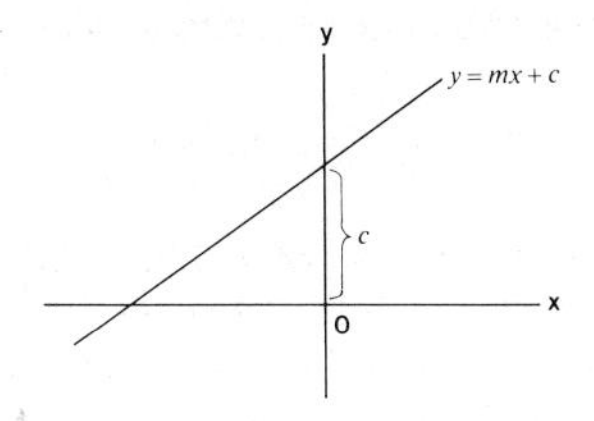

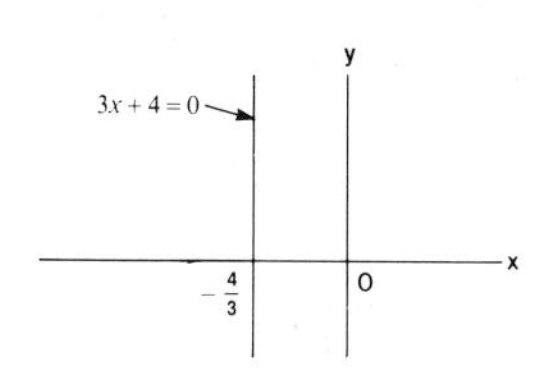

$3x + 4 = 0$ is the equation of the line.
LATIN, *linea*, line, thread; *aequus*, equal

linear programming a mathematical method for working out how best to use available resources under various (usually limiting) circumstances (sometimes abbreviated to LP).
Example A sausage manufacturer has a number of different ingredients to choose from to make up sausage meat. Each ingredient contains different percentages of fats, proteins, water, etc., and also differs in cost per kilogram. The problem of determining which ingredients can be used at the least overall cost while still making sure a given minimum level of protein/maximum level of fat is included can be solved by linear programming.

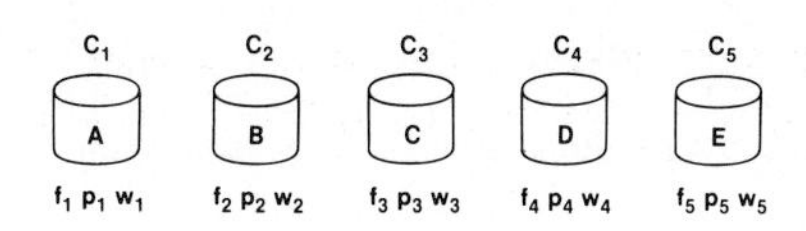

f = fat p = protein w = water
How much of A, B, C, D, E to take to make

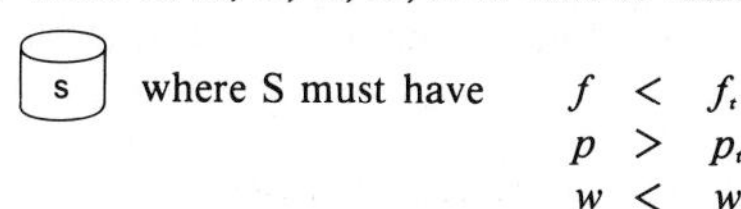

where S must have $f < f_t$, $p > p_t$, $w < w_t$

so that cost of S is a minimum.
LATIN, *linea*, line, thread; *programma*, a public notice

linear regression a way of reducing a set of results or observations to a **linear equation.** One technique used is called the **least squares method**. Another is to plot the results on graph paper and then draw in a straight line which best fits the results.
Example Suppose the volume of sales of small computers is measured at the end of each month for a period of 12 months. If y_1 is the volume sold for month t_1, y_2 for month t_2 etc., then it is possible to approximate the **relation** between y and t by an equation of the form $y = at + b$, where a and b are constants (*see* LEAST SQUARES METHOD).
LATIN, *linea*, line, thread; *regressus*, gone back

litre a unit of volume in the **SI** that is equal to 1000 cubic centimetres. In Australia the symbol is L; elsewhere it is l or *l*.
Example A litre of milk will give four children one 250 mL glass each (mL is the symbol for millilitre of which there are 1000 in a litre).
GREEK, *litra*, unit of wealth, silver coin, pound

locus the path traced out by a point which obeys a specific rule or law.
Example The locus of a point which moves so that it is always the same distance from a given fixed point is a circle.

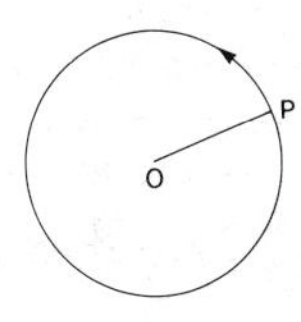

LATIN, *locus*, place

logarithm an index. A number related to another number by a special rule, which is: given a fixed number b, then the logarithm of a number N is the **power** to which the fixed number b is to be raised to produce the number N. The fixed number is called the **base**. Logarithms were used to simplify complicated multiplications and divisions when electronic calculators were not generally available. If $N = b^x$, then the logarithm of N to the base b (written $\log_b N$) is equal to x.
Example There are two types of logarithmic systems in common use. One is common logarithms where the base is 10. Tables can then be constructed so that any number can be written as a power of 10, e.g. $2 = 10^{0.3010}$. The other is **Naperian** or natural logarithms which use the number e (equal to 2.7182818...) as the base. Similar tables can be constructed so that any number can be represented as a power of e. As an

example of using logarithms to simplify calculations, suppose we have to work out

$$\frac{2369.73 \times 109.28}{3674.2}$$

without a calculator. Each number in the expression can be written as (using 10 as the base) 10^x where x is different for each number.

If $2369.73 = 10^x$, and
$109.28 = 10^y$, and
$3674.2 = 10^z$, then another way of writing the problem is

$$\frac{10^x \times 10^y}{10^z}$$

or, using the rules about indices, this can be written as 10^{x+y-z}. As long as we can find out x, y and z easily (e.g. through tables), we can then find out $(x + y - z)$, and by finding out (again through tables) what the number is whose value is 10^{x+y-z}, the calculation is completed.

GREEK, *logos*, reckoning, reason; *arithmos*, number

logarithmic function a function represented by the expression $\log_b x$ (where b represents the **base**).
Example If $y = \log_{10} x$, then y is the logarithmic function with respect to the base 10. $\text{Log}_e x$ represents the **inverse** of the growth function ($y = e^x$ is the representation of growth function).

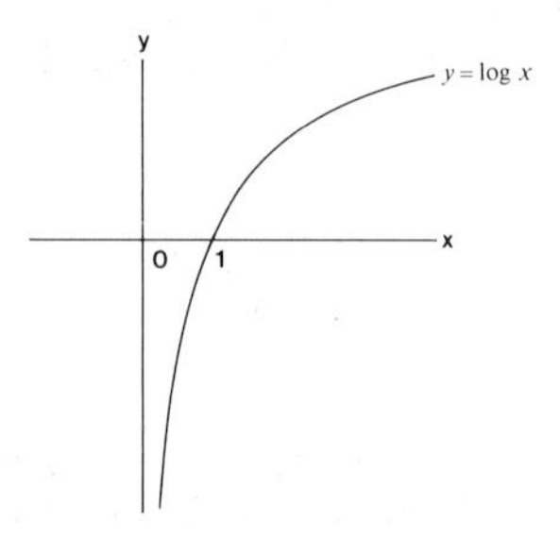

GREEK, *logos*, reckoning, reason; *arithmos*, number; LATIN, *fungi*, to perform

logic **1.** The science of reasoning. **2.** Valid reasoning, in contrast to invalid or irrational reasoning.
Examples
1. If one studies logic one can expect to learn more about the principles behind various forms of reasoning.
2. Prove, using logic, that if two sides of a triangle are equal, then the angles opposite the sides are equal.
GREEK, *logos*, reckoning, reason

long division a way of dividing one number by another by hand.
Example 3017 ÷ 37 can be written and solved as follows:

Try 37 into 3 → no go
Try 37 into 30 → no go
Try 37 into 301, this goes 8 times
Put 8 over the last digit (the 1) and subtract 8 × 37 from 301. This leaves 5. Bring down the 7 to make 57, 37 into 57 goes 1. Put 1 over the 7 and subtract 1 × 37 from 57. This leaves 20. The answer is 81 with a **remainder** of 20 (also written $81\frac{20}{37}$).
OLD ENGLISH, *lang*, long; LATIN, *dividere*, to divide

longitude one of the two angles used in navigation. It is the number of degrees east or west of a prime (or reference) **meridian**, usually that through Greenwich, England. Lines of longitude are lines which are part of great circles running from pole to pole.
Example

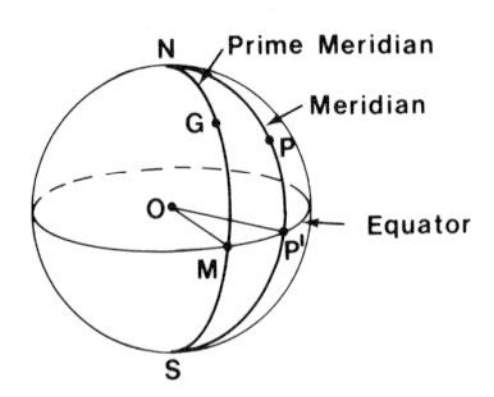

If G is Greenwich, O is the centre of the Earth, and P a point on the Earth's surface, then the longitude of P is $\angle$ MOP′°E (NGMS is the prime meridian).
LATIN, *longitudo*, length.

long multiplication a series of steps for multiplying two numbers together (each number of which is usually greater than 10) so that one number is only multiplied by one digit from the second number at a time, and the result then added together.
Example 476 × 62 can be considered as 476 × 60 + 476 × 2. This can be written in the following way, known as long multiplication.

```
    476
×    62
    952   (476 × 2)
 28 560   (476 × 60)
 29 512   (952 + 28 560)
```

LATIN, *longus*, long; *multiplicare*, to multiply

loop **1.** Part of a curve that closes back on itself. **2.** A **sequence** of computer instructions that is performed a certain number of times or is performed repeatedly until some condition is satisfied.
Examples

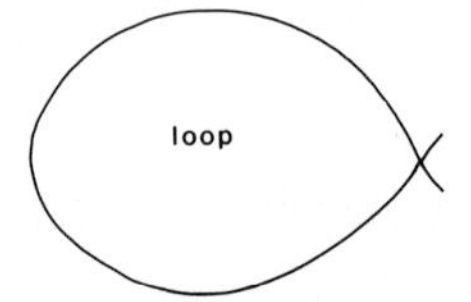

1. loop.

2. the following is a loop using the BASIC language.

```
10 FOR N = 1 to 15
20 M = 2 + N
30 NEXT N
```

MIDDLE ENGLISH, *loupe*, loop

loss the amount (usually money) one loses selling something for less than one paid for it. The opposite to **profit**.
Examples
1. Toy drums were bought at $2 per drum. They were sold for $1.50 per drum. The loss was $0.50 per drum.

$$\text{Loss (of cost price)} = \frac{50}{200} \times \frac{100}{1} = 25\%$$

2. A man bought some shares at $10 per share. One year later he sold them for $7 per share. He lost $3 per share.

$$\text{Loss (of cost price)} = \frac{3}{10} \times \frac{100}{1} = 30\%$$

$$\text{Loss (of selling price)} = \frac{3}{7} \times \frac{100}{1} = 43\%$$

OLD ENGLISH, *los*, dispersion, loss

lowest common multiple (also **LCM**) the smallest number into which two or more given numbers will divide evenly.
Example The LCM of 3, 6, and 4 is 12 and the LCM of a, ab and b^2 is ab^2 (assuming there are no factors common to a and b other than 1).
OLD NORSE, *lagr*, low; LATIN, *communis*, common; *multiplus*, many fold

Mm

M (m) the thirteenth letter in the English alphabet. As a roman numeral M stands for 1000.
GREEK, μ (mu) [12th letter of the Greek alphabet]

magic number the number which is the sum of any row or column or diagonal in a **magic square**.
Example In

8	1	6
3	5	7
4	9	2

the magic number is 15, since each row, each column and each diagonal add up to 15.
OLD PERSIAN, *magus*, sorcerer; LATIN, *numerus*, number

magic square a group of numbers in the form of a square, of which all the rows, columns, and diagonals add up to the same total. The numbers are 1, 2, 3, . . . n^2, where n is the number of rows (or columns). Magic squares were known by the ancient Egyptians and Chinese.
Example

4	9	2
3	5	7
8	1	6

Each row adds up to 15, as does each column, and each diagonal. 15 is called the magic number connected with that square. Another magic square is as follows.

16	2	3	13
5	11	10	8
9	7	6	12
4	14	15	1

OLD PERSIAN, *magus*, sorcerer; LATIN, *quadra*, square

major larger (antonym to **minor**).
Examples

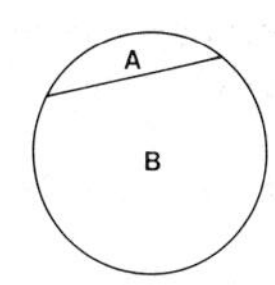

B is the major **segment** of the circle.

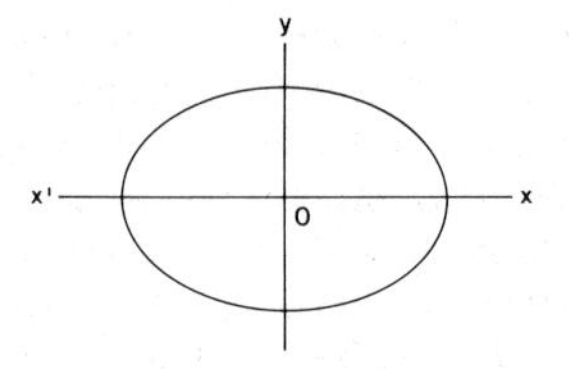

X'OX is the major axis of the ellipse shown.
LATIN, *maior*, greater

mantissa the decimal part of a common **logarithm**.
Example In $20 = 10^{1.3010}$, the $\log_{10}20 = 1.3010$, and the mantissa is 0.3010 (*see also* CHARACTERISTIC).
LATIN, *mantissa*, a makeweight

mapping the action of relating **points** or **elements** in one **region** or **set** to other points or elements in another region or set. One set of elements is called **objects**, the other **images**. A one-to-one mapping of a set onto itself is called a **transformation**. The relation can usually be expressed by some rule.
Examples

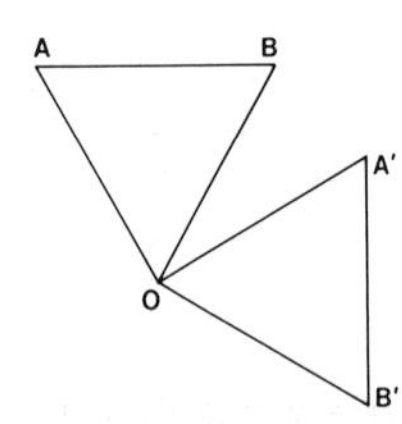

When the triangle *OAB* is rotated about *O* through an angle of 90° then *A'* is the image of *A* as a result of the given rotation, *B'* is the image of *B*, *C'* is the image of *C*. The mapping here is a transformation (a rotation) of points *A*,*B* in the plane *OAB* on to the plane *OA'B'*. Note: In $y = f(x)$, f is a mapping of x into y.

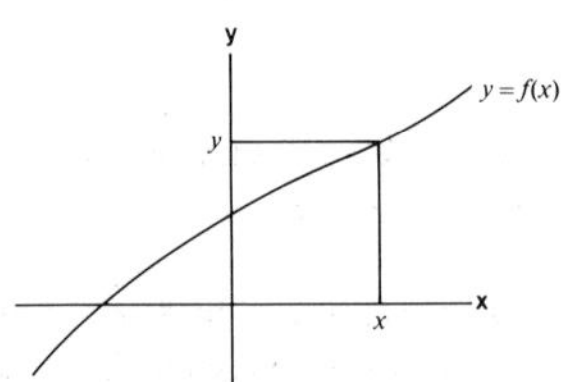

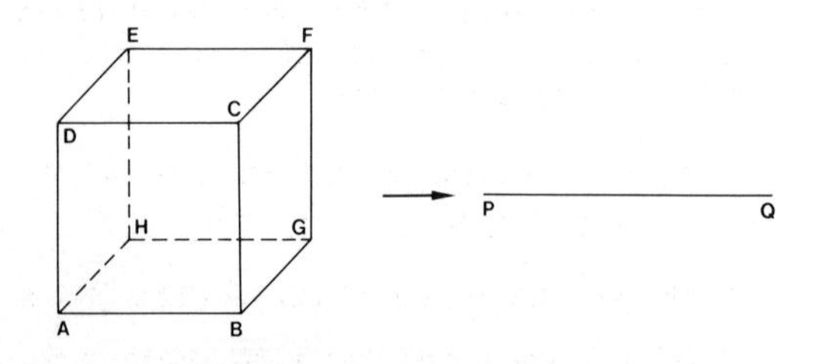

ABCDEFGH can be mapped onto *PQ*.
LATIN, *mappa*, napkin

mathematical system a collection of **principles**, **theorems** and rules which are all related, and which can be used to solve problems of a particular kind.
Examples
The French mathematician and philosopher **Descartes** invented a new mathematical system called **analytical** (or **coordinate**) **geometry**.
Sir Isaac **Newton** and a German mathematician **Leibniz** both independently discovered the mathematical system known today as **differential calculus**.
A simple mathematical system is our **number** system (0, 1, 2, 3, 4, 5, . . .). It allows us to count, compare sizes, work out many day to day problems in handling money etc.
GREEK, *mathema*, science; *sustema*, a composite whole

mathematics the study of the properties of patterns using symbols. The patterns can be number patterns, or shapes, or arrangements. The symbols can be alphabetic, numeric, or special ones to represent how patterns or parts of patterns are related. Pure mathematics deals with patterns only. Applied mathematics takes the results of pure mathematics and applies them to solving problems in life. Mathematics includes ideas of abstraction, generalisation, proof, making inferences, deductive-thinking, and drawing logical conclusions from valid assumptions.
Examples
1. In the pattern $1 + 2 + 3 + 4 + \ldots + n$ (the sum of the first n positive numbers) someone a long time ago discovered that it could be shortened to $\frac{1}{2}n(n + 1)$.
2. Another pattern whose properties have been studied is the **triangle**. A triangle with two equal sides has several properties, for example the angles opposite the equal sides are also equal. Also if a triangle is drawn with two equal angles, then the sides opposite can be shown to be equal.

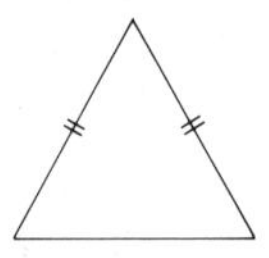

3. The following pattern can be represented by a simpler pattern using the symbols y, x, 3 and $=$, *viz.*, $y = 3x$, where x stands for the first eight positive whole numbers.

1	2	3	4	5	6	7	8
3	6	9	12	15	18	21	24

GREEK, *mathema*, science

matrix (*pl.* **matrices**) an **array** of symbols (numbers or letters) written in the form of a square or rectangle. Each symbol is called an **element**, and is often written a_{ij} where i represents the ith row, and j the jth column.

Example

If $A = \begin{pmatrix} 2 & 3 & 1 \\ 4 & 1 & 0 \\ 1 & 0 & 2 \end{pmatrix}$, then $A_{23} = 0$.

$\begin{pmatrix} 1 & 1 \\ 0 & -1 \end{pmatrix}$ is a 2×2 matrix.

If $\begin{pmatrix} x & y \\ p & q \\ a & b \end{pmatrix} = P$, then $P_{31} = a$

Note: A matrix can represent a **transformation**. For example if (x, y) is the object point, and (x', y') is the image point, then:

$\begin{pmatrix} x' \\ y' \end{pmatrix} = \begin{pmatrix} a & b \\ c & d \end{pmatrix}\begin{pmatrix} x \\ y \end{pmatrix}$ represents the mapping or transformation of $\begin{pmatrix} x \\ y \end{pmatrix}$ onto $\begin{pmatrix} x' \\ y' \end{pmatrix}$. The column $\begin{pmatrix} x \\ y \end{pmatrix}$ is known as a **vector**.

LATIN, *matrix*, womb

matrix addition the operation of adding two or more matrices together. The two matrices must have the same number of rows and columns.

Examples

If $A = \begin{pmatrix} 1 & 3 & -2 \\ 4 & 7 & 9 \\ 5 & 1 & -8 \end{pmatrix}$ and $B = \begin{pmatrix} a & b & c \\ 3 & 2 & 1 \\ x & y & z \end{pmatrix}$, then

$A + B = \begin{pmatrix} 1+a & 3+b & -2+c \\ 7 & 9 & 10 \\ 5+x & 1+y & -8+z \end{pmatrix}$

LATIN, *matrix*, womb; *addere*, add, put to

matrix multiplication the operation of multiplying two or more matrices together. The two matrices must have rows and columns equal, or the columns of the first matrix equal to the rows of the second matrix, and the rows of the first equal to the columns of the second.

Examples

If $A = \begin{pmatrix} a_{11} & a_{12} \\ a_{21} & a_{22} \end{pmatrix}$ and $B = \begin{pmatrix} b_{11} & b_{12} \\ b_{21} & b_{22} \end{pmatrix}$

then $A \bullet B = \begin{pmatrix} a_{11}b_{11} + a_{12}b_{21} & a_{11}b_{12} + a_{12}b_{22} \\ a_{21}b_{11} + a_{22}b_{21} & a_{21}b_{12} + a_{22}b_{22} \end{pmatrix}$

If $P = \begin{pmatrix} p_{11} & p_{12} \\ p_{21} & p_{22} \\ p_{31} & p_{32} \end{pmatrix}$ and $Q = \begin{pmatrix} q_{11} & q_{12} & q_{13} \\ q_{21} & q_{22} & q_{23} \end{pmatrix}$

then $P \bullet Q =$

$$\begin{pmatrix} p_{11}q_{11} + p_{12}q_{21} & p_{11}q_{12} + p_{12}q_{22} & p_{11}q_{13} + p_{12}q_{23} \\ p_{21}q_{11} + p_{22}q_{21} & p_{21}q_{12} + p_{22}q_{22} & p_{21}q_{13} + p_{22}q_{23} \\ p_{31}q_{11} + p_{32}q_{21} & p_{31}q_{12} + p_{32}q_{22} & p_{31}q_{13} + p_{32}q_{23} \end{pmatrix}$$

LATIN, *matrix*, womb; *multiplicare*, to fold many times

maximum (*pl.* **maxima**) **1.** The greatest value of a set of values. **2. Absolute** maximum (of a function). The value of the function which is greater than all other values that the function can take (in the defined **range**). In symbols, $f(a)$ is an absolute maximum of the function $f(x)$ if $f(a) \geqslant f(x)$ for every x in the given **domain**. **3. Relative** maximum (of a function). The value of the function which is greater than all other values of the function close to the place of the maximum. In symbols, $f(a)$ is a relative maximum if $f(a) \geqslant f(x)$ for values of x just smaller and just greater than a.

Examples

1.

x	−1	0	1	2	3	4	5	6	7	8
y	−6	−2	0	2	7	6	5	3	1	−9

In the above table the maximum of y is at $x = 3$ and the maximum value of y is 7.

2.

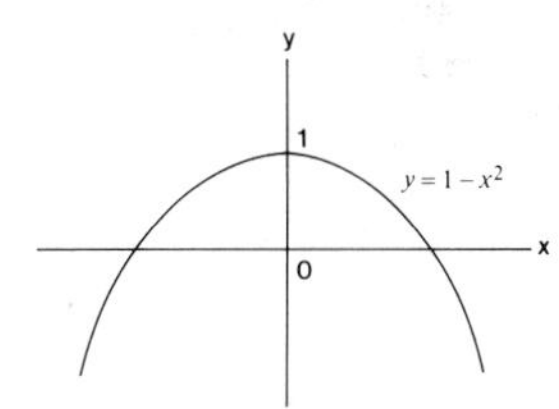

$y = f(x)$ has an absolute maximum of 1 at $x = 0$ (for *all* x).

3.

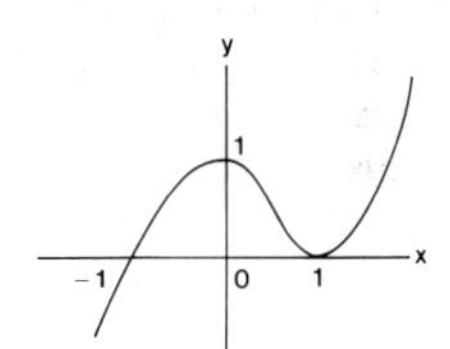

$y = f(x)$ has a relative maximum of 1 at $x = 0$.

LATIN, *maximus*, greatest

mean **1.** Arithmetic mean (of a set of **elements**): The sum of the elements divided by the number of elements (also known as **average**). **2.** Geometric mean (of two elements): The square root of the two elements multiplied together. **3.** Means: The two middle terms in a proportion.

Examples

1. The arithmetic mean of 10, 12, 15, 7, 9, and 13 is

$$\frac{10 + 12 + 15 + 7 + 9 + 13}{6} = 11$$

2. The geometric mean of 12 and 3 is $\sqrt{12 \times 3} = 6$.

3. If $\frac{a}{b} = \frac{c}{d}$ $(a:b = c:d)$, then b and c are the means (a and d are the **extremes**).

LATIN, *medius*, middle

measure **1.** The result of comparing the size of "how much" of anything with a known standard. **2.** The known standard or unit itself.
3. A device for making such a comparison.
Examples
1. There are several measures which can be made as far as a tennis ball is concerned. The **diameter** gives us one measure of the size. The volume is another measure which can be worked out from the diameter. Other measures in general are **areas, lengths, masses,** and time.
2. A square centimetre (1 cm^2) is a unit of area.

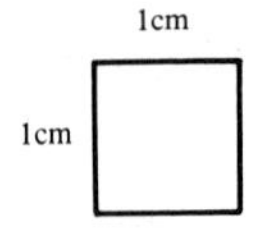

3. A tape measure is often used to work out how long a sports court is.
LATIN, *metiri*, to measure

measurement **1.** The act of finding out the size or "bigness" of anything by comparing it to a known standard or **unit**. **2.** The result found out by comparing the size of something with a known standard or unit. **3.** A way of measuring.
Examples
1. Can anyone do the measurements of the room to find out how much wallpaper we need?
2. I have the correct measurements of the room.
3. The measurement of the distance was given in metres.

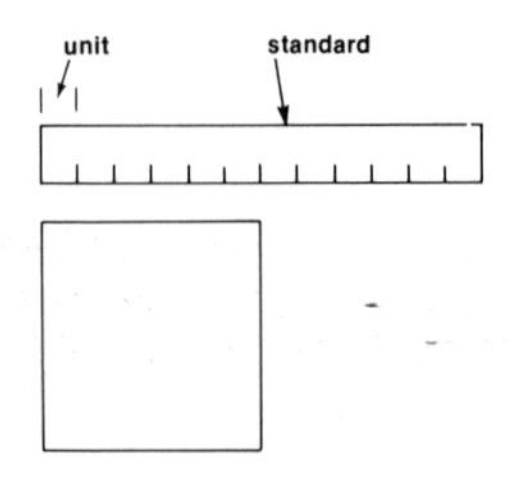

The length of the side of the square is 6 units.
LATIN, *metiri*, to measure

median **1.** The middle number in a **series** of numbers arranged in order of size. If there is an **even** number of **elements** in the series, the median is the **arithmetic mean** of the two middle terms. **2.** (of a triangle) The line drawn from the **vertex** of a triangle to the mid-point of the side opposite.
Examples
1. The median of 0, 1, 2, 2, 3 , 4, 4, 5, 7 is 3.
The median of 1, 1, 2, 3 , 4, 5 is $\frac{1}{2}(2 + 3) = 2\frac{1}{2}$.
2.

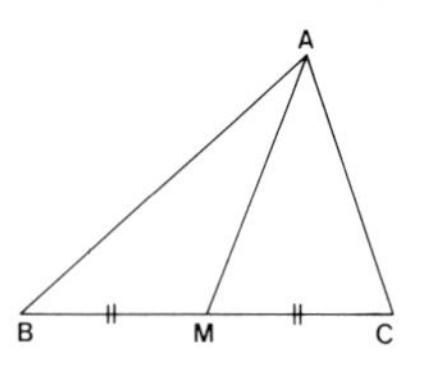

AM is a median.
LATIN, *medius*, middle

mega- prefix meaning one million.
Example A megatonne is one million tonnes.
GREEK, *megas*, great

member an element belonging to a **set**. The symbol ϵ (epsilon) means "is a member of".
Example 3 is a member of the set of counting numbers. $3 \epsilon \{1, 2, 3, 4, 5, \ldots\}$.
LATIN, *membrum*, a member

mensuration **1.** The act or process of measuring.
2. The **measurement** of geometric quantities.
Examples
1. Can we solve the problem by mensuration?
2. Mensuration such as volumes of spheres, cubes, pyramids etc. is a subject many people are taught.

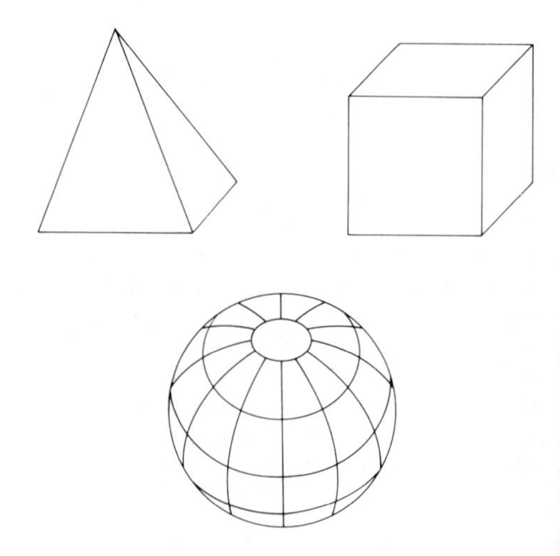

LATIN, *mensura*, measure

Mercator('s) projection a method of **mapping** points on the surface of a **sphere** on to a plane. It is used in map making to represent sections of the world (a sphere) on a plane surface. **Meridians** and lines of **latitude** become straight lines at right angles to each other, hence areas further away from the equator appear larger than areas closer to the equator.

Example

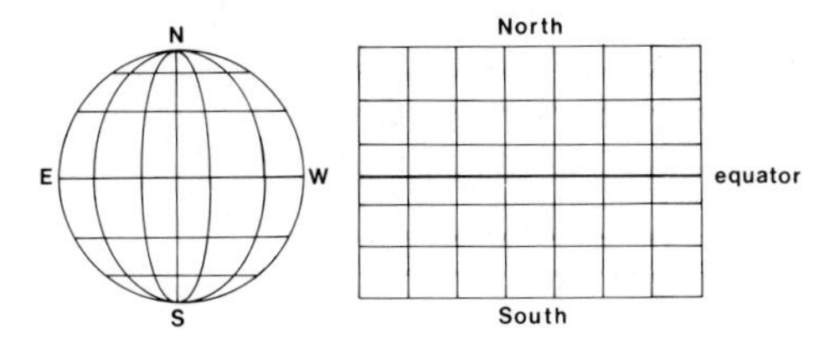

after G. Mercator, Flemish mapmaker; LATIN, *projectus*, thrown forward

meridian an **arc** of a **great circle** passing through the North and South poles of the Earth. The **prime** meridian is a meridian passing through Greenwich (England)
Example

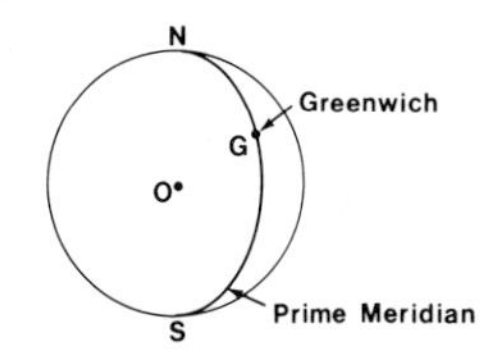

G has the longitude 0°E.
LATIN, *meridies*, midday

metre a basic unit of length in the metric system and the **SI**. It is written as m for short. It was originally defined as one ten millionth of the distance from North Pole through Paris to the equator (on a meridian), then defined as the distance between two marks on a metal alloy bar kept at the Weights & Measures Bureau in France. More recently it was defined as 1 650 763.73 wavelengths of orange-red radiation from the gas Krypton-86. In 1983 it was defined in terms of the speed of light.
Example A man can be about 1.8 metres in height.

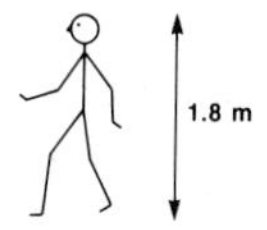

GREEK, *metron*, measure

mid-ordinate method (or rule) a way of finding the approximate **area** under a curve between two points by dividing the **interval** between the two points into a certain number of equal subdivisions and summing the rectangles formed by each subdivision's width and the height of the curve at the **midpoint** of that subdivision.
Example

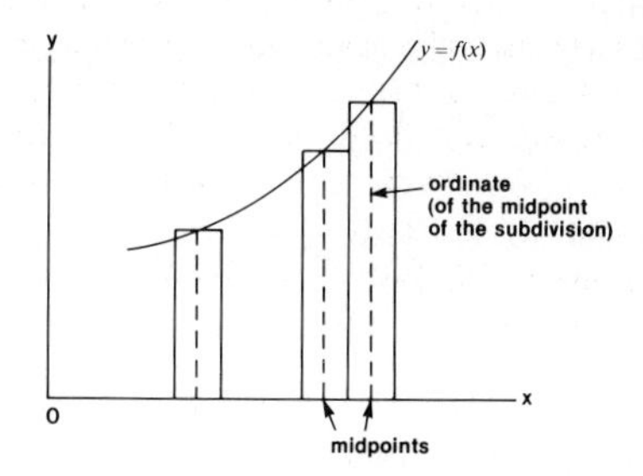

As the subdivisions are chosen smaller, so the approximation of the area is better.
LATIN, *medius*, middle; *ordinare*, to arrange in order

midpoint the point which separates a **line segment** into two equal parts.
Example M is the midpoint of the line segment XY.

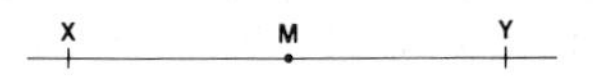

LATIN, *medius*, middle; *punctus*, pricked

midpoint formula a formula for finding the coordinates of the **midpoint** of a **line segment**. If $A(x_1, y_1)$ and $B(x_2, y_2)$ are two points of a line segment, then $P(x_m, y_m)$ is calculated as follows:

$$x_m = \frac{x_1 + x_2}{2}$$

$$y_m = \frac{y_1 + y_2}{2}$$

In three dimensions, if $A(x_1, y_1, z_1)$ and $B(x_2, y_2, z_2)$ are the two end points, then the midpoint $P(x_m, y_m, z_m)$ is similarly calculated by:

$$x_m = \frac{x_1 + x_2 + x_3}{3}$$

$$y_m = \frac{y_1 + y_2 + y_3}{3}$$

$$z_m = \frac{z_1 + z_2 + z_3}{3}$$

Example

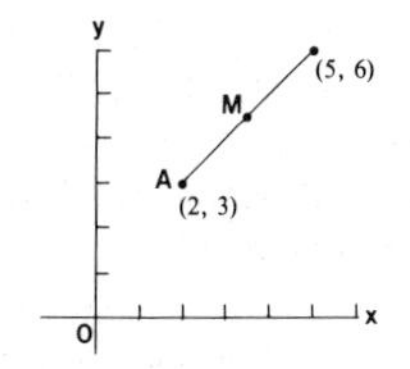

LATIN, *medius*, middle; *formula*, small form

milli- prefix meaning one thousandth.
Example A milligram (mg) is a thousandth of a **gram**.
LATIN, *mille*, thousand

millimetre one thousandth of a metre (symbol is mm).
Example In most countries rainfall is measured in mm.
LATIN, *mille*, thousand; GREEK, *metron*, measure

minimum **1.** The **least** value of a **set** of values. **2.** Absolute minimum (of a function): The value of the function which is smaller than all other values that the function can have (in the **domain** of definition). In symbols, $f(a)$ is an absolute minimum of the function $f(x)$ if $f(a) \leqslant f(x)$ for every x in the given domain. **3.** Relative minimum (of a function): The value of the function which is smaller than the other values of the function close to the value in question. In symbols, $f(a)$ is a relative minimum if $f(a) \leqslant f(x)$ for values of x just smaller and just greater than a.
Examples
1.

x	−1	0	1	2	3	4	5	6	7	8
y	−6	−2	0	2	7	6	5	3	1	−9

In the above table, the minimum value of y is at x = 8 and the minimum value of y is −9.
2.

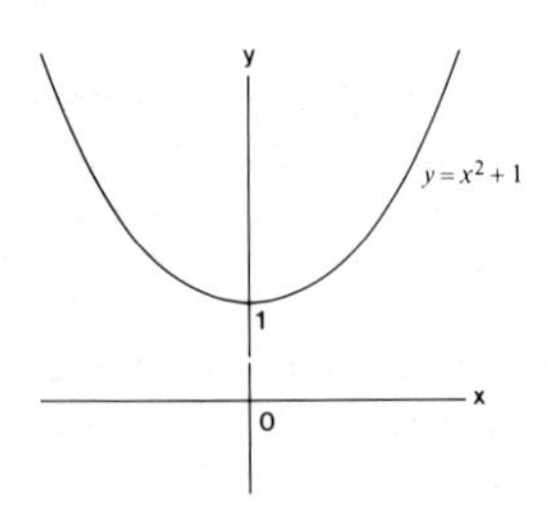

$y = f(x)$ has an absolute minimum of 1 at $x = 0$ (for all x).
3.

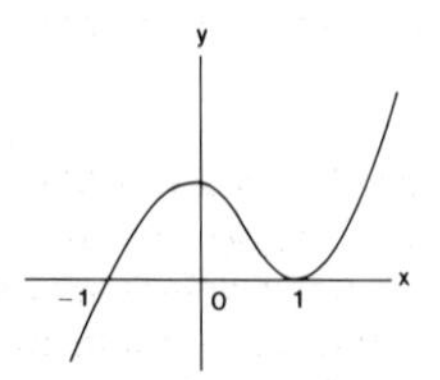

$y = (x + 1)(x - 1)^2$ has a relative minimum of 0 at $x = 1$.
LATIN, *minimus*, least

minor smaller (antonym of **major**).
Examples

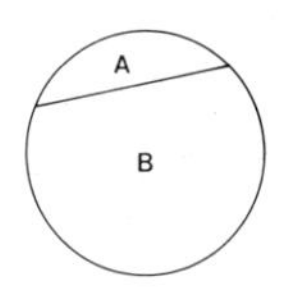

A is the minor **segment** of the circle.

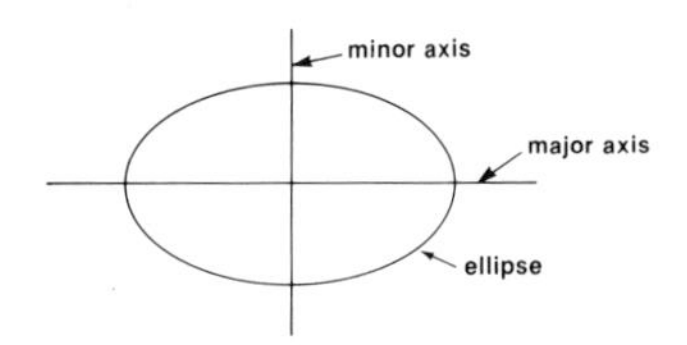

LATIN, *minor*, smaller

minus **1.** Made smaller by taking away. Sign or symbol used is −. **2. Negative** or on the negative side of a scale. **3.** The minus sign (−). **4.** A minus or negative value.
Examples
1. 12 minus 7 equals 5 (12 − 7 = 5).
2. Minus two, minus fifteen degrees (−2, −15°).
3. There should be a minus sign before the number.
4. When a bigger number is taken away from a smaller number you always get a minus.
LATIN, *minor*, smaller

minute the sixtieth part of a **degree**, and the sixtieth part of an hour (the symbol is ′).
Example There are 60 minutes in a degree, and 360 degrees in a circle.

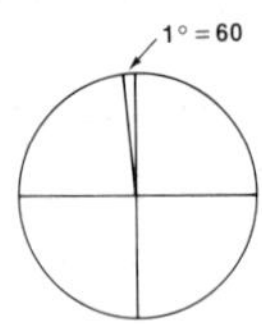

LATIN, *minutus*, small

mixed number a number written as a whole number and a **proper fraction**.
Example $^{16}\!/_5$ can be written as the mixed number $3\frac{1}{5}$, where $3\frac{1}{5}$ is short for $3 + \frac{1}{5}$. Note: A mixed **decimal** is a number containing a whole number and a decimal fraction, as 2.314.
LATIN, *miscere*, to mix; *numerus*, number

Möbius, August (1790–1868) German mathematician particularly remembered for the mathematics of the **möbius strip**—a one-sided and one-edged surface popular in topology.

möbius strip a surface with only one side. It is made by taking a long strip of paper, giving it a twist and then joining the ends together.

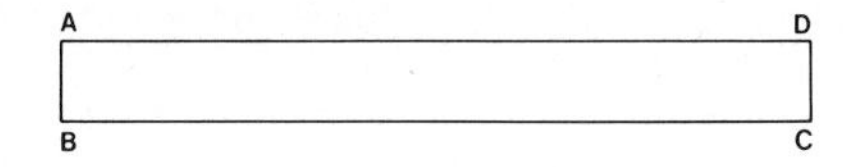

A is joined to *C* and *B* to *D*.

Example

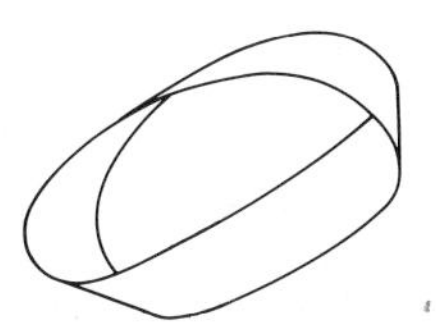

A möbius strip only has one edge, as can be shown if you take a pencil and follow the edge starting at any point.
after Möbius, German mathematician; MIDDLE LOW GERMAN, *strippe*, strap, thong

mode the **element** or thing which occurs most often in a given **set**.
Example The price of one apple from each of 10 shops was 4¢, 5¢, 5¢, 5¢, 4¢, 6¢, 5¢, 3¢, 8¢, 7¢. The mode is 5¢, because it occurs the most often (four times) in the set. Note: if both 4¢ and 5¢ occurred most often, both are modes.
LATIN, *modus*, measure, manner, size.

model a mathematical pattern whose known properties are applied to some aspect of life in order to study it better.
Example The following sales figures were recorded over a period of months as follows:

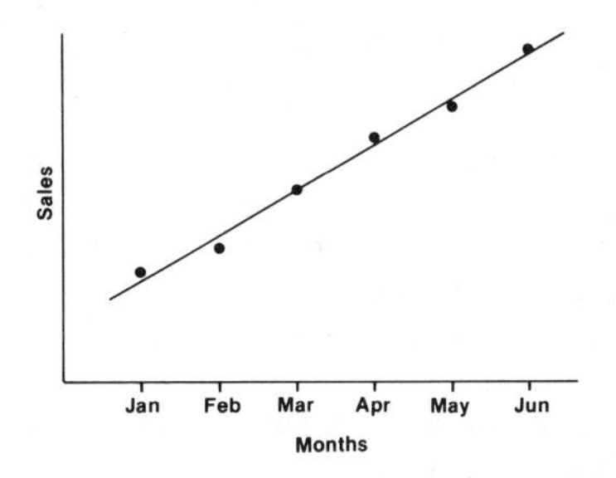

One pattern which could be applied to this situation is a straight line. This is one model which can then be used to find out more about sales over a period of time.

LATIN, *modus*, measure, manner, size

modulus **1.** The value of a number disregarding its sign. The symbol used is | | (*see* ABSOLUTE VALUE). **2.** (of a complex number) The number formed by taking the **square root** of the sum of the squares of the **real** and **imaginary** parts of a complex number. The symbol used is | | (*see* ABSOLUTE VALUE). **3.** A number which is used as a **divisor** in a special kind of arithmetic. Symbol is modulo or mod. **4.** The number which is used to change a **logarithm** in one system to a logarithm in another system.
Examples
1. $|-7.1| = 7.1$.
2. $|3 - 2i| = \sqrt{3^2 + (-2)^2} = \sqrt{13}$.
3. In the so-called **clock arithmetic**, numbers divided by the given modulus are equal if their remainders are equal. For example, 25 divided by 12 gives a remainder of 1, since $25 = 2 \times 12 + 1$, as $13 \div 12$ gives a remainder of 1. Both have the same remainder, and from the point of view of a clock with hours running from 1 to 12 only, each time the hour hand travels around once it comes back to pointing to 1, 2, 3, etc. up to 12. Therefore, 25 on the clock face is 1, and so is 13. This can be written $25 \equiv 13 \pmod{12}$.
4. If $N = 10^p$, then $\log_{10}N = p$, but $\log_e N = p\log_e 10$. $\log_e 10$ is the modulus for changing from the decimal or common logarithms to the so-called **natural** or **Naperian logarithms**.
LATIN, *modus*, measure, manner, size

modulus (or modular) arithmetic (also **clock arithmetic**) a special kind of arithmetic which has rules for addition and subtraction that follow the rules for adding/subtracting hours on a clock, or similar.
Example A traditional clock has 12 positions for 12 hours marked on a dial. As soon as the hour hand passes 12 we start counting from the beginning, i.e. 7 o'clock + 6 hours = 1 o'clock. This is written 7 + 6 = 1 mod 12 in modulus or clock arithmetic. Similarly 4 + 13 = 5 mod 12 (4 o'clock + 13 hours is 5 o'clock). In general, we can construct a similar system so that any number (e.g. 8) is the divisor (e.g. 7 + 4 = 3 mod 8).
LATIN, *modus*, measure, manner, size; *arithmos*, number

monotonic referring to a **variable** or **function** whose values either increase consistently, or decrease consistently (usually within a particular **range**).
Examples A function whose values increase within a particular **interval**, is called a monotonic increasing function (in that interval). If the function's values decrease within a particular interval, it is called monotonic decreasing.

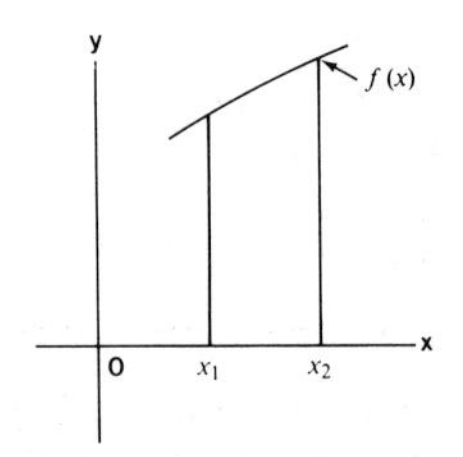

$f(x)$ is monotonic increasing between x_0 and x_1, i.e. as x increases from x_0 to x_1, $f(x)$ increases also. This can be expressed more precisely as follows: $f(x)$ is monotonic increasing in the interval x_0 to x_1 if and only if for every pair of real numbers p, q where $x_0 \leqslant p < q \leqslant x_1$, $f(q) > f(p)$.

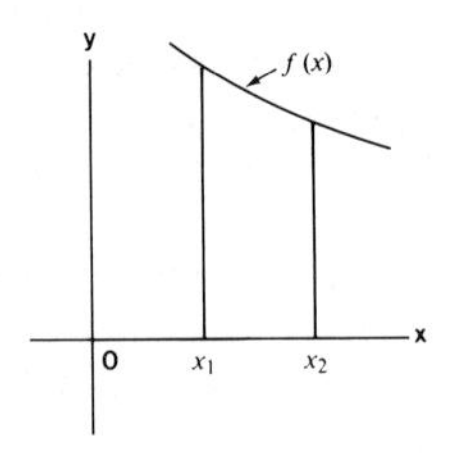

$f(x)$ is monotonic decreasing between x_0 and x_1, i.e., as x increases from x_0 to x_1, $f(x)$ decreases.
GREEK, *monotonus*, having one tone

motion geometry a way of studying geometry using movement of points, lines, and planes (also known as **transformation geometry**).
Examples
1. A point which moves so that it is always a fixed distance from a given point, generates a circle.

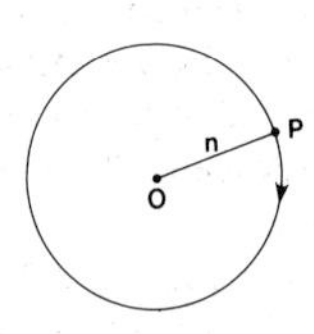

2. A straight line moved perpendicular to itself a distance equal to itself, will form a square.

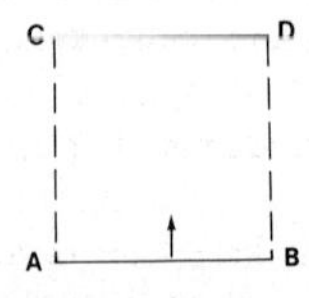

LATIN, *movere*, to move; *geometria*, a measuring of land

moving average (statistics) an **average** of a **set** of values calculated over a number of given time periods. The average "moves" because it is recalculated each time a new period is finished.
Example The following table shows sales over 18 months.

Year	Month	Sales ($'000)	12 Month Moving Average
1984	Jan.	250	There is no moving average for the first six months since we do not have sales earlier than January, 1984. In this example the number of given time periods is 12 months. We have put the first average (Jan. to Dec.) opposite the July position. It could also be put opposite Dec. We could also use averages calculated over three or five years. Each time a new period is completed, we use the results for this period and drop off the result from the first time period in the calculation.
	Feb.	260	
	Mar.	320	
	Apr.	402	
	May	410	
	June	503	
	July	550	
	Aug.	475	
	Sept.	380	
	Oct.	310	
	Nov.	295	
	Dec.	270	
1985	Jan.	265	
	Feb.	298	
	Mar.	335	
	Apr.	412	
	May	440	
	June	485	

First average =
$$\frac{250+260+320+402+\ldots+380+310+295+270}{12}$$

Next average =
$$\frac{260+320+402+410+\ldots+310+295+270+265}{12}$$

and so on.
LATIN, *movere*, to move; FRENCH, *avarie*, damage to ship or cargo

multiple a **number** which is made up of two or more **factors** other than 1.
Example 6 is a multiple of 2 and 3.
$3a^2b$ is a multiple of 3, a^2 and b, also of a, ab, $3a$, etc.
Note: multiples of any number (e.g. 5) are whole numbers times the given number (e.g. 2×5, 3×5, 4×5, etc.).
LATIN, *multiplus*, manyfold

multiplicand, *see* MULTIPLICATION

multiplication (originally referring to integers) **1.** The process of adding a quantity to itself a certain number of times. **2.** (in general mathematics) A **binary** operation which is **closed** and **associative**. The symbols used for the multiplication operation are • or ×.

Examples
1. $3 \times 5 = 5 + 5 + 5$ ($= 15$)
Note: The number to be added to itself is called the **multiplicand** and the number saying how many times it has to be added to itself is called the **multiplier**.
2. **Real** numbers (e.g., $\frac{1}{2}$, $\frac{3}{7}$) can be multiplied, e.g. $\frac{1}{2} \times \frac{3}{7} = \frac{3}{14}$ (**closure** illustration). Also $\frac{1}{2} \times (\frac{3}{7} \times \frac{1}{5}) = (\frac{1}{2} \times \frac{3}{7}) \times \frac{1}{5}$ (**associative** illustration).
LATIN, *multiplicare*, to fold many times

multiplier, *see* MULTIPLICATION

multiplicative inverse the inverse of an element '*a*' is that element which, when combined with '*a*', produces the identity element for multiplication.
Example $\frac{1}{4} \times 4 = 1$, and this means $\frac{1}{4}$ is the multiplicative inverse of 4, and also that 4 is the multiplicative inverse of $\frac{1}{4}$. Note: The multiplicative inverse of any number (except 0) is the **reciprocal** of that number (i.e. 1 divided by the given number).
LATIN, *multiplicare*, to fold many times; *invertere*, to invert, turn upside down

multiplication table a table drawn up so that one can easily find the result of multiplying any two numbers from 1 to 12 (usually) together.
Example

	1	2	3	4 ...
1	1	2	3	4
2	2	4	6	8
3	3	6	9	12
4	4	8	12	16

i.e. $1 \times 1 = 1$, $1 \times 2 = 2$ etc.
$2 \times 1 = 2$, $2 \times 2 = 4$ etc.

LATIN, *multiplicare*, to fold many times; *tabula*, a board

mutually exclusive referring to **events** (in **probability**) or to arrangements which cannot occur at the same time.
Examples
1. Since a head and a tail cannot turn up from a single coin when thrown, the two events are mutually exclusive.
2. Two people (*P* and *Q*) can only be seated on two chairs out of a row of five chairs in any one arrangements. *P* occupying the second chair and *Q* the fourth excludes the possibility of *P* or *Q* occupying any other two chairs at the same time.
LATIN, *mutuus*, exchanged, borrowed; *exclusus*, shut out

Nn

N (n) the fourteenth letter of the English alphabet. In mathematics it is used as the symbol for an indefinite number.
Examples
$\sum_{r=1}^{n} r^2$ is a shorthand way of writing the following:

$1 + 2^2 + 3^2 + 4^2 + \ldots + n^2$, where n is not known exactly (Σ, Greek letter sigma, is used to represent **sum**).
x^n means $x \times x \times x \times \ldots x$ where x is multiplied by itself n times where n is not given a definite value.
$x^{\frac{1}{n}}$ or $\sqrt[n]{x}$ is the nth root of x (this means that $x^{\frac{1}{n}} \times x^{\frac{1}{n}} \times x^{\frac{1}{n}} \ldots x^{\frac{1}{n}} = x$, where $x^{\frac{1}{n}}$ is multiplied by itself n times.
GREEK, ν (nu) [thirteenth letter of the Greek alphabet]

nano- prefix meaning one thousand millionth (10^{-9}). The symbol is n.
example 1 nanosecond (ns) = 0.000000001 seconds.
GREEK, *nannos*, a dwarf

Naperian logarithm, *see* NATURAL LOGARITHMS

Napier, John (1550–1617) a Scottish mathematician particularly remembered for the invention of **logarithms** and the invention of a device, called Napier's bones, for mechanically computing **products** and **quotients** of numbers.

natural logarithms **logarithms** which are worked out with reference to the **base** e ($e = 2.718\ldots$). They are also called Naperian logarithms as they were invented by John **Napier** (ordinary or common logarithms are worked out to the base 10). The symbol for natural logarithm is $\log_e$.
Example If $N = e^2$, then $\log_e N = 2$. Examples such as $\log_e 7$ are usually worked out by looking up special log tables, or by calculation.
LATIN, *natura*, nature; GREEK, *logos*, reckoning; *arithmos*, number

natural numbers the positive whole numbers. Also known as the counting numbers.
Example 1, 2, 3, 4, 5, ...
LATIN, *natura*, nature; *numerus*, number

nautical mile the length of an **arc** on the surface of the earth which **subtends** one sixtieth of a **degree** (equal to 1 minute) at the centre of the earth.

Example

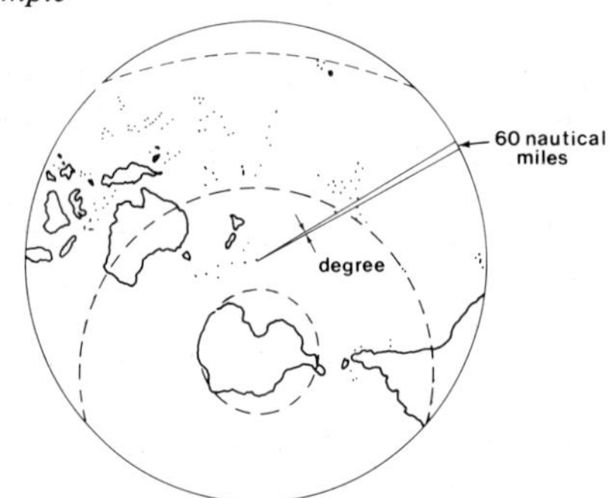

$1' = 1$ nautical mile.
$360° = 21\,600$ nautical miles.
LATIN, *nauta*, a sailor; *mille*, a thousand

negative numbers numbers used to count or measure in the opposite direction or sense to the **positive numbers**. The sign used to show a number which is negative is $-$.
Example -1.3, -4.24, $-a$.
Note: A negative number which when added to a positive number gives zero is called the **inverse** of the positive number.
LATIN, *negare*, to deny or negate; *numerus*, number

net 1. A **network**. **2.** A plane shape which when folded along definite lines becomes a solid. **3.** (adj.) Remaining after deductions.
Examples
1. This is the net of a cube.

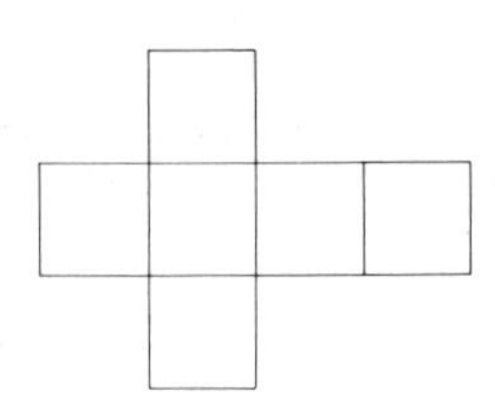

2. The net weight was 2 kilograms (i.e. the weight of the contents after subtracting the weight of the containing and packing).
OLD ENGLISH, *nett*, net

network a group of connected lines.
Example

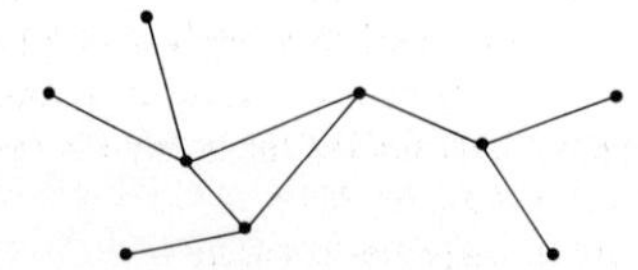

Note: The lines are called **arcs**, the points at which lines meet are called **nodes**, and the areas bounded by the lines are called **regions**.
OLD ENGLISH, *net*, net; *weorc*, act, deed, work

Newton, Isaac (1642–1727) English mathematician of great distinction, jointly responsible with, though independent of, Gottfried **Leibniz** for the invention of the **calculus**. Also noted for many discoveries in physics including the composition of white light, law of universal gravitation, and discoveries in hydrostatics and hydrodynamics.

newton the basic unit of force in the **SI** and MKS (*M*etre *K*ilogram *S*econd) systems. It is equal to the force required to accelerate a mass of 1 kilogram at the rate of 1 metre per second.
Example When a newton moves through a distance of 1 metre, one **joule** of work is done!
after Isaac Newton, English physicist

Newton's method a way of finding a solution (approximate) to an **equation** by starting with an approximate value and then improving its accuracy. It uses the basic idea of differential **calculus**. If $f(x) = 0$ is the equation and if at $x = x_0$, $f(x_0) \doteqdot 0$, then

$$x_1 = x_0 - \frac{f(x_0)}{f'(x_0)}$$

is a better approximation than x_0. Repeating this gives

$$x_2 = x_1 - \frac{f(x_1)}{f'(x_1)}$$

and on so.
In graph form:

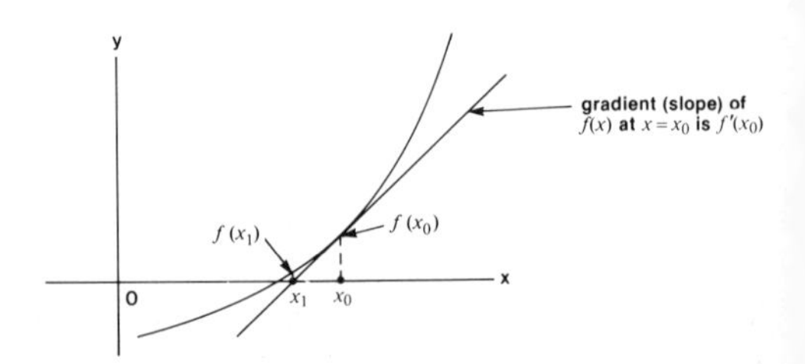

Examples
If $f(x) = x^2 + 2x - 5$, $f'(x) = 2x + 2$
then $f(2) = 2^2 + 4 - 5 = 3$
$f(1.5) = (1.5)^2 + 3 - 5 = 0.25$
Choose $x_0 = 1.5$

$$x_1 = 1.5 - \frac{f(1.5)}{f'(1.5)} = 1.5 - \frac{0.25}{5} = 1.45$$

$$x_2 = 1.45 - \frac{f(1.45)}{f'(1.45)} \text{ etc.}$$

after Isaac Newton; GREEK, *methodos*, a going after, pursuit

node the point at the beginning or end of a line or **arc** in a **network**.

Example

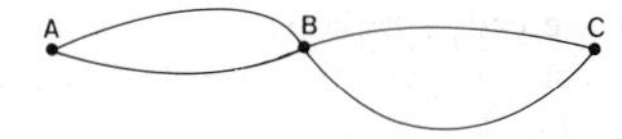

A, *B*, *C* are nodes.
LATIN, *nodus*, a knob, a knot

Noether, Emmy (Amalie) (1882–1935) a German mathematician, especially remembered for her contributions to the study of modern **algebra**.

nomogram (also **nomograph**) a device for working out the value of a variable using three scaled lines. The three lines represent three related quantities, two of which will be known, and the third one to be worked out. It uses the laws of **logarithms**.
Example Electricians use the formula:
amperes $= \frac{\text{volts}}{\text{ohms}}$. The three variables, amperes, volts, and ohms can be represented by three lines scaled as follows:

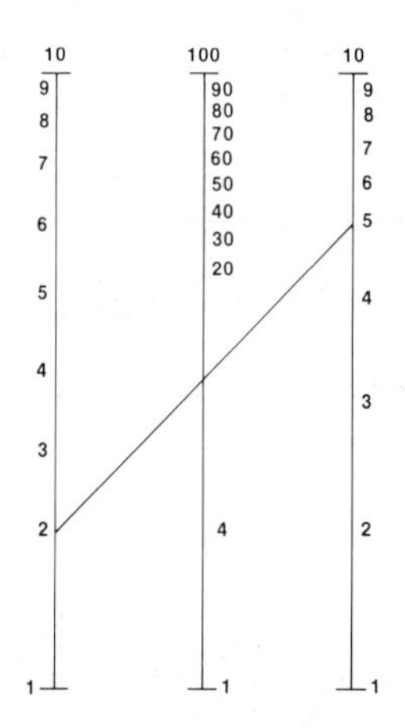

To use the nomogram, place a ruler or straight edge on the two known values, and read off the third value from where the line cuts the third scale. As in the example, if we know the current is 2 amperes, and the resistance 5 ohms, then the voltage is 10 volts.
GREEK, *nomos*, usage, law; *gramma*, letter, line

nonagon a **polygon** figure having nine straight sides and nine angles.
Example A nonagon can have all or some **vertices** pointing outwards.

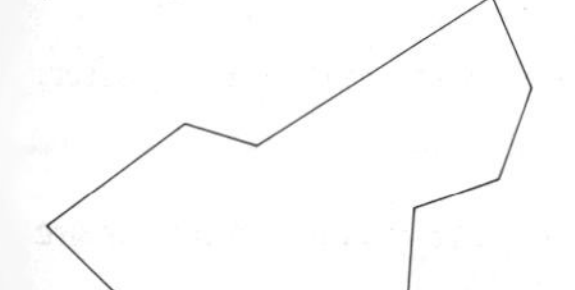

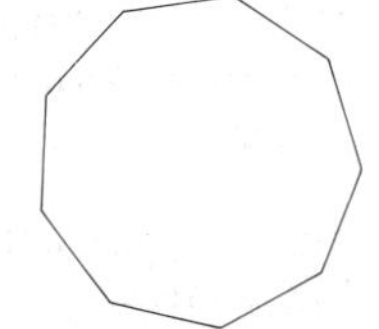

LATIN, *nonus*, nine; GREEK, *gonos*, angled

non-euclidean geometry any of the several **geometries** not based on **Euclid's** assumptions. One assumption in particular has been varied to generate two distinct geometries. This is the **assumption** that through a point outside a line, only one **parallel** line can be drawn. By assuming either that two or more lines can be drawn through that point, which will not intersect that given line, or that no lines can be drawn which will not intersect the original line, two different geometries can be constructed.
Example Riemannian geometry (developed by G. Riemann, a nineteenth century German mathematician) is one non-euclidean geometry which assumes no parallel lines. Another non-euclidean geometry is that of Lobachevsky, also developed independently by a Hungarian mathematician Bolyai, which assumes that through any point there are an **infinite** number of lines which can be drawn parallel to a given line away from the point.

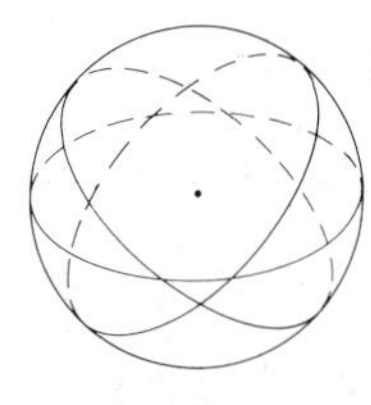

On a sphere, all lines which are parts of **great circles** intersect, so that the study of these elements is an example of Riemannian geometry.
after Euclid, Greek mathematician; GREEK, *geometria*, measuring land

normal a straight line perpendicular to another line or **plane**, especially a line or plane **tangent** to a curve or surface. Graphic symbol is ∟.
Examples

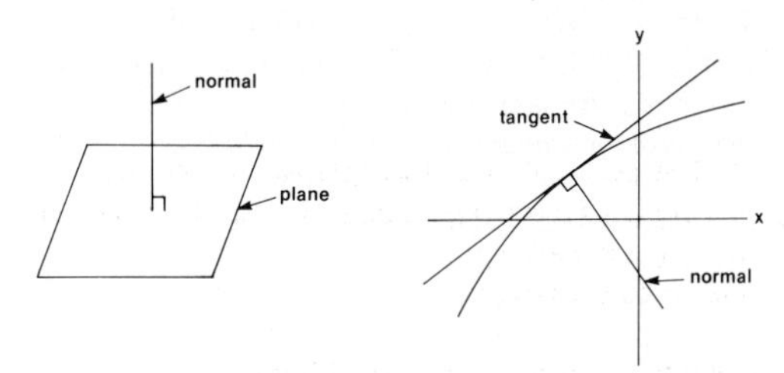

Note: In two-dimensional space (i.e. a plane) the **gradient** of the tangent to a curve multiplied by the gradient of the normal = −1. In three-dimensional space the **direction ratios** of the normal to the plane $ax + by + cz = d$ are (a, b, c).
LATIN, *norma*, carpenter's square

normal distribution (statistics) a set of numbers grouped together according to how often each occurs so that they fit the "bell" shape shown below. "Normal" is used in the sense of standard.
Example Many scientific observations of a single measurement when plotted on a graph show this shape. Its equation is: $y = \frac{1}{\sqrt{2\pi}\sigma} e^{-(x-\bar{x})^2/(2\sigma^2)}$
where σ is the **standard deviation**
$\bar{x}$ is the arithmetic mean
π, e are mathematical constants.

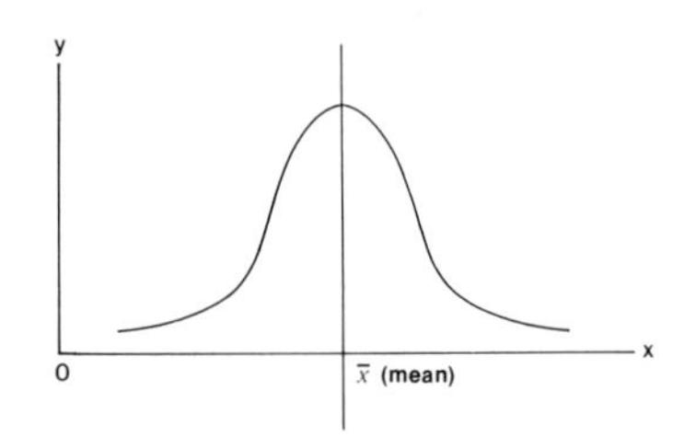

LATIN, *norma*, carpenter's square; *distribuere*, to assign or grant parts of

notation a special way of representing figures, numbers, quantities or other facts or values.
Example Instead of writing $3 \times 3 \times 3 \times 3 \times 3$, we use an index notation, such as 3^5, where 5 is a figure called the **index**. In **logarithms**, an example of the notation used is $\log 0.02 = \bar{2}.3010$, where $\bar{2}$ means -2. In the **decimal system** we use position notation, where the particular position of a figure to the right or left of the decimal point indicates which power of ten is understood to be multiplied by the figure in that position: for example, in 237.21, the first 2 represents 2×10^2, the 3 represents 3×10^1, the 7 represents 7×10^0, the second 2 represents 2×10^{-1}, and the 1 represents 1×10^{-2}.
LATIN, *nota*, a note

nought the symbol 0, which stands for zero.
Example 0.7 is read nought point seven, and means no units seven tenths (7×10^{-1}).
OLD ENGLISH, *nawiht*, no creature or thing

null hypothesis (statistics) a statement (to be proved true) that says there is no really significant difference between two samples of things, and that any difference that does occur is only the result of errors due to the **random** way the samples were obtained.
Example When people perform a number of experiments only changing certain things, they can use the null hypothesis to see if the difference in results is likely to be due to chance or not. This can help them to determine how significant the differences are (*see* STANDARD ERROR OF THE MEAN).
LATIN, *nullus*, not any; GREEK, *hupothesis*, foundation

null set, *see* EMPTY SET

number 1. An idea of quantity, or how many units (of something). 2. A **symbol** used to represent how many (of something).
Examples
1. The number of circles in the following diagram is 5.

2. The number five is represented by the symbol 5.
There are many different kinds of numbers, including **natural** numbers, **rational** numbers, **real** numbers and **complex** numbers.
LATIN, *numerus*, number

number line a straight line on which numbers are marked. All **real numbers** can be represented as points on the line.
Example

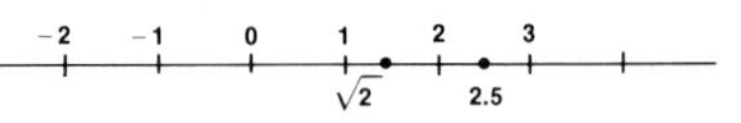

LATIN, *numerus*, number; *linea*, line

number pattern a way in which **numbers** can be arranged.
Example A famous pattern is **Pascal's triangle**.

1
1 1
1 2 1
1 3 3 1
1 4 6 4 1

One of the things mathematicians do is study number patterns to find out rules about them.
LATIN, *numerus*, number; MIDDLE ENGLISH, *patron*, model, design

number sentence a statement which shows a connection between two or more **numbers**.
Example $2 + 7 = 9$, $24 \div 3 = 8$ are number sentences. So also are $5 > 3$, and $2 < 9$ (where $>$ means "is greater than", and $<$ means "is less than").
LATIN, *numerus*, number; *sententia*, a way of thinking

number system a way of representing **numbers**
Examples
Our most common number system comes from the Arabs. The arabic system used the following ten symbols for counting: 0, 1, 2, 3, 4, 5, 6, 7, 8, 9. The Arabs also used the position of these symbols to mean something. For example 12 means one lot of 10 plus 2. Also 257 means two lots of 100 plus five lots of 10 plus 7. This system is now called the hindu-arabic system.

Another number system is that used by the Romans. They used symbols such as I, II, III, IV, V and so on. The arabic system is easier to use in calculations than the roman. Just imagine a hand calculator working with roman numerals!
LATIN, *numerus*, number; GREEK, *sustema*, organized whole

numeral the symbol or group of symbols which represents a **number**.
Example The number two can be represented by the symbol 2 (in the **decimal** system), and 10 (in the **binary** system), II (in the Roman system), also 4/2, 6/3 etc.
LATIN, *numerus*, number

numeration **1.** A way of numbering things. **2.** The art of expressing any **number** in words that is already given in figures.
Examples
1. A million is 1 000 000. A billion is 1000 million (US), or 1 million million (British).
2. 34.26 is said as thirty four point two six.
LATIN, *numerus*, number

numerator the **number** written above the line in a **fraction**.
Example 2 is the numerator of the fraction 2/3, a is the numerator of the number $\frac{a}{b}$.
LATIN, *numerus*, number

Oo

object **1.** The point or set of points to which a **transformation** is applied. **2.** Any **member** of a set to be **mapped** by a given **operation**.
Examples
1.

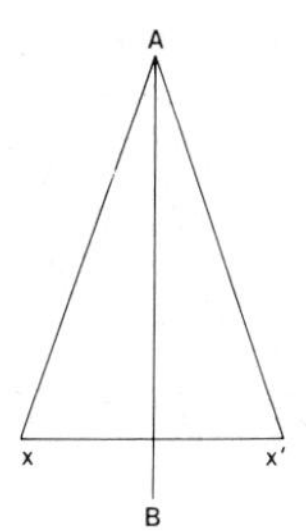

The line AX is the object. The line AX' is the **image** as a result of a **reflection** about the line AB.
2. If for every x, there corresponds (a **mapping**) $f(x)$, then for a particular x (= 3) the mapping is $f(3)$. 3 is called the object, and $f(3)$ the **image** or corresponding functional value.

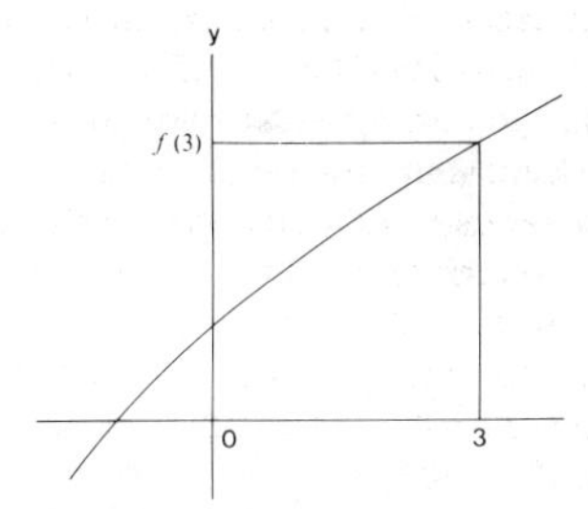

LATIN, *objectus*, thrown against

oblate flattened.
Example The earth, which is slightly flattened at the poles, is an oblate **spheroid**.

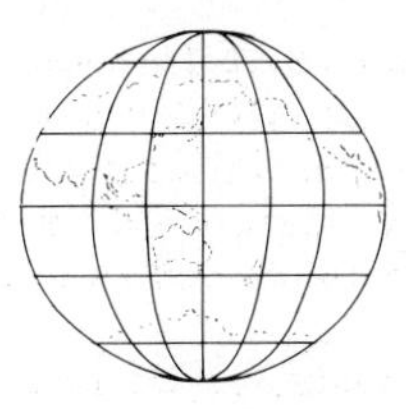

LATIN, *oblatus*, brought to or offered

oblique slanting.
Example An oblique line is neither **perpendicular** nor **horizontal**.

LATIN, *obliquus*, inclined

obtuse referring to an angle which is greater than 90°, but less than 180°.
Example

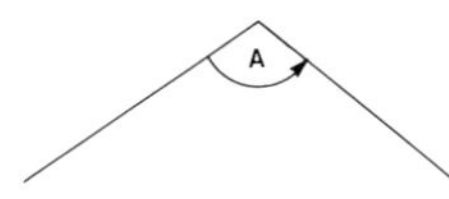

A is an obtuse angle. An obtuse triangle has an obtuse angle in it.
LATIN, *obtusus*, blunted

octa- a prefix meaning eight.
Example An octad is a group of eight units or figures.

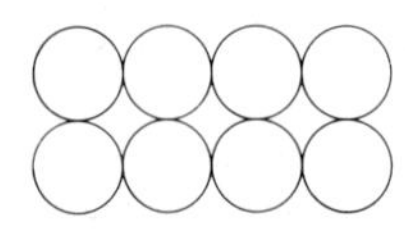

GREEK, *okto*, eight

octagon a figure having eight straight sides.
Example

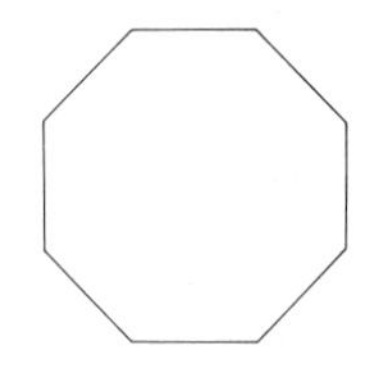

GREEK, *oktagonos*, having eight angles

octahedron a solid shape having eight triangular faces.
Example

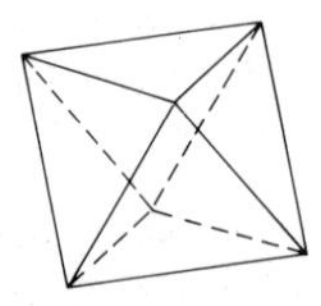

A **regular** octahedron is one in which the faces are **equilateral** triangles.
GREEK, *oktaedron*, eight faces

octal system a way of counting in groups of eight (compared with ten in the **decimal system**). The eight symbols used are 0, 1, 2, 3, 4, 5, 6, 7.
Example

Decimal	*Octal*
1	1
2	2
3	3
4	4
5	5
6	6
7	7
8	10
9	11
10	12
11	13
12	14
13	15
14	16
15	17
16	20
17	21
18	22
19	23
20	24

GREEK, *okta*, eight; *sustema*, an organized whole

odd function A function $f(x)$ which has the property that $f(x) = -f(-x)$, and in particular $f(0) = -f(-0) = 0$.
Examples $y = x$, $y = x^3$, $f(x) = x^5 + x^3$, etc., are all odd functions.

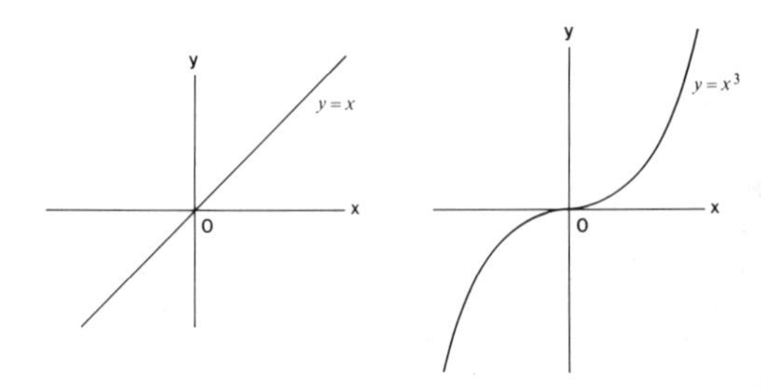

OLD NORSE, *oddi*, triangle, point; LATIN, *functio*, activity, performance

odd number a whole number that cannot be divided by 2 without leaving a remainder of 1.
Example 1, 3, 5, 7 are all odd numbers, as is any number ending in 1, 3, 5, 7, or 9.
OLD NORSE, *oddi*, triangle, point

odds (probability) the **ratio** expressing the likelihood of an event or outcome taking place. Odds originally meant the difference in advantage between one competitor's position and the position of another. A weaker opponent could be given odds (or points) to

equalize the chances of winning in a contest. Hence, **odds on**, better than 1:2 that an event will happen.
Example What are the odds of a six turning up on the throw of a die? (The odds are 1:6.)
OLD NORSE, *oddi*, triangle, point

ogive a cumulative **frequency distribution** curve derived from a cumulative **frequency polygon** (*see* CUMULATIVE FREQUENCY).
Example

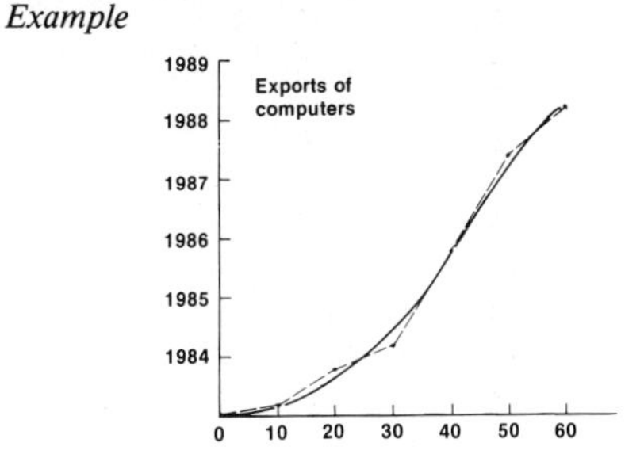

OLD FRENCH, *augive*, a pointed arch

on (probability) indicating that the **ratio** *a:b* in the phrase *a:b* on is taken to mean put up *a* to win *b*.
Example The favourite is 2:1 on. This means you would have to outlay two units to win one unit. Compare this with the phrase 2:1. This means you outlay one unit to win two.
OLD ENGLISH, *on*, on

one the number representing a single thing. Symbol used to represent one is 1.
Example One when added to itself makes two. 1 + 1 = 2.
OLD ENGLISH, *an*, one

one-dimensional referring to measurement in one direction only.
Example A line is one-dimensional in that any point can be located by a single measurement from a given point of reference (called the origin).

OLD ENGLISH, *an*, one; LATIN, *dimensio*, a measuring

one-to-one correspondence a relationship between two **sets** of **elements** such that every **member** of the first set is paired with a unique element in the second set, and every member of the second set is likewise paired with a member in the first set.

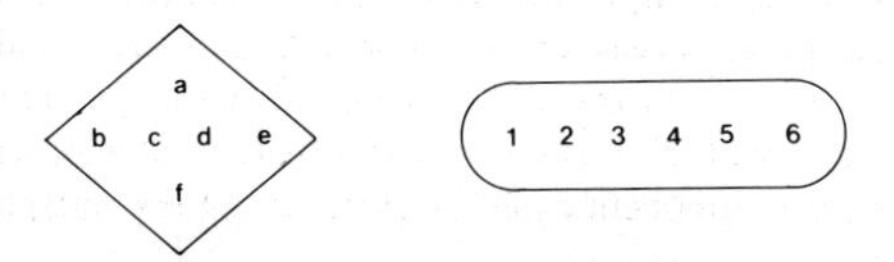

Examples
1. The sets $\{x\}$ and $\{\frac{3x}{4} + 1\}$ are in a one-to-one correspondence.
2. The sets {1, 2, 3, 4, 5, . . .} and {2, 4, 6, 8, 10, . . .} are in a one-to-one correspondence.

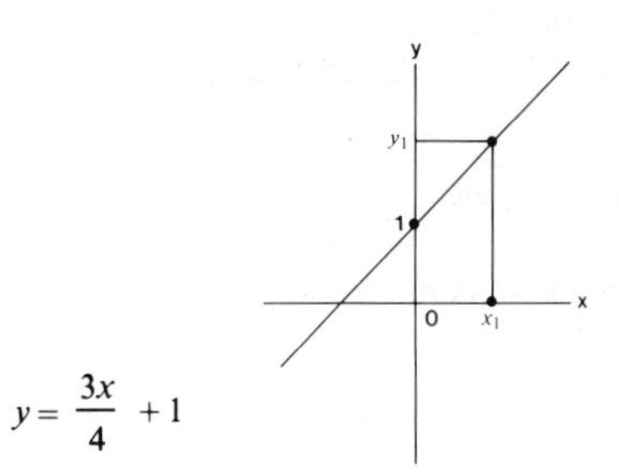

$$y = \frac{3x}{4} + 1$$

OLD ENGLISH, *an*, one; LATIN, *correspondere*, to respond together

open number sentence a number sentence with a number (or numbers) to be filled in. A frame such as □ or △ is usually used to show a number is to be put in to make the sentence true.
Examples
7 + □ = 13 is an open number sentence. The number which makes the number sentence true is 6, since 7 + 6 = 13.
6 − □ > 2 is another open number sentence. Any number less than 4 will make the sentence true.
OLD ENGLISH, *open*, open; LATIN, *numerus*, number; *sententia*, a way of thinking

open sentence a **number sentence** which contains one or more unknown quantities.
Examples
The number sentence $x + 3 > -1$ contains x as an unknown quantity, and is therefore an open sentence.
$x + 3y = 4$ is also an open sentence which contains two unknown quantities, x and y.
OLD ENGLISH, *open*, open; LATIN, *sententia*, a way of thinking

operation **1.** The action of combining **elements** in a **set**. The result of an operation on one or more elements is usually another element in the set. **2. Order** of operations. The rule when simplifying expressions is: do *B*rackets first, then *O*f, then *D*ivide or *M*ultiply, then *A* ddition or *S*ubstraction. (This can be remembered as BODMAS.)
Examples
1. + is the symbol for a binary operation (addition) on numbers such as 3 + 2 = 5, 1¼ + 2½ = 3¾, −2 + 2½ = ½. ×, ÷, − are other kinds of operations.
2. (19 − 12) ÷ 6 is simplified by working out the brackets first, that is, 19 − 12 = 7 then dividing 7 by 6 = ⅞.

$$7 \times (9 - 3) \div 2 = 7 \times 6 \div 2$$
$$= 42 \div 2 \text{ or } 7 \times 3$$
$$= 21$$

$(a - 2)^2 + 6 \div 2a - 3$ simplifies as follows:

$(a - 2)(a - 2) + \dfrac{6}{2a} - 3$, which is different from

$(a - 2)^2 + 6 \div (2a - 3)$, which simplifies to

$(a - 2)(a - 2) + \dfrac{6}{2a - 3}$.

LATIN, *operari*, to work

operator a symbol indicating an action to be performed on something.
Examples
1. i is an operator which represents a rotation of 90°. $i3 = 3i$ changes a point located at 3 on one number line to a point 3 on a line at right angles to the first line.

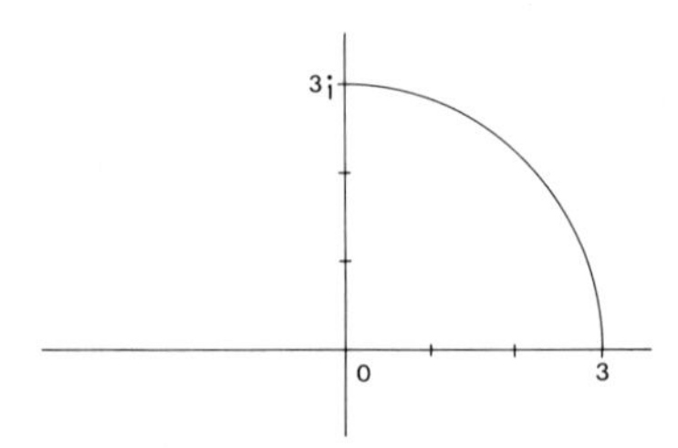

2. If we define Z to mean "double the value of", then Z is an operator. $Z3 = 6$, $Z12 = 24$ etc.
3. Other more common operators are $+$, $\times$, $\surd$ etc.
LATIN, *operari*, to work, labour

order **1.** The arrangement or **sequence** of things. **2a.** The smallest **power** to which an **element** in a group is raised to give the identity. **2b. Order of rotational symmetry**: the number of positions a shape appears the same in one rotation. **3.** The size of a **matrix**. If a matrix has three rows and four columns, its order is said to be 3×4. **4.** The number of lines or arcs that meet at the **node**. **5.** Order of operations (*see* OPERATION).
Examples
1. In an ordered pair, the arrangement of the two numbers is important.
2a. If T is an operator meaning turn upside down, then TT or T^2 would mean "turn upside down twice", which results in the original position represented by the identity.

2b. The shape such as ▯ has an order of rotation of 2 (since there are two positions where it appears the same in one complete rotation).

3. The order of the matrix $\begin{pmatrix} 2 & 6 \\ 3 & 4 \\ 1 & 5 \end{pmatrix}$ is 3×2.

4.

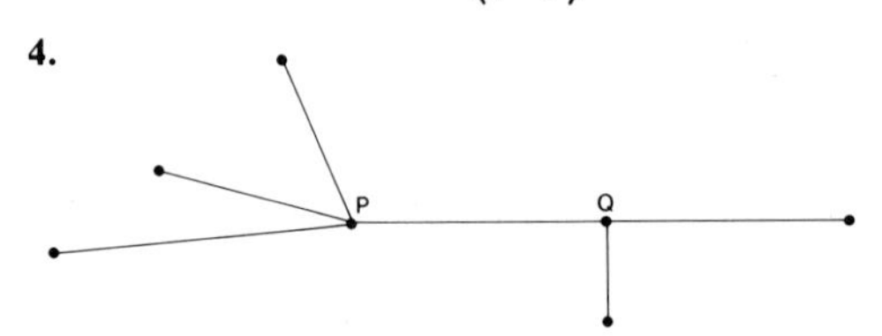

The order of P is 4.
LATIN, *ordo*, a straight row

ordered pair any two **numbers** where the **order** is important. Usually used to locate a point in a plane with reference to two intersecting lines.
Example In a frame of reference an ordered pair (3,4) means a point whose location is found 3 units to the right from the y-axis, 4 units upright from x-axis.

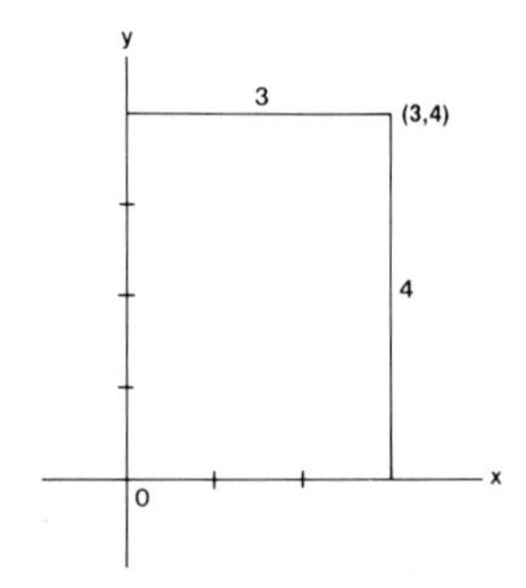

LATIN, *ordo*, a straight row; *par*, equal

ordinal number a **number** indicating a position in a **sequence**.
Example First, second, fifth, forty-third (*see also* CARDINAL NUMBER).

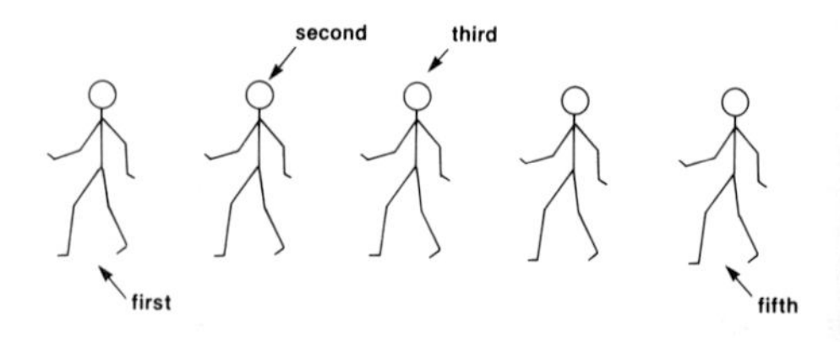

LATIN, *ordo*, a straight row; *numerus*, number

ordinate the distance of a **point** from the x-axis. It is the second number in an **ordered pair** representing the position of a point.

Example 4 in the pair of coordinates (3,4).

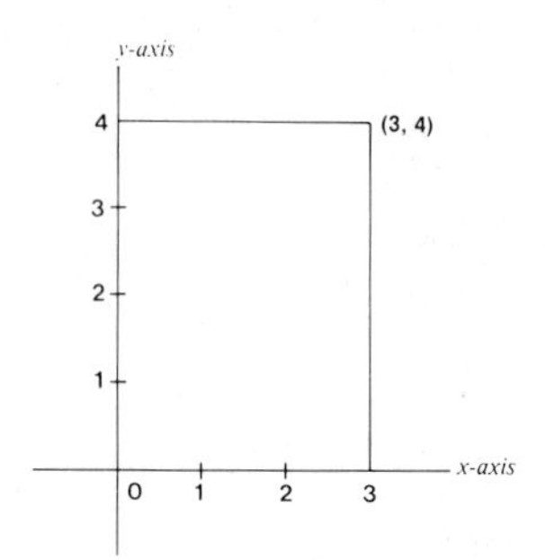

LATIN, *ordinare*, to arrange in order

origin a fixed point in a coordinate system to which all other points are referred. In a two-dimensional system, the origin is the intersection of the x and y axes. The coordinates of the origin are (0, 0). In a three-dimensional system, the origin is at (0, 0, 0). The origin can also refer to the centre of an **enlargement** transformation.
Example

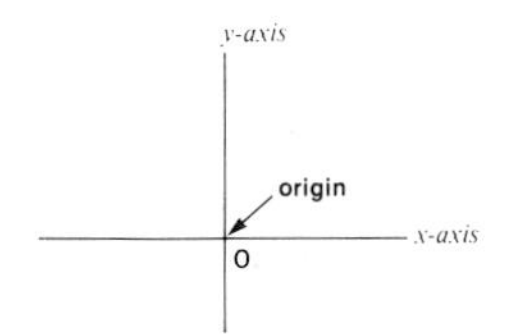

LATIN, *oriri*, to rise

orthocentre the point of intersection of the three **altitudes** of a triangle.
Examples

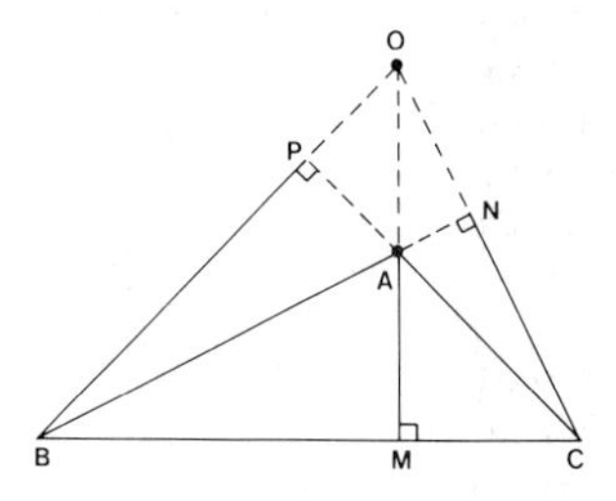

In the triangle ABC, the altitudes are AM, BP, CN. These three lines when extended meet in O. O is the orthocentre of triangle ABC, and in this case lies outside the triangle ABC.

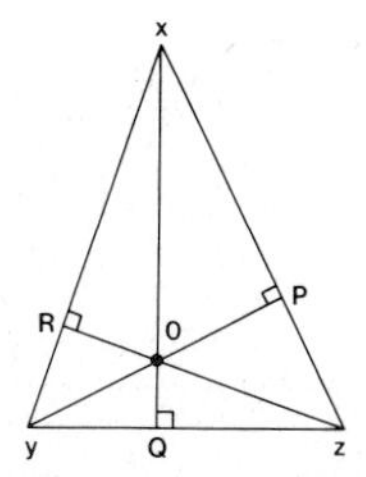

In the triangle XYZ, the altitudes are XP, YQ, RZ. These three lines meet at O, which is the orthocentre of triangle XYZ. In this case O lies inside the triangle.
GREEK, *orthos*, straight, correct, upright; *centrum*, centre

Pp

palindromic number a number which reads the same backwards as forward. A palindrome is a word, phrase or verse which reads the same backwards as forwards.
Example 2112, 131, 95459.
GREEK, *palindromos*, running back again

parabola a curve drawn so that any point on it is always the same distance from a given fixed point and a given fixed line. The point is called the **focus**, and the fixed line the **directrix**. It is represented by the general equation $y = ax^2 + bx + c$. A parabola can also be formed by a plane cutting a cone parallel to the slant edge (*see* CONIC SECTION).
Examples

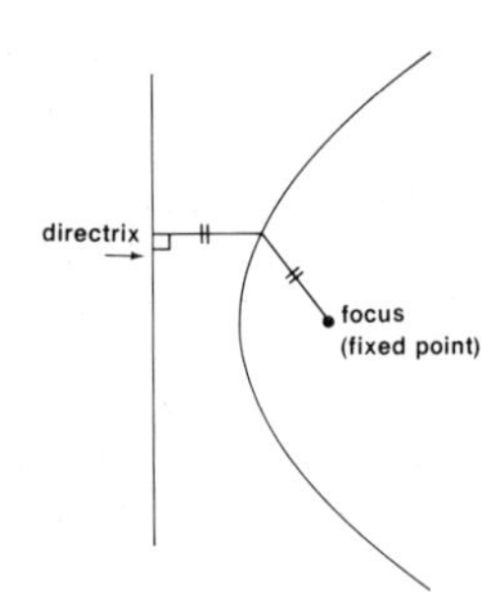

Equation is $y^2 = 4ax$.

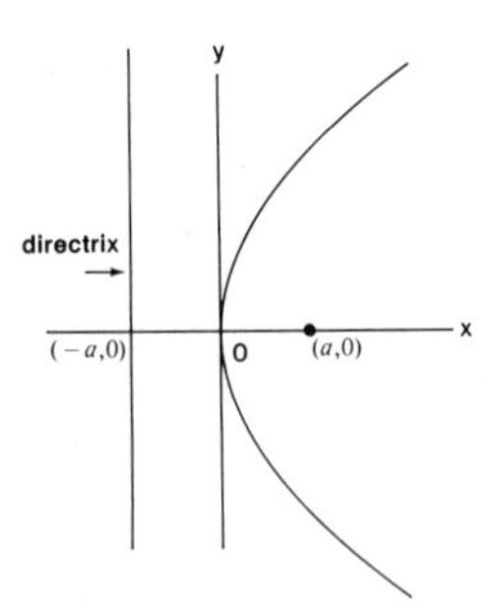

GREEK, *parabole*, placing next to

paraboloid (geometry) a surface obtained by rotating a **parabola** about its principal axis.
Example

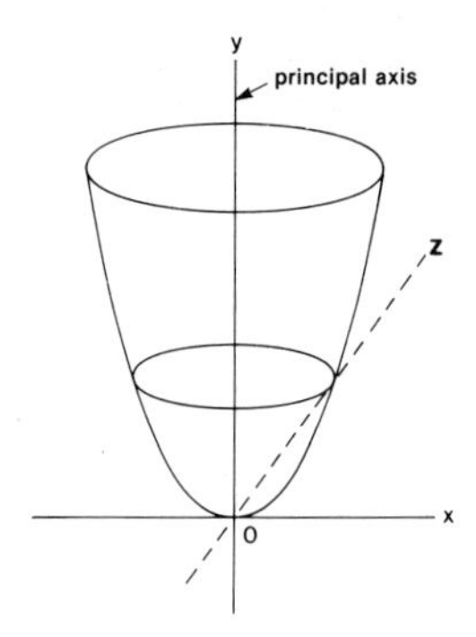

Its equation is $\frac{x^2}{a^2} + \frac{z^2}{c^2} = 2by$.

GREEK, *parabole*, placing next to; *eidos*, form or shape

parallel referring to two or more lines in a **plane** which do not meet no matter how far they are extended. Two parallel lines are always the same distance apart. Symbol is $||$.
Example A pair of railway lines.

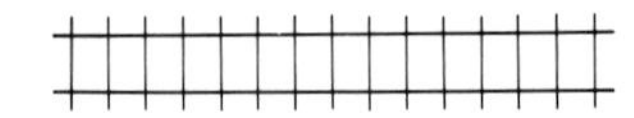

GREEK, *parallelos*, beside one another

parallelogram a four sided figure whose opposite sides are **parallel**. It has the following properties: opposite sides and angles are equal; and the **diagonals** bisect each other.
Example

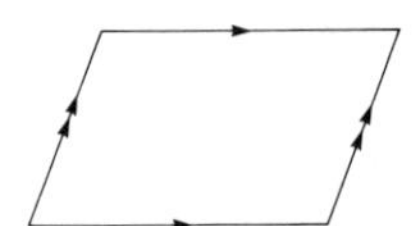

The area of a parallelogram is base × perpendicular distance between the **base** line and the line parallel to it.
GREEK, *parallelos*, beside one another; *gramma*, a writing, letter

parameter a **variable** or a **constant** (which can take on different values) in a mathematical **expression**. The value taken by the variable or "constant" restricts or determines the form of the expression. A parameter often has a physical or graphical meaning.
Examples
1. In the equation $y = x + c$, c is a parameter which

determines where the line as represented by $y = x + c$ cuts the y-axis. If c is limited to whole numbers, then $y = x + c$ represents a family of parallel lines as below.

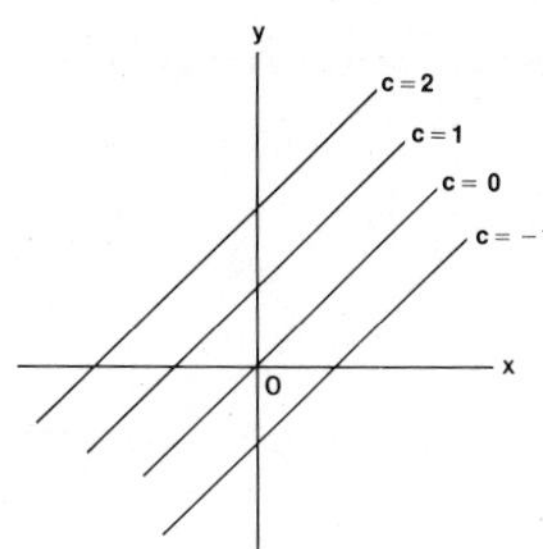

2. In the equation $y = mx + 1$, m is a parameter which determines the slope of the line passing through the point (0,1). $y = mx + 1$ is a family of lines with different slopes passing through the point (0,1).

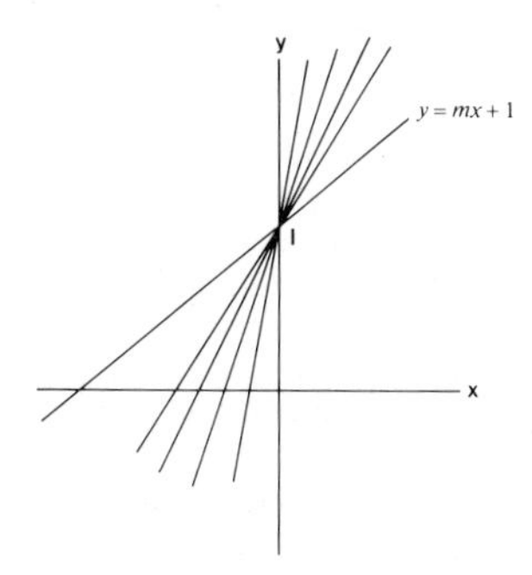

3. For the parabola $x^2 = 4ay$, any point on it can be represented using a parameter t by the equations $x = 2at$, $y = at^2$.
GREEK, *para*, alongside; *metron*, measure

partial fractions a set of **fractions** whose **algebraic** sum is a given fraction.
Example The fraction $\frac{1}{x^2 - 1}$ can be split into $\frac{1}{2(x - 1)} - \frac{1}{2(x + 1)}$ using the following technique:

Let $\frac{1}{x^2 - 1} = \frac{A}{x - 1} + \frac{B}{x + 1}$

Then $1 = A(x + 1) + B(x - 1)$

$1 = x(A + B) + A - B$

Equating coefficients, we have

$$A + B = 0$$
$$A - B = 1$$
$$2A = 1$$
$$A = \tfrac{1}{2},\ B = -\tfrac{1}{2}$$

Partial fractions can often be worked on more easily than the original fraction. For example,

$$\int \frac{1}{x^2 - 1}\,dx = \int \frac{1}{2(x - 1)} - \frac{1}{2(x + 1)}\,dx$$
$$= \tfrac{1}{2}\log(x - 1) - \tfrac{1}{2}\log(x + 1)$$
$$= \tfrac{1}{2}\log\frac{(x - 1)}{(x + 1)}$$

LATIN, *pars*, part; *fractio*, a breaking

Pascal, Blaise (1623–1662) French mathematician of great genius. Pascal's studies include work on the mathematics of **conics**, **binomial expansions** $[(a + b)^n]$ where he introduced **Pascal's triangle** for determining coefficients, and the geometry of the **cycloid**.

Pascal's triangle a triangular pattern of **numbers** as follows:

```
        1
       1 1
      1 2 1
     1 3 3 1
    1 4 6 4 1
   .         .
   .         .
```

The pattern is continued by adding each pair of numbers in a row and putting the sum underneath. It is used to determine the **coefficient** of the terms of the **expression** $(x + y)^n$ where n represents the row plus 1.
Example $(x + y)^4 = x^4 + 4x^3y + 6x^2y^2 + 4xy^3 + y^4$, which is the row 1 4 6 4 1 in Pascal's triangle.
after Blaise Pascal, French mathematician; LATIN, *triangulus*, three angled

pattern a **set** of **points**, **lines** or **numbers** put in some special arrangement. There is usually some rule or rules connecting the points, lines or numbers.
Example

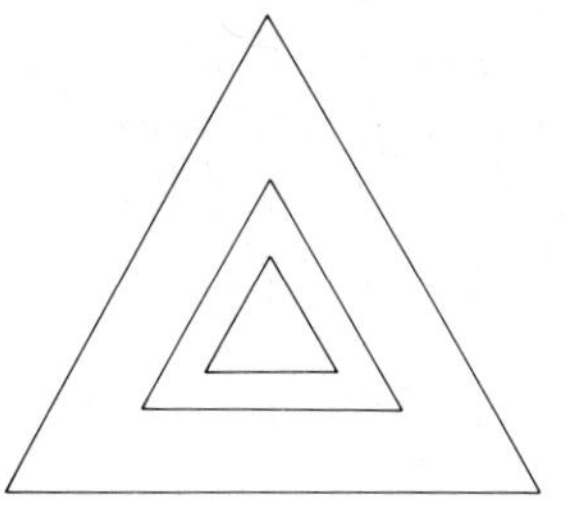

This pattern consists of 3 triangles. The first inner triangle is drawn similar to the outer but with its side half the length of the larger triangle. The second triangle is drawn in like fashion.

This pattern consists of dots equally spaced from each other in the shape of a rectangle.

1, 4, 9, 16, 25, 36, 49, 64, . . .

This pattern of numbers is made up of the squares of whole numbers 1, 2, 3, 4, 5, 6, 7, 8, . . .
MIDDLE ENGLISH, *patron*, model, design

penta- prefix meaning five.
Example A pentad is a group of five; **pentagon**.
GREEK, *penta*, five

pentagon a five sided figure having five angles. A **regular** pentagon has all sides and angles equal.
Examples

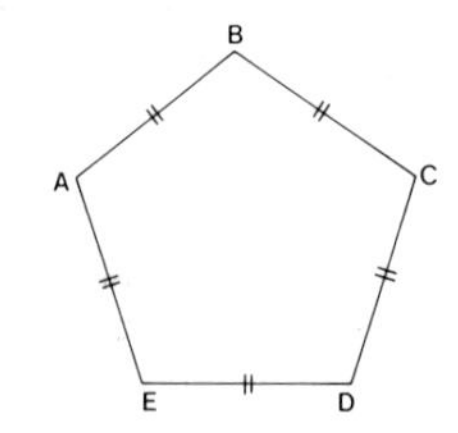

ABCDE is a regular pentagon.

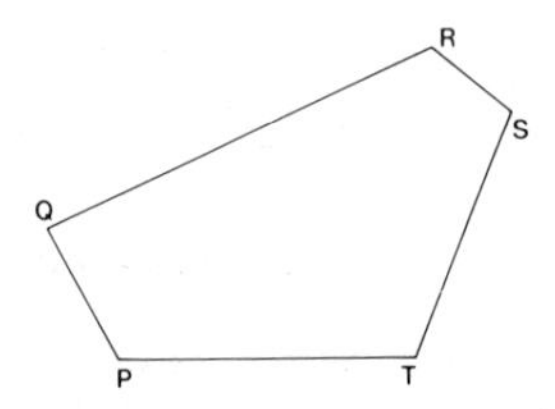

PQRST is an ordinary pentagon.
The sum of the interior angles of a pentagon is six right angles.
GREEK, *pentagonon*, five angles

percentage **1.** A **fraction** written as a part of one hundred. **2.** The amount calculated by applying a per cent (written as percent in the United States) to a given quantity. Per cent is the rate expressed as a fraction with a denominator of one hundred.
Examples
1. $\frac{1}{2} = \frac{50}{100}$ and is written as 50 per cent. The symbol for per cent is %, which represents 100 written differently.
2. 20% of $100 was the percentage the salesman received as his commission (i.e. $20).
LATIN, *per centum*, by the hundred

perfect number a whole **number** which equals the sum of all its **factors** except itself.
Examples
$6 = 1 + 2 + 3$
$28 = 1 + 2 + 4 + 7 + 14$
LATIN, *perficere*, to complete; *numerus*, number

perfect square a **number** or **expression** which can be written as the **product** of two equal factors. Note: The **square root** of a perfect square is one of the two equal factors.
Examples
1. $9 = 3 \times 3 = 3^2$
2. $256 = 16 \times 16 = 16^2$
3. $x^2 + 2x + 1 = (x + 1)^2$
LATIN, *perficere*, to complete; *quadra*, square

perimeter the distance around the boundary of a figure.
Example The perimeter of a square is four times the length of one side. The perimeter of a circle is its **circumference**. $C = 2\pi r$, where π is pi and r is the radius.

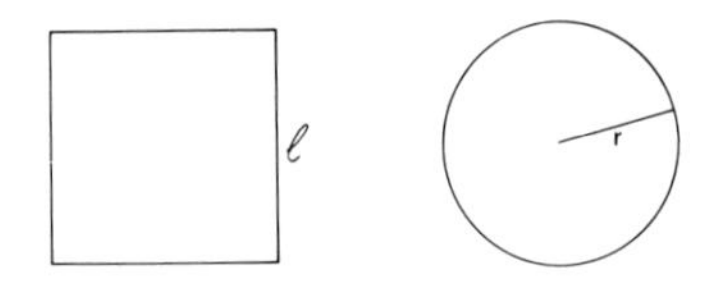

GREEK, *peri*, around; *metron*, measure

period the smallest interval of x in which a **function** $f(x)$ takes on all its values and then repeats for each interval.
Example

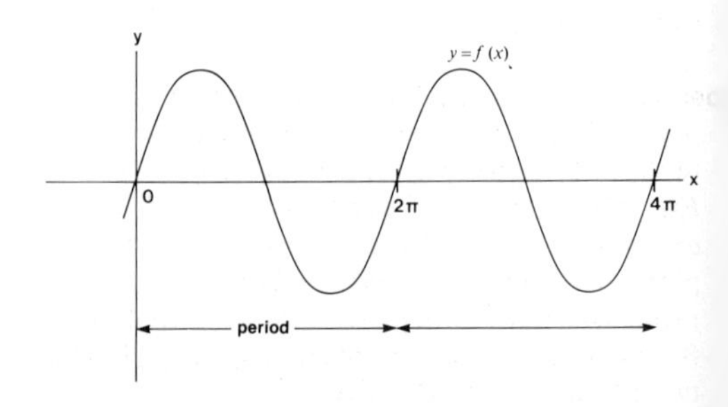

In the diagram $f(x)$ is a function which repeats every 2π units.
GREEK, *periodos*, a circuit, period

periodic decimal a repeating **decimal**. Symbol used to show the repeating digits is ¨
Example 2.23232323..., where 23 repeats forever. This is written $2.\dot{2}\dot{3}$. A periodic decimal always represents a fraction, e.g. $2.232323\ldots = \frac{221}{99}$.
Another example is $1.1111\ldots = 1.\dot{1}$.
GREEK, *periodos*, a circuit, period

periodic function a **function** which repeats its values after a certain **interval**.
Example $y = \sin x$, $y = \cos x$, $y = \tan x$, etc., are all periodic functions. In the case of $y = \sin x$, y repeats its values after the interval $0 \rightarrow 2\pi$.

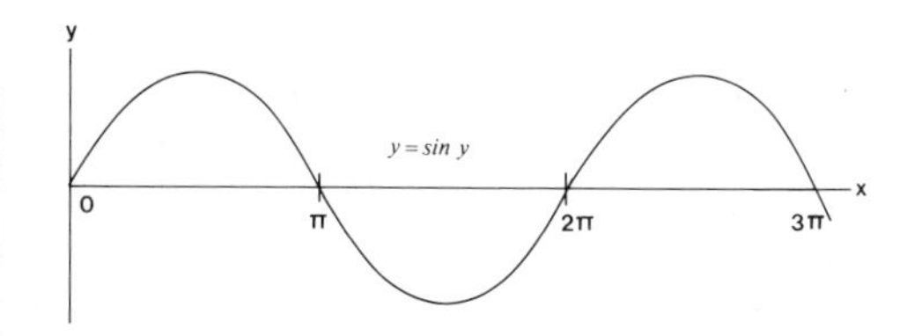

GREEK, *periodos*, a circuit, period

perpendicular **1.** At right angles or 90° to. Symbol used is ⊥. **2.** A line at right angles to another.
Examples

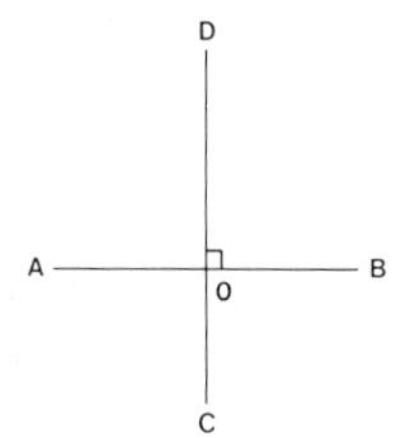

1. AB is perpendicular to CD.
2. CD is a perpendicular to AB.
LATIN, *perpendiculum*, a plumb line

permutation any of the total **number** of arrangements possible in a group of **elements**. A permutation refers to the order in which the elements are grouped.
Example a, b, c can be arranged in six ways: *abc*, *acb*, *cab*, *bac*, *bca*, *cba*, but there are only two permutations of a and b: *ab*, *ba*. The total number of permutations of n elements is $n!$ Note: The formula for finding the number of permutations of m elements taken from a group of n elements is

$$^{n}P_{m} = P^{n}_{m} = \frac{n!}{(n-m)!},$$

where ! means **factorial**.
LATIN, *permutare*, to change thoroughly

peta a prefix meaning one thousand million million (10^{15}). The symbol is P.
Example If you were to travel 100 petametres (Pm), you would end up far out in space, a long way beyond our solar system.
GREEK, perhaps from *penta*, five

pi the **ratio** of the **circumference** of a circle to its diameter. The symbol for pi is π. Its value is the same for all circles and equal to approximately 3.14159 (to five decimal places). $\frac{22}{7}$ is a good approximation but not the exact value.
Example If the diameter of a circle is known, then the circumference is πd. The area of a circle is πr^2.
GREEK, Π, π (pi) [16th letter of the Greek alphabet—and first letter in *perimetros*, perimeter]

pico- a prefix meaning one million millionth (10^{-12}). The symbol is p.
Example A picosecond (ps) is 0.000000000001 seconds. A beam of light travels less than a centimetre in one picosecond.
SPANISH, *pico*, very small

pictogram a special kind of **graph** or chart using easily understood symbols to show information.
Example

Boy | Number of apples eaten
Bill
Jack
Bob
= apple

LATIN, *pingere*, to paint; *gramma*, a writing, letter

pie chart a chart in the form of a circle or pie with **sectors** showing **fractions** of the whole. It is used to represent information.
Example

A B C D E

Company A has 20% of the market.
Company B has 15% of the market.
Company C has 25% of the market.
Company D has 5% of the market.
Company E has 35% of the market.
MIDDLE ENGLISH, *pie*, pie; LATIN, *charta*, paper

place value the value given to the position a **digit** has in relation to neighbouring digits. A point is usually used to separate the whole number from the **fraction** part. To the left of the point, each position represents a power of the **base** number used to count with. To the right, each position represents a particular fraction of the base number (*see* DECIMAL SYSTEM). It is a short-hand way of writing numbers.
Examples In the decimal system a number written 543.26 means $5 \times 10^2 + 4 \times 10 + 3 \times 1 + 2 \times \frac{1}{10} + 6 \times \frac{1}{10^2}$. To the left of the point, the first position represents units, therefore a 3 in that position indicates 3 units; the second position represents tens, therefore a 4 in that position indicates 4 tens (or forty); the third position represents hundreds, therefore a 5 in that position indicates 5 hundreds (or 500). The place value of 5 in the number 543.26 is hundreds. To the right of the point the first position represents tenths, therefore a 2 in that position indicates 2 tenths; the second position represents hundredths, and therefore a 6 in that position shows 6 hundredths; and so on.
LATIN, *platea*, an open space; *valere*, to be worth

plan a drawing or diagram which is the result of projecting a solid object onto a **plane** surface.
Examples
1. This is a plan of a tetrahedron.

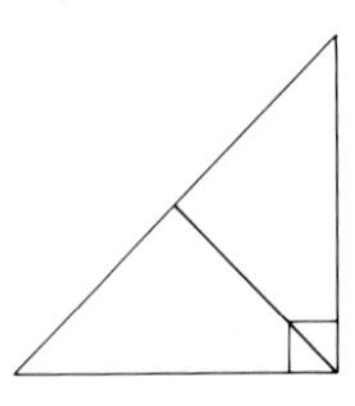

2. The plan of two cubes of sides 3 cm and 2 cm respectively is:

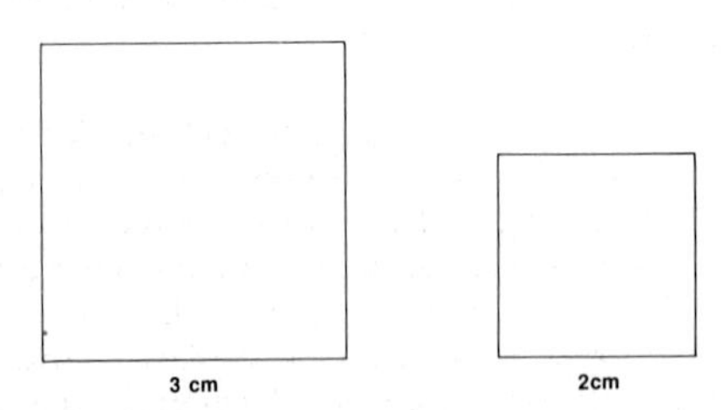

LATIN, *planus*, flat

plane **1.** A surface such that every line joining any two points in it lies wholly within it. **2.** Referring to a plane.
Examples
1.

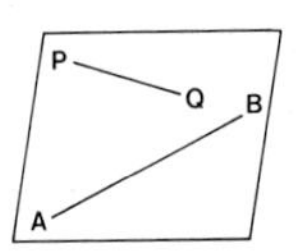

AB and *PQ* lie within the plane. A plane can be represented in **three dimensions** by the general equation $ax + by + cz = d$ in **coordinate** goemetry. (also written $S: ax + by + cz = d$, where S stands for surface and : stands for described by the equation.
2. A plane section is a figure formed by a plane cutting a solid.

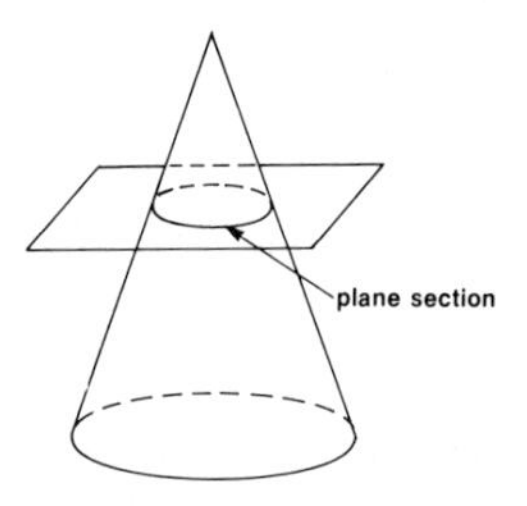

A plane shape is one whose points are all in one plane, such as a circle, triangle, rectangle etc.
LATIN, *planus*, flat

planimeter an instrument which allows one to find the area of a plane figure by tracing its pointer along the **perimeter** of the figure.
Example

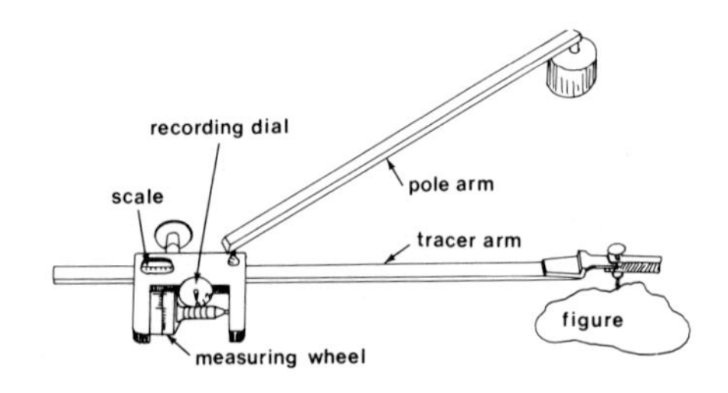

LATIN, *planus*, flat; GREEK, *metron*, measure

Platonic solids regular solids.
Examples A **regular tetrahedron** and a **regular dodecahedron** are two types of Platonic solids. (There are five in total, the others being a **cube, octahedron** and **icosahedron**.)
After Plato, a Greek philosopher; LATIN, *solidus*, solid

plus to be added. Symbol used in +, and this can also mean that a quantity is **positive**.

Examples
8 + 4 = 12.
+3 means 3 positive units, or 3 units to the right of the origin along an *axis*.

0 +1 +2 +3

LATIN, *plus*, more

point 1. A dot on a surface. It has position, but no size. 2. The dot representing the position between the **units** and the **fractions** in a particular counting system.
Examples
1.

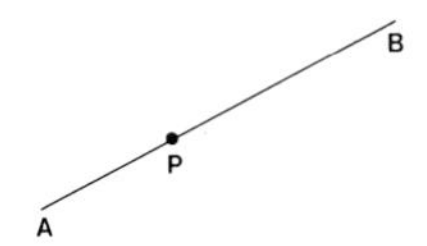

P is a point on the line *AB* 2 cm from *A*.
2. 459.8 (here, it is a **decimal** point, but it could be used in any **base** system).
LATIN, *punctus*, pricked

point of inflection (also **inflexion**) a point on a curve where the **gradients** either side of the point are both positive or both negative. If the gradient is zero at the point, it is called a **horizontal** point of inflection. In general, a point of inflection is a point $x = a$ where $f''(a) = 0$ and the sign of the values of $f''(x)$ at a point just greater than $x = a$ is different to the sign of the value of $f''(x)$ at a point just smaller than $x = a$.
Examples
1.

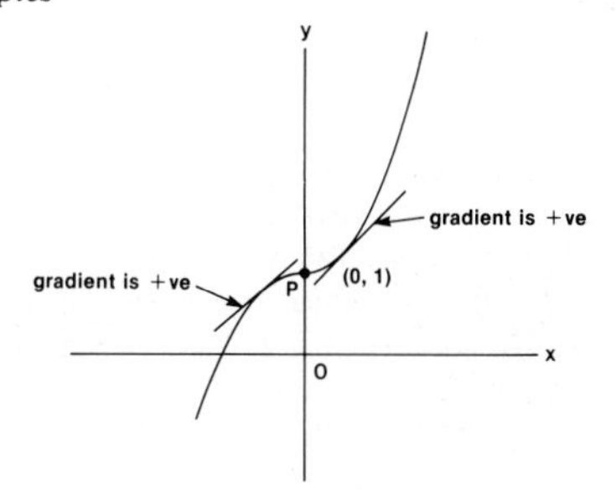

P is a horizontal point of inflection on the curve $y = x^3 + 1$. If the second **derivative** of the particular function is zero for the same value of *x* that the first derivative is zero, then the point is a horizontal point of inflection, that is, $\frac{dy}{dx} = 3x^2 = 0$ when $x = 0$; $\frac{d^2y}{dx^2}$ (second derivative) $= 6x = 0$ when $x = 0$, therefore the point $(x = 0, y = 1)$ is a horizontal point of inflection (*see also* INFLEXIONAL TANGENT).

2. The curve of $y = \tan x$ is shown below.

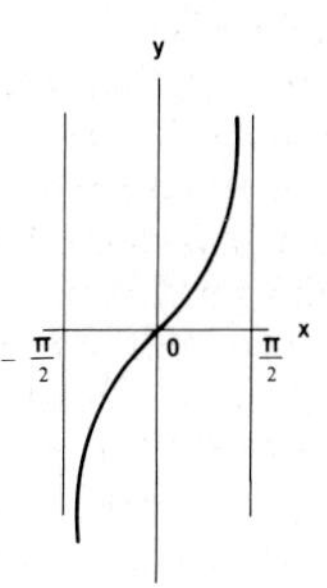

$P(x = 0, y = 0)$ is a point of inflection since $\frac{d^2y}{dx^2} = (2 \tan x.\sec^2 x)$ is negative for $x < 0$, and positive for $x > 0$.
LATIN, *punctus*, pricked; *inflectere*, to bend

point symmetry **symmetry** of an **object** about a point so that the **image** resulting from a **rotation** of 180° about the point has the same position as the object.
Example

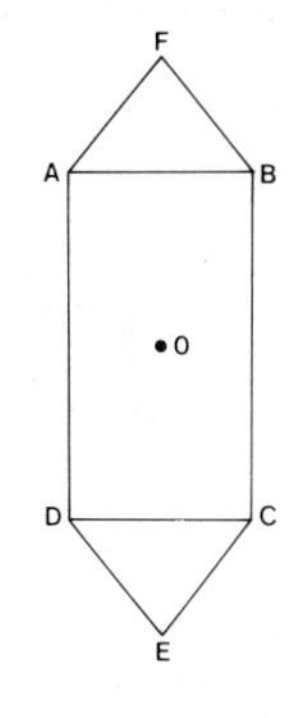

AFBCED has point symmetry about 0.
LATIN, *punctus*, pricked; GREEK, *summetros*, of like measure

poisson distribution (statistics) a **distribution** of the **probability** of unlikely outcomes of events occurring in a large number of independent **events**. It differs from the **binomial distribution** in that the Poisson distribution describes situations where the total number of outcomes is not known (as it is with the binomial distribution). If *p* is the average number of outcomes of an event, then the probability of the outcomes is given by the successive terms of the following: e^{-p}, pe^{-p}, $(p^2/2!)e^{-p}$, $(p^3/3!)e^{-p}$, etc.
Example Over a period of 50 years the following table of shark deaths was made.

No. of deaths in one year	0	1	2	3	4	5	6	7
No. of years this number of deaths occured	10	16	11	7	4	1	1	0

The total number of deaths was 86. The average number of deaths per year is 1.72. According to the poisson distribution, the probability of 0 deaths is $e^{-1.72}$, therefore the number of years for which 0 is likely is $50 \times e^{-1.7} = 9.1$ approximately (compare this to the actual number of years, 10). Similarly, for one death, the probability is $1.72 \times e^{-1.72}$, etc.
after Poisson, French mathematician

polar coordinates coordinates of a point given as the distance (or **radius**) from an origin (or **pole**) and an angle from the horizontal axis (polar **axis**).
Example

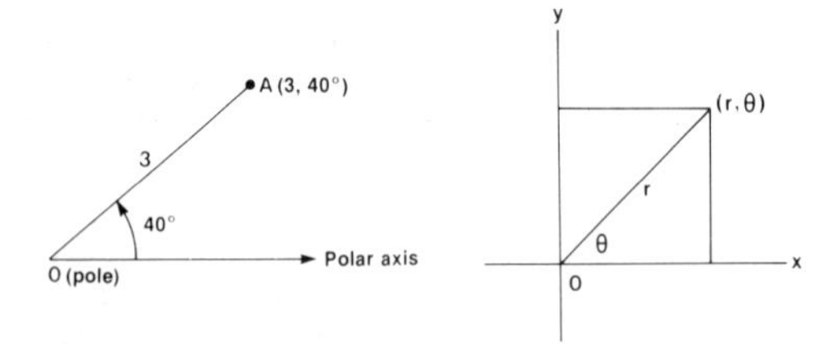

Note: Polar coordinates can be converted to **cartesian** coordinates by using the equations $x = r\cos\theta$, $y = r\sin\theta$.
LATIN, *polus*, pole; *coordinatio*, arrangement in some order

polar form the representation of a **complex** number (z) using **polar coordinate** notation rather than **cartesian** coordinate notation: z is expressed in the form $r(\cos\theta + i\sin\theta)$ instead of $x + iy$.
Example

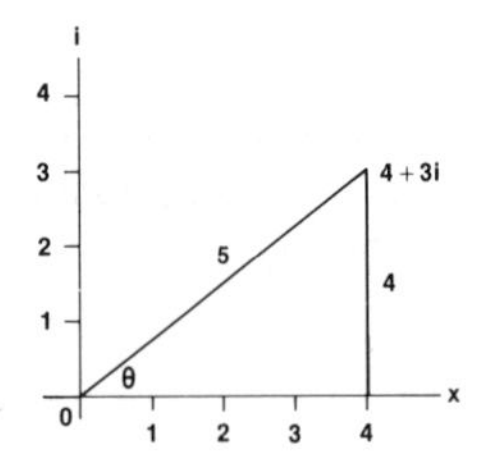

$z = 4 + 3i$ can be represented as $5(\cos\theta + i\sin\theta)$ where $\cos\theta = 4/5$, $\sin\theta = 3/5$.
LATIN, *polus*, pole; *forma*, form, shape

polygon a plane figure bounded by straight lines only.
Example a triangle is a three sided polygon, a **quadrilateral** is a four sided polygon, etc. (*see* PENTAGON, HEXAGON).

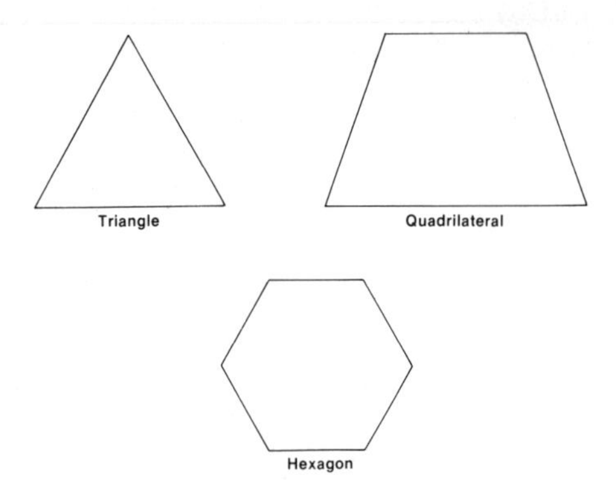

Note: The sum of the **interior** angles of any **convex** polygon which has n sides is $(2n - 4)$ right angles.
GREEK, *polugonos*, many angles

polyhedron (*pl.* **polyhedra**) a solid made up of only plane faces. The plane faces are all **polygons**.
Examples **Cube**; **prism**; **pyramid**; **octahedron**. A regular polyhedron has all sides equal, for example, cube, regular octahedron.

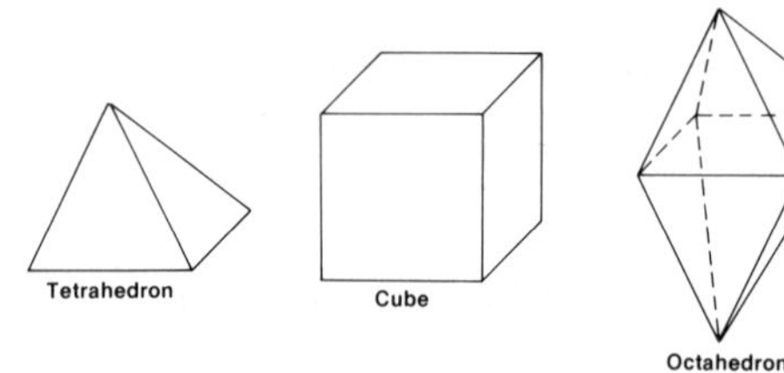

GREEK, *poluedros*, having many sides or seats

polynomial **1.** An **expression** of two or more **terms**. **2.** An **algebraic function** of two or more terms added together, where each term is made up of a constant and one or more **variables** raised, in general, to a positive whole number power. The general form of a polynomial in one variable is

$$a_0x^n + a_1x^{n-1} + a_2x^{n-2} + \ldots + a_{n-1} + a_n$$

where $a_0 \neq 0$, and $a_1, a_2, a_3, \ldots a_n$ are all constants, and n is a positive whole number. The highest value of n is called the **degree** of the polynomial.
Examples
1. $x + y + z^2$ is a polynomial.
2a. $2x^3 + 3x^2y + 4xy^2 - 5y^3$ is a polynomial of degree 3 in 2 variables.
b. $x^4 + 3x^2 - 2x^2 + x - 6$ is a polynomial of degree 4 in a single variable.
GREEK, *poly*, many; LATIN, *nomen*, name, part

polyominoes shapes made by equal squares which join along a complete side.
Examples A domino is a polyomino made from 2

squares. A tromino is a polyonimo made from 3 squares.
GREEK, *poly-*, many; LATIN, *domino*, a half mask

population (statistics) the set of all individuals, or items or things from which a sample is taken.
Examples
The country town has a population of 10 000 adults. We asked 500 people as a sample their opinion of living in the town.
A sample of 100 cereal packets was taken at random from the production area and weighed to see how accurate the weights were. The population here was all the cereal packets produced in a given time period (e.g. a day's production).

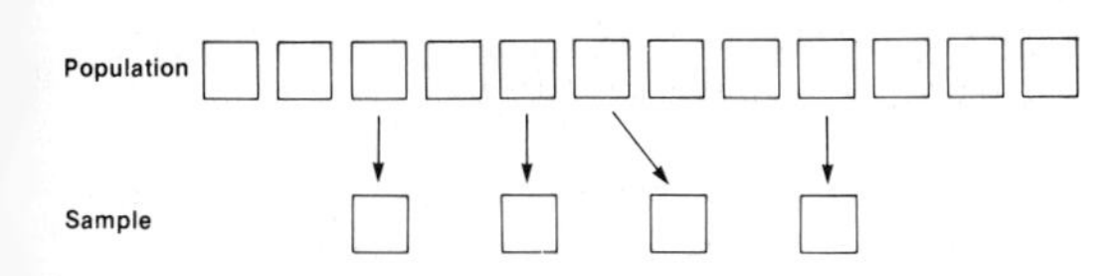

LATIN, *populus*, people

positive greater than **zero**.
Example A positive number such as 4, 0.23, or 712.8 lies to the right of zero on the number line.

LATIN, *positivus*, arbitrarily laid down, dogmatic

positive definite a function $f(x)$ is positive definite if for all values of x, $f(x) > 0$. Note: $f(x)$ can be shown to be greater than 0 if the **absolute minimum** of $f(x)$ is greater than 0.
Example $f(x) = x^2 + 1$. Since the minimum value of $x^2 + 1$ is 1 (when $x = 0$), then $f(x) > 0$ for all x, and is positive definite. The procedures of **calculus** can be used to find the minimum, as follows:
$\frac{df}{dx} = 2x = 0$ when $x = 0$. There is a **turning point** at $x = 0$. Since $\frac{d^2f}{dx^2} = 2$, (i.e. positive) the turning point is a minimum.

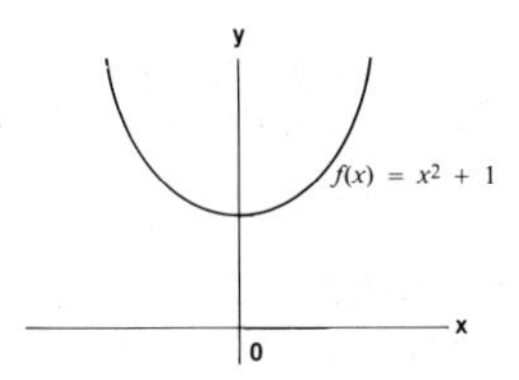

LATIN, *positivus*, arbitrarily laid down, dogmatic; *definitus*, determined, defined

postulate an **assumption** used (in combination with other postulates) as the basis for constructing various **mathematical systems**.
Example There is only one straight line which can be drawn through two points.

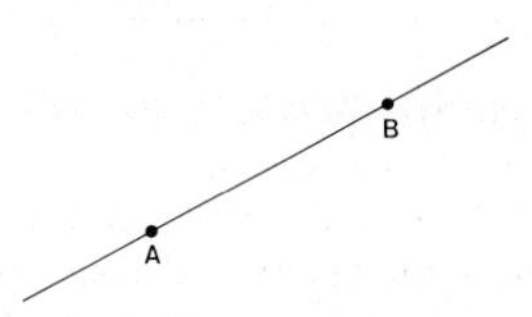

LATIN, *postulare*, to request, demand

power the **number** obtained by multiplying a given number by itself a certain number of times. (The number of times the given number is multiplied is called power n.) The symbol used for power (or **exponent**) is written as a superscript.
Examples
$8 = 2 \times 2 \times 2 = 2^3$. 8 is a power of 2. It is two to the power 3, or two **cubed** (3 is the superscript).
$9 = 3 \times 3 = 3^2$. 9 is a power of 3. It is three to the power 2 or three **squared**.
LATIN, *posse*, to be able

prime **1.** A prime number, or a number which can only be divided without remainder by itself and one. **2.** Relating to such a number.
Example 2, 3, 5, 7, 11, 13, 17, 19 are all prime numbers. A prime factor of 12 is 3. Another prime factor is 2. Note: 1 is not regarded as a prime number by most mathematicians. Numbers which are not prime are **composite**.
LATIN, *primus*, first

primitive **1.** Any form in **geometry** or **algebra** from which another form is derived. **2.** A **function** $F(x)$ whose **derivative** is $F'(x) = f(x)$. $F(x) + c$ (where c is any constant) is also known as the indefinite **integral** of $f(x)$.
Examples
1. A triangle is derived from these line segments cutting each other. ABC is a geometric figure or form derived from the primitive line.

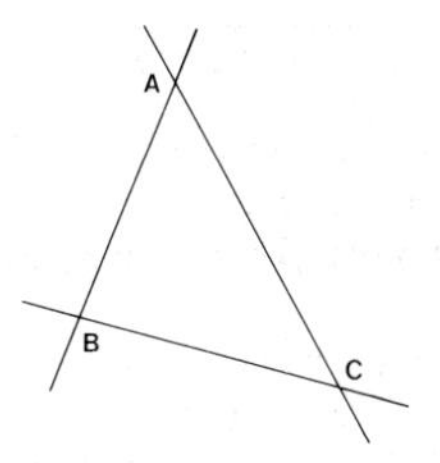

2. The function x^2 is said to be a primitive of the function $2x$, since $2x$ is the derivative of x^2. $x^2 + c$ is the indefinite integral of $2x$ (c is any constant).
LATIN, *primitivus*, first of its kind

principle a basic truth, law or something taken to be this. Principles form the building blocks on which one can construct (**mathematical**) systems, **theorems**, etc.
Examples
One of the principles of **coordinate goemetry** is that **points**, **lines**, **planes**, and other shapes can be represented by **algebraic** equations with reference to fixed axes. For example, a straight line can be represented by the **equation** $y = mx + c$ (m represents the slope or **gradient**, and c the point on the y-axis where the line cuts it).

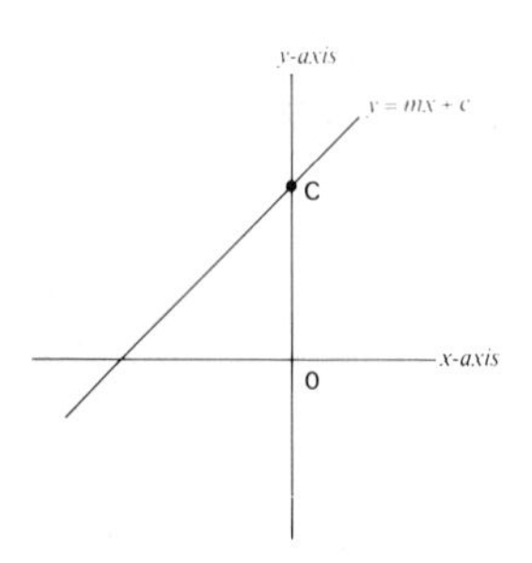

A circle can be represented by $x^2 + y^2 = r^2$.

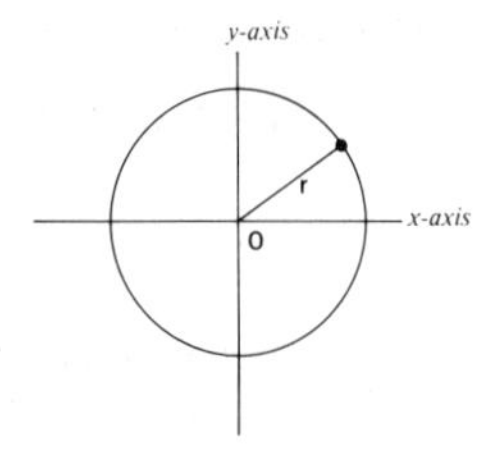

Another principle often used to prove certain mathematical patterns is **induction**.
LATIN, *princeps*, first

prism a solid shape which has a uniform cross-sectional area.
Examples

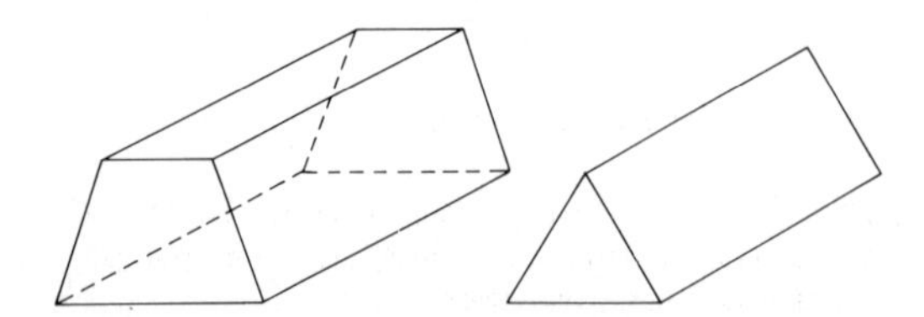

Note: Prisms are named according to the shapes of an end, so that there are triangular, pentagonal prisms etc. A prism whose end is at **right angles** to the sides is called a right prism. For the volume of any right prism, $V = A \times h$, where A = cross-sectional area, h = height of the prism.
GREEK, *prisma*, a thing sawed

probability **1.** A measure of how likely an **event** is to occur. **2.** The **ratio** of the number of certain specified outcomes or events to the total number of possible outcomes. Probabilities lie between 0 and 1.
Example The probability of a six turning up from a thrown die is $\frac{1}{6}$, since there are six possible outcomes, and only one way in which a six can turn up.

LATIN, *probabilis*, provable, able to be tested

product the result obtained when two or more **numbers** are multiplied together.
Examples
35 is the product of 5 and 7.
60 is the product of 3 and 4 and 5.
LATIN, *producere*, to produce

product rule a rule used in **calculus** to work out the result of **differentiating** the product of two **algebraic functions**. The rule is: given $y = u(x).v(x)$, then $dy/dx = u\, dv/dx + v\, du/dx$ (often written $y' = uv' + u'v$).
Example If $y = x(x - 1)^3$,

$$\frac{dy}{dx} = 3x(x - 1)^2 + (x - 1)^3$$
$$= (x - 1)^2 (3x + x - 1)$$
$$= (x - 1)^2 (4x - 1)$$

LATIN, *producere*, to produce; *regula*, rule

profit a quantity (usually money) one receives after deducting the cost price from the selling price. Sometimes abbreviated as: $P = SP - CP$ (selling price − cost price). It can also be expressed as a percentage of cost or selling price.
Examples
1. A bicycle is bought secondhand for \$50. Next month it is sold for \$70. Profit = \$20 (\$70 − \$50).

$$\text{Profit \% (of cost price)} = \frac{\$20}{\$50} \times \frac{100}{1} = 40\%.$$

2. A greengrocer buys apples at 5¢ each. He sells them at 8¢ each. His profit on each apple is 3¢ (8¢ − 5¢). Profit % (of selling price) = $\frac{3}{8} \times 100 = 37.5\%$.
LATIN, *proficere*, to advance, profit

progression a **series** of **numbers** or quantities, each one related to the previous one by some rule.
Example The following group of numbers is an **arithmetic progression** (AP) 1, 3, 5, 7, 9, ...,51. Each number is two bigger than the previous number. a, ar, ar^2, ar^3, ar^4, ...is an example of a **geometric progression** (GP). Each quantity is related to the previous by being r times the previous value. Progressions occur in real life as in the following example: At what time between 3 p.m. and 4 p.m. is the minute hand over the hour hand? Since the hour hand has moved through five minutes in each hour, by 3.15 p.m. the minute hand has moved through 15 minutes, the hour hand has moved $15 \times 5/60 = 15/12$ minutes. When the minute hand has moved through to 3.15$^{15}/_{12}$ p.m., the hour hand has now moved through $15/12 \times 1/12 = 15/144$. These movements continue, so that the final position is the sum of $15 + 15/12 + 15/12^2 + \ldots$ (sum of a GP.) Another kind of progression is the **harmonic**, where the **reciprocals** of each **term** in the harmonic progression form an arithmetic progression, for example, 1, $1/3$, $1/5$, $1/7$, $1/9$ ...
LATIN, *progredi*, to go forward

projection the action of transforming a set of points (on a line, figure or solid) to another set of points according to specified geometrical rules. The initial set of points is called the object, the final set the image.
Example

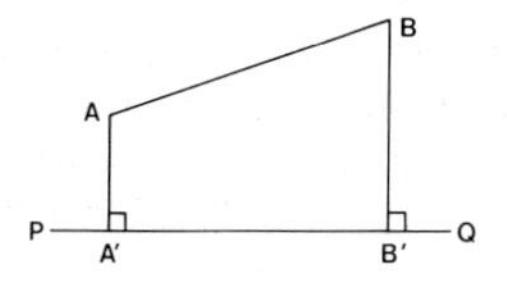

$A'B'$ is the perpendicular projection of AB on to the line PQ.

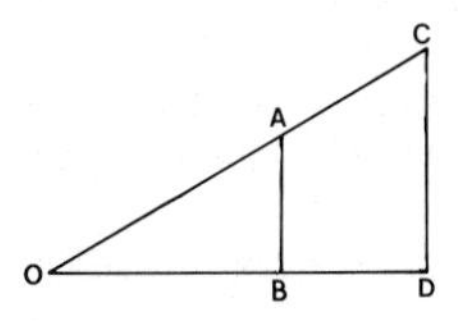

CD is a projection of AB from O.
LATIN, *proicere*, to throw forth

pronumeral a symbol standing for a numeral.
Example In the following, @ and δ are pronumerals.

$8 \times @ = 24$ $\quad$ $6 \times 42 = 42 \times \delta$
$12 \div \delta = 2$ $\quad$ $13 + @ = 12 + 4$

LATIN, *pro*, on behalf of; *numerus*, number

proper fraction a **fraction** whose **numerator** is less than its **denominator**.
Example $3/4$, $2/4$, $4/7$ are proper fractions.
LATIN, *proprius*, one's own, personal; *fractio*, a breaking

proportion a **relation** between four quantities a, b, c and d such that $\frac{a}{b} = \frac{c}{d}$. Also written as $a{:}b = c{:}d$.
Hence, **proportional**, having the same **ratio**—symbol is $\propto$. Proportional can be used as a noun also, as in **mean proportional** (also known as a **geometric mean**), **third proportional** and **fourth proportional**.
Examples
1. The distance (D) travelled by a wheel is proportional to the number (n) of revolutions made. $D \propto n$ is the same as writing $\frac{D}{n}$ = a constant.
2. The distance up the incline is proportional to the perpendicular height.

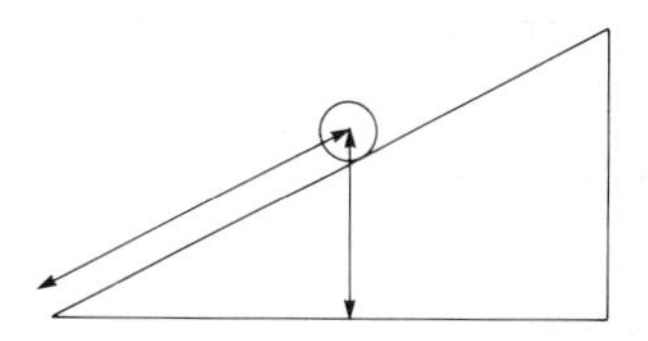

3. In the equation $\frac{a}{b} = \frac{b}{c}$, b is the mean proportional, and c is the third proportional. In the equation $\frac{1}{3} = \frac{4}{12}$, 12 is the fourth proportional to 1, 3, 4 and 12.
LATIN, *proportione*, for its (his) share

protractor a flat, circular or semicircular instrument, graduated around the circular edge, used to measure or mark off angles.
Example

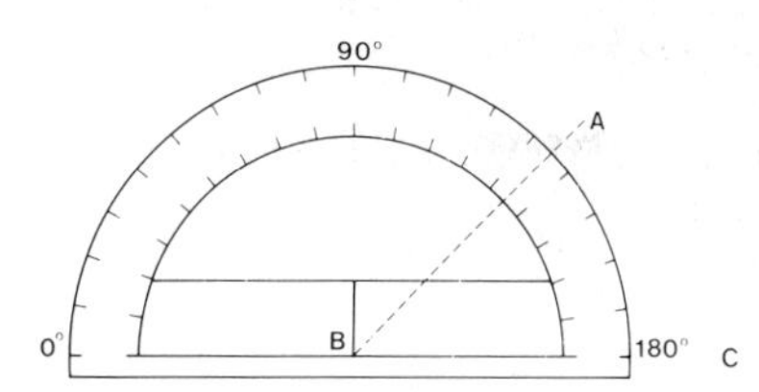

LATIN, *protrahere*, prolong, defer

pyramid a solid shape with flat faces that has a **polygon** as a base, and whose sides are **triangles** meeting at a common **vertex**.

Example

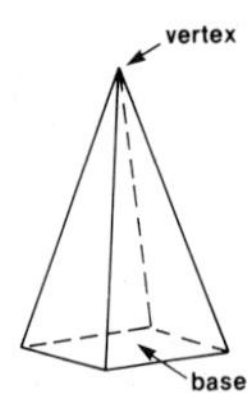

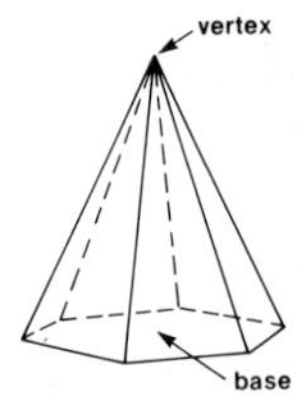

Note: Volume of any pyramid = ⅓ × base area × **perpendicular** height of vertex from base.
LATIN, *pyramis*, pyramid

Pythagoras (585–500 BC) a famous Greek mathematician who founded his school at Croton, a Greek settlement in Southern Italy. Here he formed a religious and philosophical brotherhood, the Pythagoreans. Discoveries attributed to Pythagoras and his followers include the mathematics of triangular, square and polygonal numbers, the invention of perfect numbers and the possible proof of **Pythagoras' theorem**.

Pythagoras' theorem in a right angled triangle, the **square** on the **hypotenuse** (longest side) is equal to the sum of the squares of the other two sides.
Examples

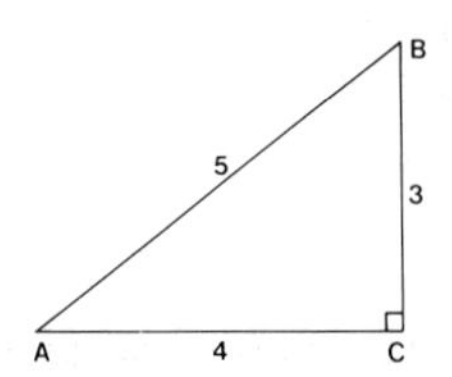

$AB^2 = AC^2 + BC^2$, i.e. $5^2 = 4^2 + 3^2$

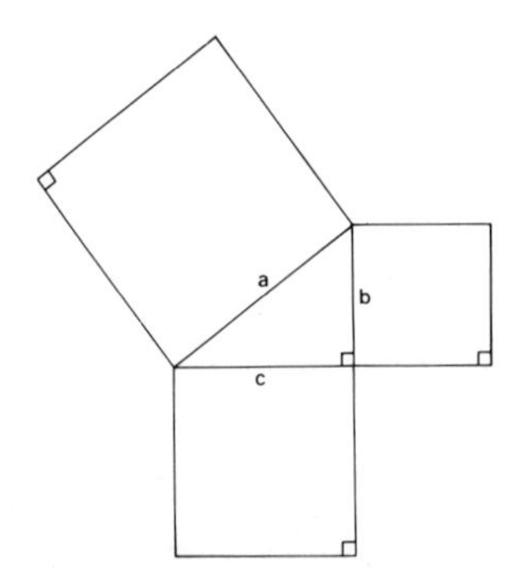

In general, $a^2 = b^2 + c^2$
after Pythagoras, Greek mathematician; GREEK, *theorein*, to observe, look at

Qq

quadrant 1. A quarter of a **circle**. 2. Any of the four parts into which a **plane** is divided by a pair of **axes** cutting each other at **right angles**.
Examples
1.

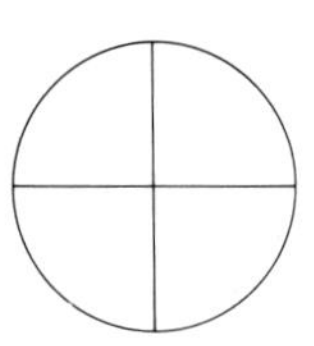

2.

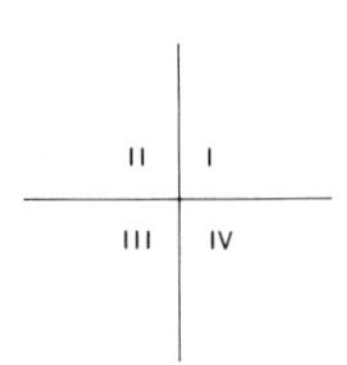

LATIN, *quadrans*, fourth part, quarter

quadratic involving the second **power** (and no higher power) of an unknown quantity or **variable**.
Example A **quadratic equation** is an equation which involves unknown **variables** to the second power. $x^2 + 10x + 7$ is a quadratic expression.
LATIN, *quadratus*, squared

quadratic equation an **equation** whose unknown is to the second **power** only. The general form of the quadratic equation is $ax^2 + bx + c = 0$.
Example $x^2 + 2x + 3 = 0$, $x^2 = 5$ are quadratic equations. The quadratic formula is one of the methods used to solve a quadratic equation. It is:

$$x = \frac{-b + \text{or} - \sqrt{b^2 - 4ac}}{2a}$$

Other methods are **factorisation** and **completion of the square**.
LATIN, *quadratus*, squared; *aequus*, equal

quadrature the process of finding a square whose area is the same as that of the area bounded by a curve. The ancient Greeks were very interested in problems of this nature.

Example Finding the area of a circle was one ancient problem to which quadrature applied.
LATIN, *quadratus*, squared

quadrilateral a plane figure having four sides and four angles.
Example

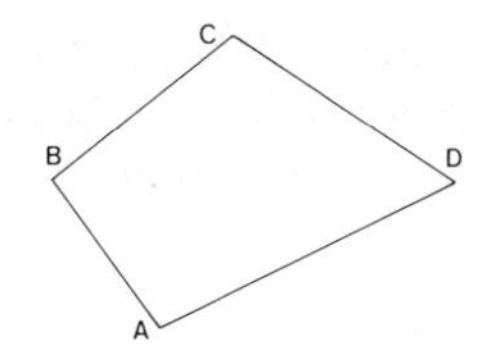

ABCD is a quadrilateral. The sum of the **interior** angles of any **convex** quadrilateral is four right angles (360°).
LATIN, *quadri*, four; *latus*, side

quartile the value of a **data element** at the end of one of the four equal parts of a data set which has been laid out on a proper **cumulative frequency**.
Example Twenty families were asked how many children they had. The results were presented as follows:

Cumulative frequency	1	5	10	15	18	19	20
Frequency (no. of families)	1	4	5	5	3	1	1
No. of children per family	1	2	3	4	5	6	7

The first quartile is at two children since one quarter of 20 is 5, and the value of the data at the fifth family is two children. The next quartile is at three children since the second quarter of 20 is 10 and the corresponding number of children is three. The next quartile is at four children since the third quarter of 20 is 15 and the corresponding number of children is four.

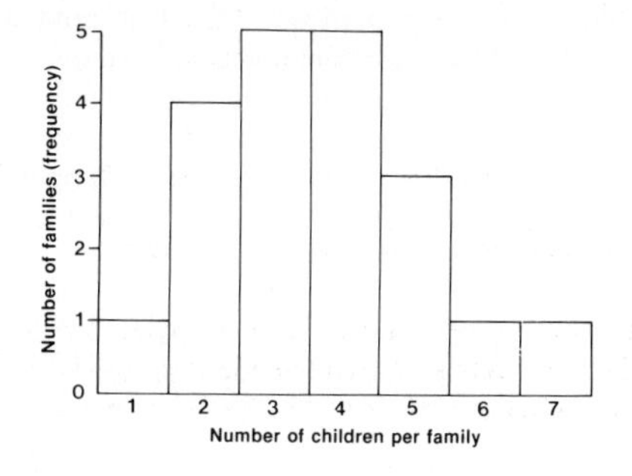

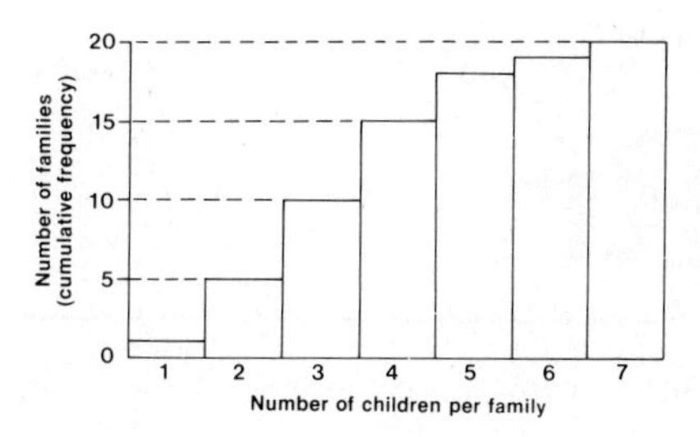

LATIN, *quartus*, fourth

quotient the result obtained by dividing one **number** by another (also called the **answer** to a **division**).
Example The quotient of 12 divided by 3 is 4.
LATIN, *quotiens*, how many times

quotient rule a rule used in **calculus** to work out the result of **differentiating** the **quotient** of two **algebraic functions**. The rule is: given $y = u(x)/v(x)$,

then $\dfrac{dy}{dx} = \dfrac{1}{v} du/dx - \dfrac{u}{v^2} dv/dx$

$= (v\dfrac{du}{dx} - u\dfrac{dv}{dx}) / v^2$

(often written $y' = (vu' - uv')/v^2$

Example If $y = \dfrac{x}{x+4}$,

$$\frac{dy}{dx} = \frac{(x+4)\cdot 1 - x(1)}{(x+4)^2}$$

$$= \frac{x+4-x}{(x+4)^2}$$

$$= \frac{4}{(x+4)^2}$$

LATIN, *quotiens*, how many times; *regula*, rule

Rr

radian the **unit** of plane angle (symbol is rad or ^{c}) and a supplementary unit in the **SI**. One radian is equivalent to $\frac{180}{\pi}$ degrees, or approximately 57 degrees. The unit of solid angle is called the steradian (symbol is sr). A radian is the angle subtended by an arc equal in length to the radius of a circle.

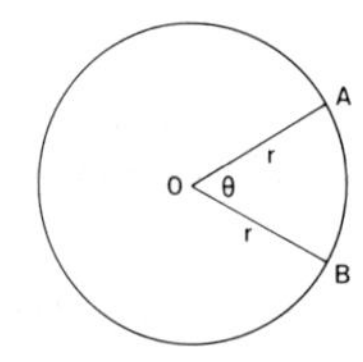

$\angle AOB = 1^c = \theta$ if arc $AB = OA = OB = r$.

Example $90° = \frac{\pi}{2}$ (radians are understood if no degree symbol is used).
LATIN, *radius*, rod, ray

radical **1.** Relating to or forming the **root** of a **number** or quantity. **2.** The root so formed. The radical sign is $\sqrt{\ }$.
Example $\sqrt{7}$, $\sqrt{x+1}$ are radicals.
LATIN, *radix*, root

radius the distance from the centre of a circle or sphere to a point on the **circumference** of the circle or sphere.
Example

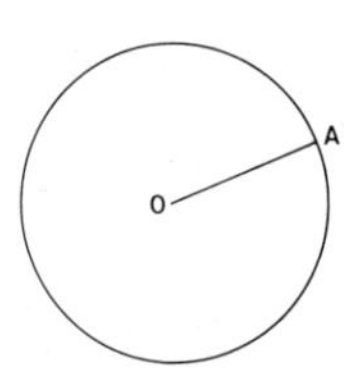

OA is the radius.
LATIN, *radius*, rod, ray

radius vector a line joining the **origin** to a point in a **polar** or **spherical** coordinate system.
Example

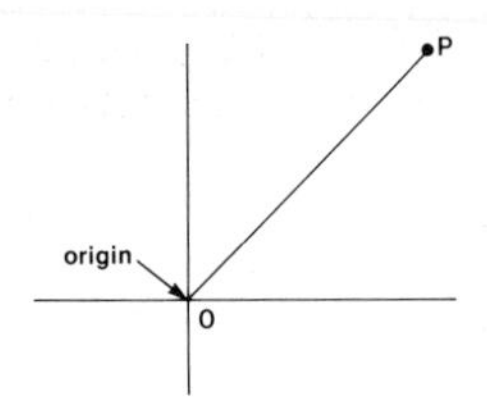

OP is the radius vector of the point P.
LATIN, *radius*, rod, ray; *vector*, carrier

radix a number or quantity which is made the **base** of a system of counting.
Example Two is the radix, or base of the **binary** system, 10 is the radix or base of the **decimal** system.

0, 1

There are two distinct symbols in the binary base. These are 0, 1.

0, 1, 2, 3, 4,
5, 6, 7, 8, 9

There are 10 distinct symbols in the decimal base. These are 0, 1, 2, 3, 4, 5, 6, 7, 8, 9.
LATIN, *radix*, root

random **1.** Haphazard, having no special pattern. **2.** (statistics) **a.** Of or relating to something that occurs by chance. **b.** Of or relating to a set of things (**samples**) taken from a larger **set** (**population**) so that each **member** of the set has an equal chance to be taken.
Examples
1. Lightning struck several trees *at random* during the storm.
2a. The boy put his hand into the bag of sweets and made a random selection.
b. So that everyone in the lottery has an equal chance, a marble is picked out *at random*.
OLD FRENCH, *randir*, to run

random number a **number** selected at **random**. In **statistics**, tables of random numbers are used to select **samples** from a given **population**.
Example If we built 10 sided dice each side numbered 0 to 9 we could use this to build up a set of random numbers. The following is a small selection from a table of random numbers:

20717 61996 61472 09504
47619 39890 17028 98528

If we wanted a sample of 20 from a group containing 99 members we can take the following members as selected from the table using two digits at a time. For example: 20, 71, 76, 19, 96, 61, 47, 20, 95, 04, 47, 61, 93, 98, 90, 17, 02, 89, 85, 28.
OLD FRENCH, *randir*, to run; LATIN, *numerus*, number

random variable (statistics) a **variable** whose values are determined from the outcome of a **random event** or experiment.
Example The sum of the two sides facing upwards of 2 dice is a random variable. It can take on values dependent on the outcome of any throw.

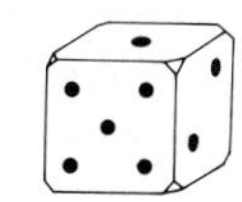

OLD FRENCH, *randir*, to run; *variare*, to vary

range 1. The **set** of values that a given **variable** or **function** can take on. 2. The difference between the highest and lowest values in a collection of statistical data.
Examples
1. In the equation $y = x^2$, the range of y is from 0 to 16 for the values of x from -4 to 4 (-4 to 4 is known as the **domain**).

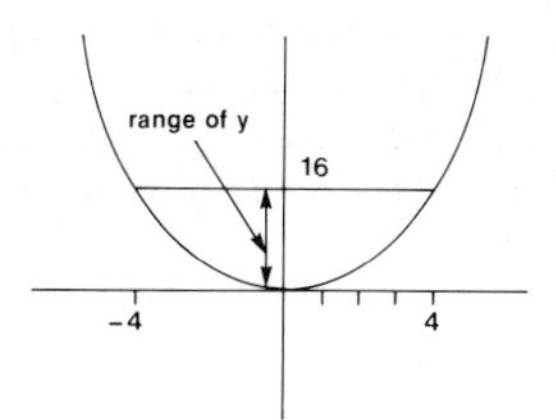

2. In a group of 10 students, the highest height was two metres, the smallest height was 1.5 metres, and the range was therefore 0.5 metres.
OLD FRENCH, *rang*, rank

rate one quantity compared to another of a different kind. Symbol used is per or /.
Example He drove the car at the rate of 80 km/h. The birth rate (number of babies being born per year) is going down.
LATIN, *rata*, judged, calculated

ratio one quantity compared to another of similar kind. Usually written either as a **fraction** or using the symbol :.

Examples
The ratio of games won to games lost was 4:5.
The ratio of men to women on the island was 1:2.

$a:b = \frac{a}{b} = \frac{2}{1}.$

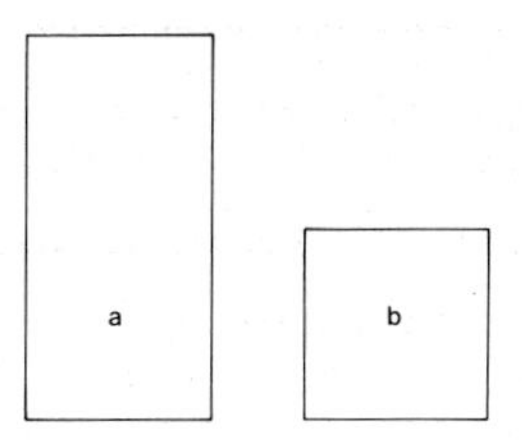

LATIN, *ratio*, reckoning, reason

rational number a **number** which can be expressed as the **ratio** of two **integers**, providing the second number is not zero.
Example $\frac{m}{n}$ where m is any integer, and n is any integer not equal to 0, is the general form of a rational number; $\frac{3}{4}, \frac{741}{1076}, \frac{-2}{41}$ are all rational numbers.
Note: integers are a **subset** of rational numbers.
LATIN, *ratio*, reckoning, reason; *numerus*, number

rationalization the **operation** of removing **radicals** from the **denominator** in an **expression** containing radicals without changing the value of the expression.
Examples
1. The rationalization of the expression $\frac{24}{\sqrt{18}}$ can be done by multiplying both top and bottom numbers by $\sqrt{18}$, as $\frac{24}{\sqrt{18}} \cdot \frac{\sqrt{18}}{\sqrt{18}} = \frac{24\sqrt{18}}{18}$ (since $\sqrt{18} \times \sqrt{18} = 18$).

Therefore $\frac{24}{\sqrt{18}} = \frac{4\sqrt{18}}{3} = \frac{4 \cdot 3\sqrt{2}}{3} = 4\sqrt{2}$

2. To rationalize $\frac{1}{\sqrt{5} - 1}$ multiply top and bottom by $\sqrt{5} + 1$, i.e. $\frac{1}{\sqrt{5} - 1} = \frac{1}{\sqrt{5} - 1} \cdot \frac{\sqrt{5} + 1}{\sqrt{5} + 1}$

$$= \frac{\sqrt{5} + 1}{5 + \sqrt{5} - \sqrt{5} - 1}$$

$$= \frac{\sqrt{5} + 1}{4}.$$

The answer has a rational denominator, which is easier for calculation purposes.
LATIN, *rationalis*, reasonable

ray a straight line coming out from one point.
Example

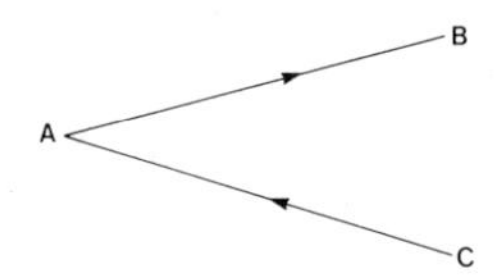

AB and *AC* are rays, emanating from *A*.
LATIN, *radius*, rod, ray

real number a **number** which can be written as a **decimal**. It can also be put on a number line. Real numbers include both **rational** and **irrational** numbers (*see also* COMPLEX NUMBER).
Example 2.371, $\sqrt{2} = 1.414\ldots$, $3/11 = 0.2727\ldots$, $\pi = 3.1419\ldots$

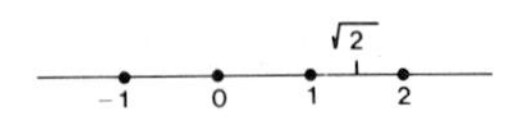

LATIN, *realis*, actual, real; *numerus*, number

reciprocal the **number** or **expression** obtained by dividing the given number or expression into 1.
Example The reciprocal of 3 is ⅓. The reciprocal of $x^2 + 1$ is $\frac{1}{x^2 + 1}$.
LATIN, *reciprocus*, alternating, returning

rectangle a four-sided figure with all angles 90°.
Example

ABCD is a rectangle. It has all the properties of a **parallelogram** as well as **adjacent** angles being equal and **diagonals** equal in length.
LATIN, *rectus*, right; *angulus*, angle

recurring decimal a **decimal** that repeats one or a group of numbers.
Example 1.2323232323... (the group 23 repeats itself). 1.6666666666... (the digit 6 repeats itself). Recurring decimals are often written with dots over the group of repeating numbers. $1.\dot{2}\dot{3} = 1.2323232323\ldots$ Also known as repeating decimals or **periodic decimals**. A recurring decimal always represents a **fraction**.
LATIN, *recurre*, to run back

re-entrant pointing inward. A re-entrant angle is an angle in a **polygon** whose value is greater than 180°.
Example

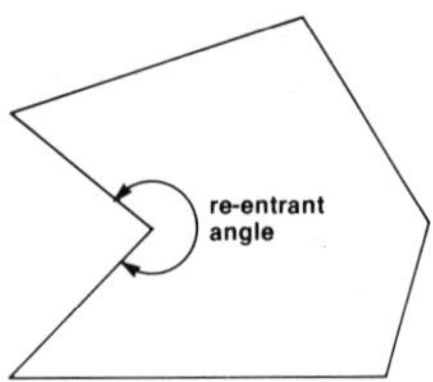

LATIN, *reintrare*, to enter again

reflect to transform a point or curve to another point or curve so that the **corresponding** point or points (on a curve) are **equidistant** from a given line. Hence, **reflection**.
Examples

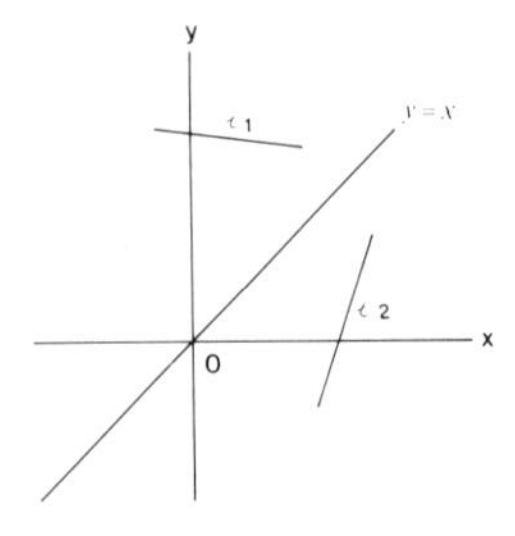

Line 1 (l_1) is reflected about $y = x$ to make line 2 (l_2). This reflection about $y = x$ makes an **inverse** function of the function (a line l_1 in this case) you started with.

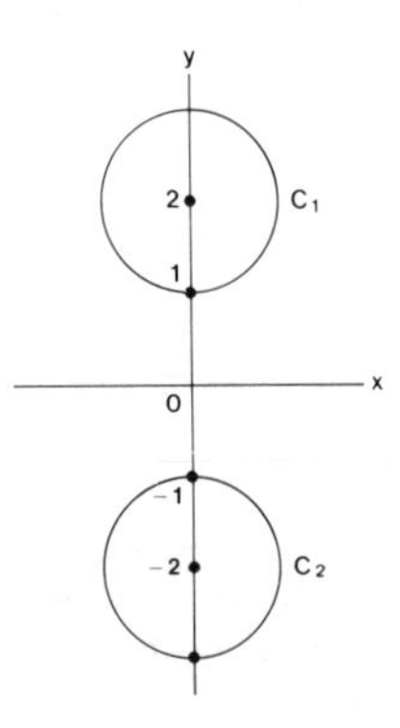

Circle 1 (C_1) is reflected about $y = 0$ to give circle 2 (C_2). The line of reflection is always an axis of **symmetry** of the **object** together with its **image**.
LATIN, *reflectere*, to bend back

reflex angle an angle greater than 180° but less than 360°.

Examples

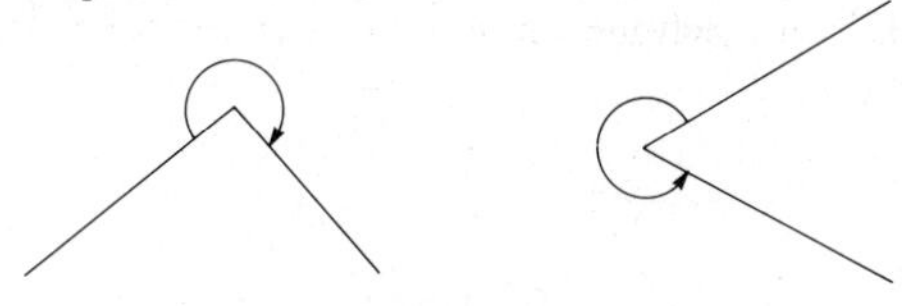

LATIN, *reflexus*, a bending back; *angulus*, angle

reflexive (of a relation) characterising a reflection in which an element is related to itself.
Examples In the relation "Joe is as heavy as Joe", 'is as heavy as' is reflexive. However "Joe is heavier than Joe" is not a reflexive relation as it is a false statement.
LATIN, *reflexus*, bent back

region an area or part of a surface.
Examples

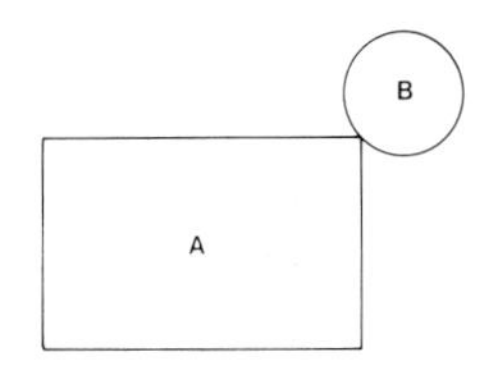

Rectangle A and circle B are two regions in the plane.

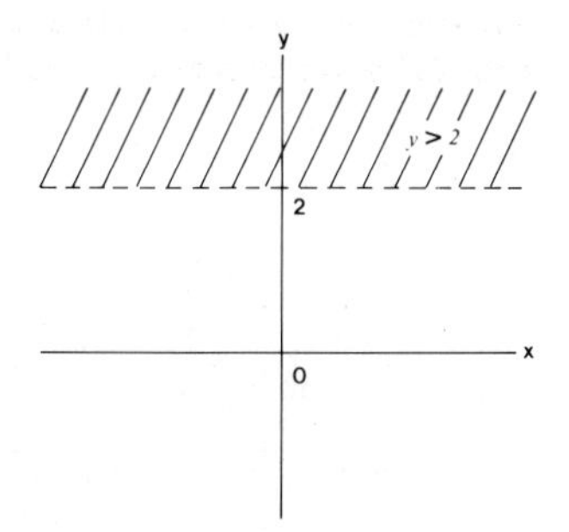

The part (area) above the line $y = 2$ is the region described by $y > 2$ (not including the line $y = 2$).
LATIN, *regio*, region, boundary

regression (statistics) a way of finding a line or curve that best fits a given set of values.
Example If we are given the following table:

Sales	10	11	13	15	17	20	22	25	29	32	37	42
Months	1	2	3	4	5	6	7	8	9	10	11	12

We can put these on a graph as below

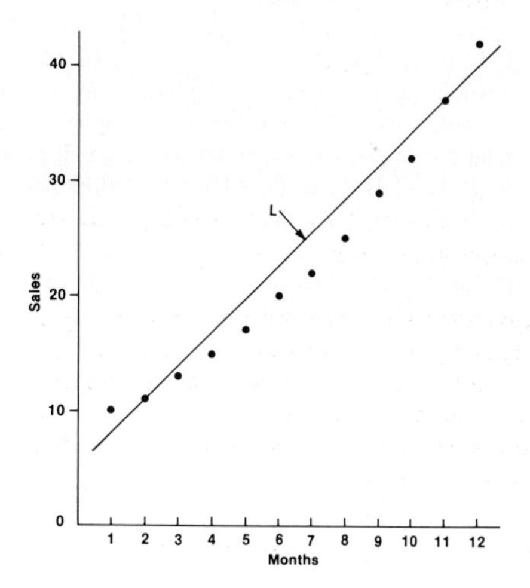

Regression allows us to work out what straight line or curve best fits. It we choose a line to best fit the values, it would look like L. There is a way to find out the formula of this line (*see* LEAST SQUARES METHOD).
LATIN, *regressus*, gone back

regular having equal sides and equal **angles** (referring to **plane** figures), or identical regular faces (in solids). Also referring to **tessellation** patterns.
Examples
1. A regular **polygon** has all sides equal and all angles equal. A regular **pentagon** has all five sides equal and all angles equal to 108°.

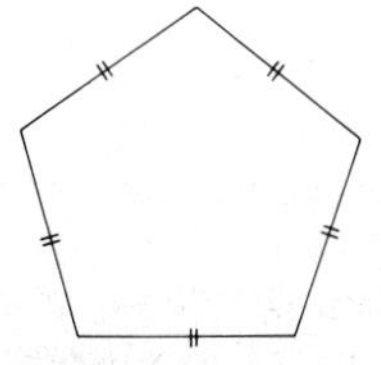

2. A regular **polyhedron** has identical polygons as faces. A **cube** is a regular polyhedron of six faces (each face is a square).

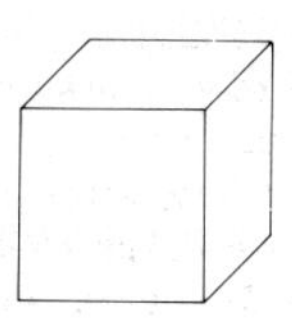

LATIN, *regula*, a rule

relation **1.** A way of connecting sets of things. **2.** Any set of **ordered pairs**.
Examples
1. In the statement 3 is a factor of 6, "is a factor" is a relation. In $y > 5$, $>$ is a relation connecting the variable y and the number 5. Examples of special relations are **equations, functions** and **mappings**.
2. $\{(1,2), (2,3), (2,4), (3,-1)\}$ is a relation made up of four ordered pairs. Note $\{1,2,3\}$ is the set of first members in the pairs (and is called the **domain**), and $\{2,3,4,-1\}$ is the set of second members in the pairs (and is called the **range**).

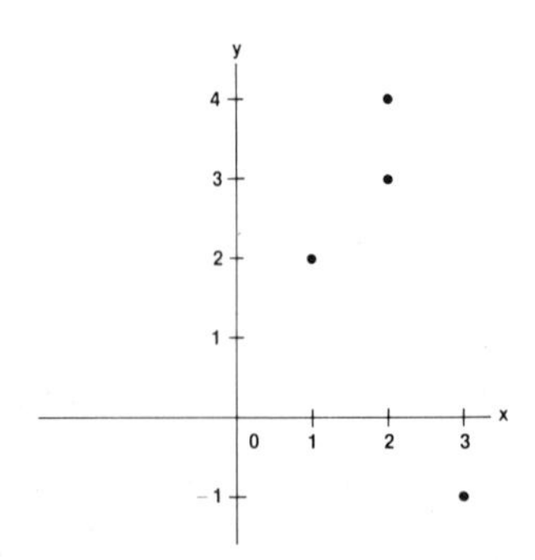

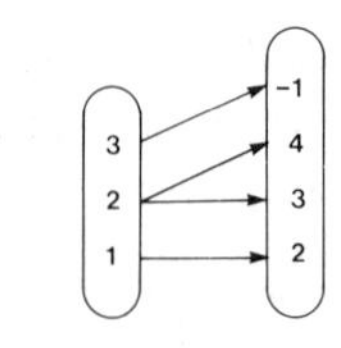

Two ways of drawing the relation
$(1,2), (2,3), (2,4), (3,-1)$

LATIN, *relatus*, carried back

relatively prime describing two or more **numbers** which only have 1 as a **common factor** between them.
Examples
3 and 8 are relatively prime (3 is actually a prime number in its own right, but 8 is not).
16 and 25 are relatively prime (even though both are made up of other **factors**).
LATIN, *relativus*, having relation or reference; *primus*, first

remainder **1.** Something left over after other parts have been taken away. **2.** In subtraction, the difference between the two given quantities. **3.** In division, the part left over after a particular quantity (called the **dividend**) has been divided by another quantity (called the **divisor**) a certain number of times.

Examples
1.

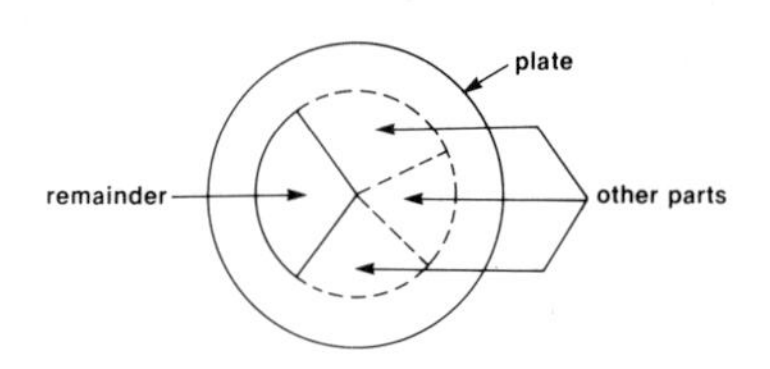

2.

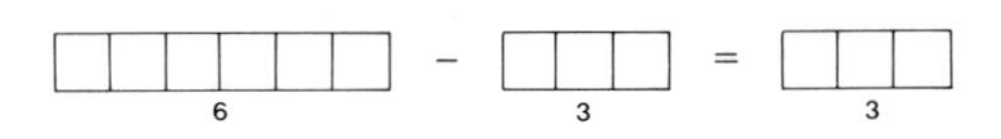

3a.

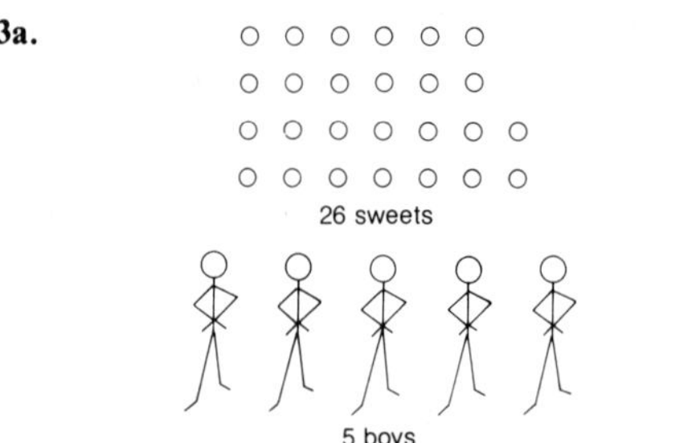

26 sweets divided amongst 5 boys = 5 with 1 remainder.

b. $(x^2 - 2x - 3) \div (x + 1) = (x - 3) +$ remainder 0.
OLD FRENCH, *remaindre*, to remain

remainder theorem if a **polynomial** $P(x)$ is divided by $(x - a)$, the remainder is $P(a)$. Since $P(x) = (x - a)Q(x) + R$, then when $x = a$, $P(a) = 0 + R$, or $R = P(a)$.
Example If $P(x) = x^3 + 2x - 1$ is divided by $(x + 1)$, then the remainder $R = P(-1)$
$= -1 -2 -1$
$= -4$.

OLD FRENCH, *remaindre*, to remain; GREEK, *theorema*, speculation, proposition

repeating decimal *see* RECURRING DECIMAL

respectively one by one in the order just mentioned.
Examples
The **coordinates** of the points A, B, C are respectively (2,3), (3,4), (−1,2). This means that A has the coordinates (2,3); B has the coordinates (3,4); and C has the coordinates (−1,2).
The ages of the four boys P, Q, R, S were respectively 9, 13, 11, 16 (that is P was 9 years old; Q was 13 years old; R was 11 years old; and S was 16 years old).
LATIN, *respicere*, to look back

resultant the sum of two **vectors**.
Example

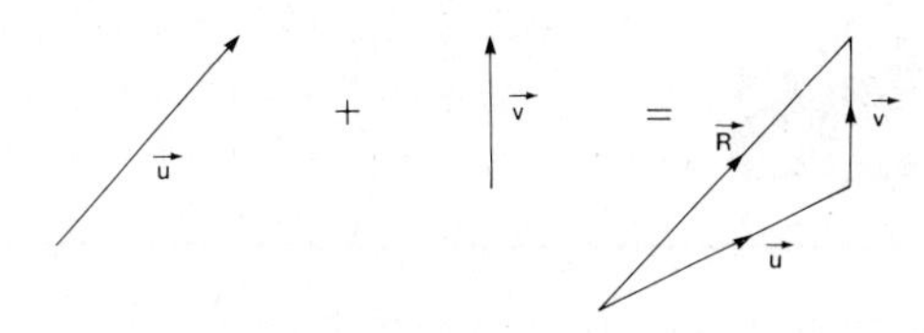

Vector R is the resultant vector u and vector v ($\vec{u}$ is sometimes used as an abbreviation for vector u).
LATIN, *resultare*, to leap back, rebound

revolution **1.** The action of moving around an orbit or circular course. **2.** A single **rotation** about a centre. **3.** The return of a point or period of time; a cycle.
Examples
1. Planet Earth revolves around the Sun once a year; each revolution is one year.

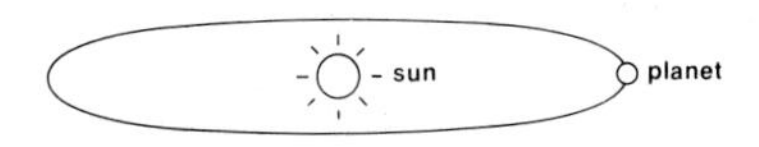

2. A coin is rolled along a table top so that the point A is now in contact with the table again. The coin has completed one revolution.

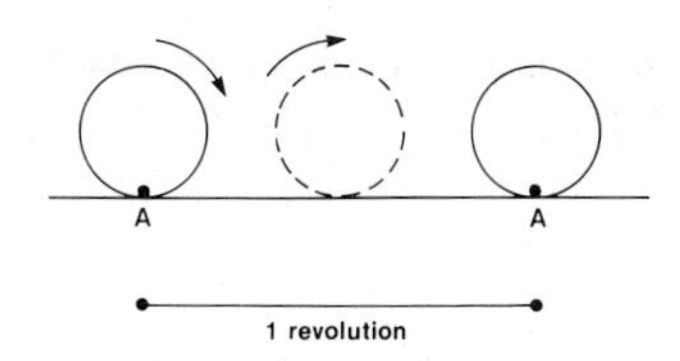

3. The number of degrees in one full revolution is 360.
LATIN, *revolvere*, to revolve, roll back

rhombus a four sided figure with opposite sides parallel and four equal sides. It has the same properties as a **parallelogram** plus the diagonals bisect at right angles and the angles through which they pass.
Example

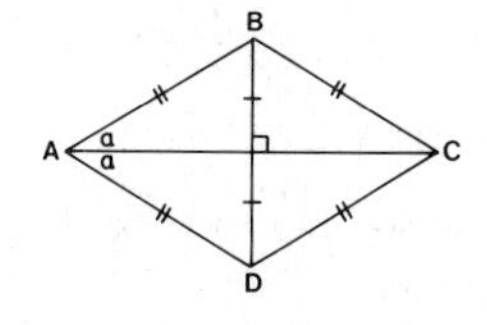

$ABCD$ is a rhombus
GREEK, *rhombos*, spinning top

right angle half a straight angle. It is an angle of 90°. Symbol often used is ∟. Hence, **right angled** (adj.).
Example

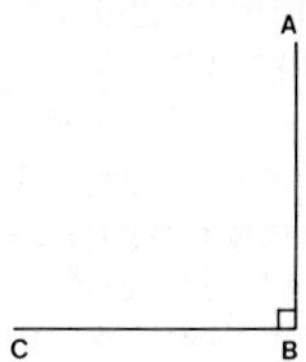

$\angle ABC$ is a right angle. The walls of most houses are at right angles to the floor.

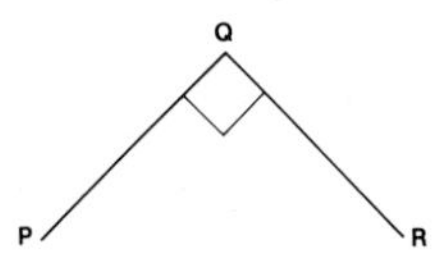

$\angle PQR$ is a right angle.
OLD ENGLISH, *riht*, right, just; LATIN, *angulus*, angle

root **1.** A **number** which is related to another number in the following way: when the (root) number is multiplied by itself an indicated number of times, it equals the specified number. **2.** (of a polynomial) A number which makes an **expression** (in one **variable**) equal to zero when the variable is substituted for by the number. Also called a solution to an **algebraic equation**.
Examples
1. $\sqrt{3}$ is called the "square root" of 3, since $\sqrt{3} \times \sqrt{3} = 3$. Also $\sqrt[3]{n} \times \sqrt[3]{n} \times \sqrt[3]{n} = n$ ($\sqrt[3]{n}$ is called the "cube root" of n).
2. The expression $x^2 + x - 2$ has two roots, since putting $x = 1$ or $x = -2$ both make the expression equal zero. Another way of saying this is $x = 1, -2$ are solutions or roots to the algebraic equation $x^2 + x - 2 = 0$.
OLD NORSE, *rot*, branch, root

rotation a **transformation** of a point or **object** so that the point or set of points turns through the same angle about the same centre.
Example

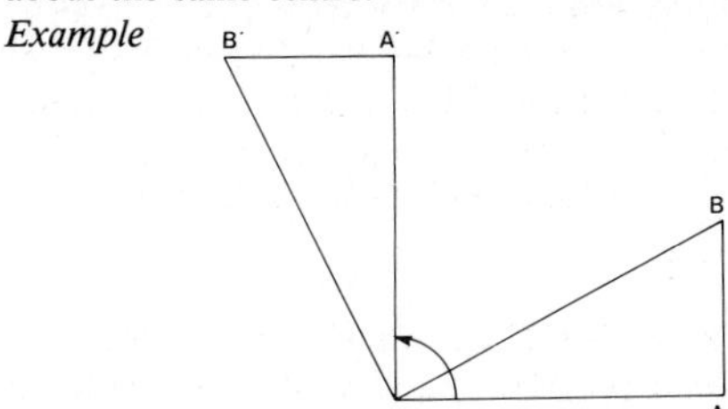

AB is rotated about through 90° to $A'B'$.
LATIN, *rota*, a wheel

rotational symmetry the property of a figure or solid which, after a **rotation**, causes each point to be mapped onto a corresponding point so that the figure or solid appears the same.
Example

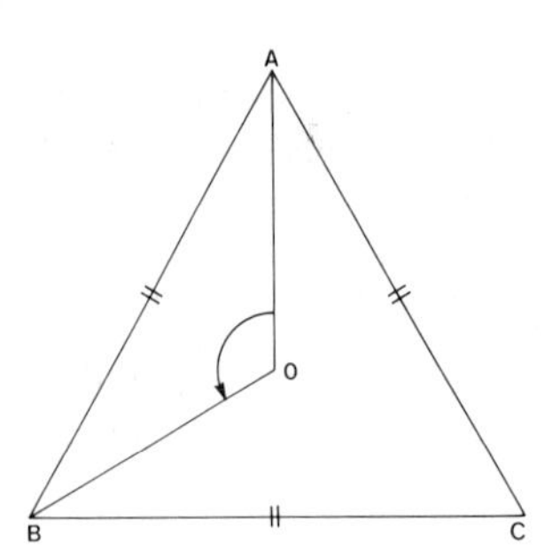

Rotating ΔABC about its centre O 120° means A moves to B, B moves to C,C moves to A, but the shape is the same.
LATIN, *rota*, a wheel; GREEK, *summetros*, of like measure

round **1.** To round *up* a **number** is to write it in a form to the specified nearest digit or digits above. **2.** To round *down* is to write a number in a form to the nearest digit or digits downwards. **3.** To round *off* is to write a number with fewer non-zero digits. One rounds up or down depending whether the digit to the right of the specified nearest digit is 5 or greater.
Examples
1. 1473 rounded up to the nearest one hundred is 1500. 1473 rounded up to the nearest ten is 1480.
2. 1473 rounded down to the nearest one hundred is 1400. (Also 1473 rounded down to the nearest tens is 1470.)
3a. $3.61 rounded off to the nearest dollar is $4. $3.61 to the nearest ten cents is $3.60 (since the next digit is 1, which is less than 5).
b. 1473 rounded off to 3 **significant figures** is 1470.
LATIN, *rotundus*, round

row a group of **numbers** or letters written in a single line across a page.
Example 1 3 4 5 is a row. The following **matrix** has two rows in it. $\begin{pmatrix} 2 & 1 \\ 0 & 3 \end{pmatrix}$
OLD ENGLISH, *raew*, a line, row

Ss

sample (statistics) a group of things (or individuals) taken from a larger group (called the **population**) and used to calculate the characteristics of the larger group.
Example Selecting 100 people at random in a small town of population 2000 and measuring their height, we can estimate the average height of the whole 2000 using statistical methods.
OLD FRENCH, *essample*, example

sample space (statistics) all **samples**, which make up a given set.
Example
Given a population:

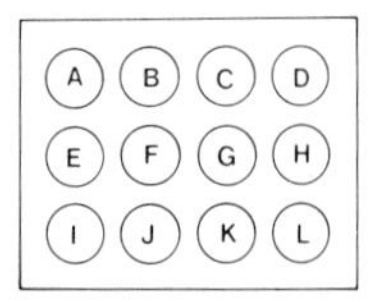

then the sample space would contain such samples as the following:

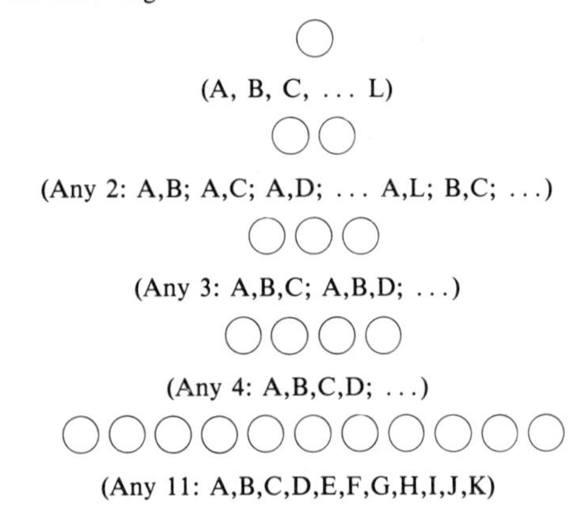

OLD FRENCH, *essample*, example; LATIN, *spatium*, space

sampling (statistics) the process of selecting a representative group of things (or individuals) from a larger group of the same things.
Example Sampling involves four things: (1) working out how big a sample; (2) type of sample; (3) means

of measuring the sample characteristics; and (4) selecting the sample.
OLD FRENCH, *essample*, example

scalar a quantity which has magnitude but no direction.
Example 3, 5.7 are scalar quantities, but the following line segment AB has both magnitude (length) and direction, and is not scalar (*see* VECTOR).

LATIN, *scala*, a ladder

scalar product (also **dot product**) the **scalar** result which occurs when two **vectors** are multiplied together. The scalar product of x and y where x and y are vectors is xy cos θ where x and y are the magnitudes of x and y and θ is the angle between the vectors. It is written $x \cdot y$.
Example A force F is applied to a box on a table at an angle θ to the direction of travel. The actual work done if the box moves s metres is Fscos θ. This can be represented as the scalar product of F and s, that is, $F \cdot s$, where F is the force vector and s is the **displacement** vector and F and s are the magnitudes of F and s respectively.
LATIN, *scala*, ladder; *producere*, to produce

scale **1.** (on a graph) The relationship between a unit length on a graph or map representing some object or length and the actual object or length. **2.** A set of marks on a line at fixed intervals used as a reference for measuring. **3.** A special way of representing figures, numbers, or quantities based on a fixed "counting number" such as 10, or 2, or 60. The position of figures relative to each other has meaning (*see* DECIMAL SYSTEM, BINARY SYSTEM).
Examples

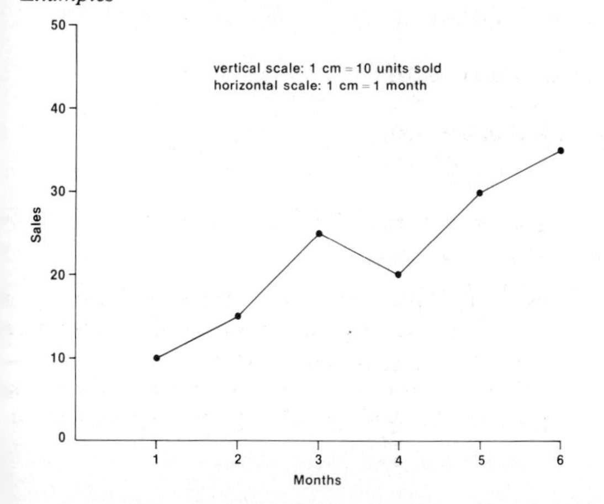

2.

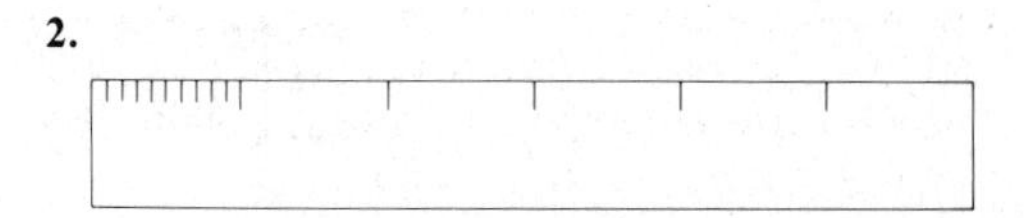

A ruler is an example of scale.

3. The denary or decimal scale used 10 as a fixed counting number, and 543 in this system means $5 \times 100 + 4 \times 10 + 3$.
LATIN, *scala*, ladder

scale drawing a drawing, plan or map of something which is in direct proportion to the object concerned.
Examples
A map of an area is often drawn such that each centimetre of the drawing represents a certain number of metres or kilometres (*see* SCALE).

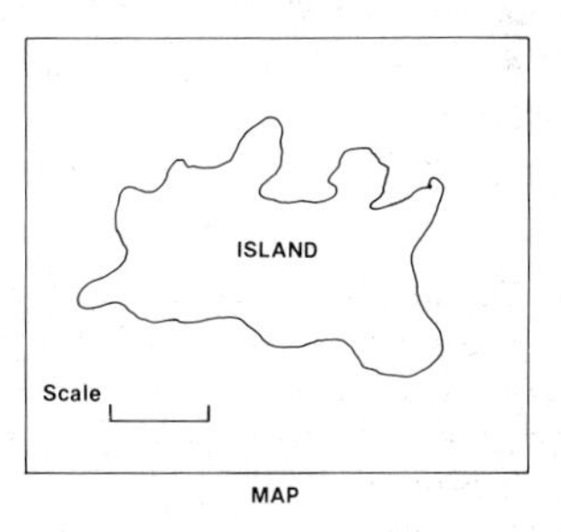

Scale: 1cm represents 1 kilometre.

The architect showed the home buyers a scale drawing of the top floor of the house.

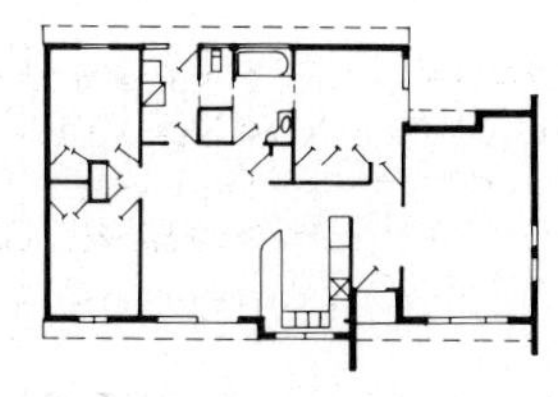

LATIN, *scala*, ladder; OLD ENGLISH, *dragan*, to drag, draw

scale factor the **ratio** of the length of one side on one figure to the length of the corresponding side on another figure, which is **similar** to the first.
Example

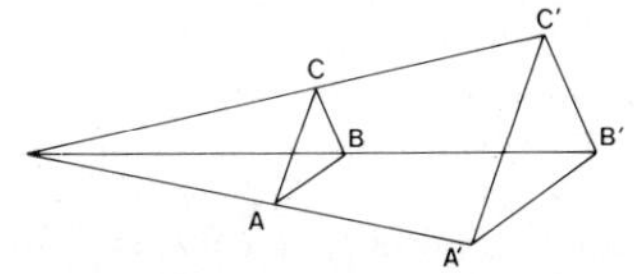

The linear scale factor between triangle ABC and $A'B'C'$ is $\frac{A'B'}{AB}$ (or $\frac{B'C'}{BC}$ or $\frac{C'A'}{CA}$). Scale factors can be positive, negative, or fractional.
LATIN, *scala*, ladder

scalene triangle a triangle whose sides are all different in length.
Example

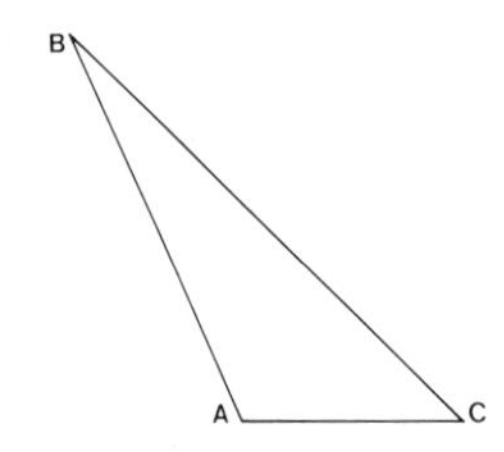

ABC is a scalene triangle, since $AB \neq AC \neq BC$.
GREEK, *skalenos*, unequal, odd

secant **1.** (geometry) A straight line that cuts a circle in two points or passes through two points on any curve. **2.** (trigonometry) The ratio of the longest side of a right angled triangle to the side adjacent to the angle in question.
Examples

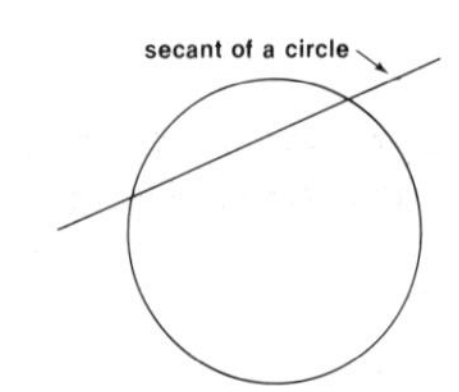

2.

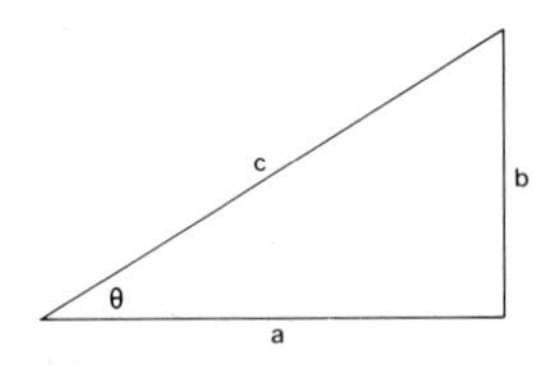

$$\text{secant of } \theta \text{ (written sec } \theta) = \frac{c}{a}$$

LATIN, *secare*, to cut

second derivative the **function** obtained from another function by **differentiating** twice (*see also* DERIVATIVE).
Example If $y = 4x^3 + 2x + 6$, then
$\frac{dy}{dx} = 12x^2 + 2$ is the first **derivative**, and
$\frac{d^2y}{dx^2} = 24x$, which is the second derivative of the function y.
LATIN, *secundus*, second; *derivare*, to draw off, derive

sector a part of a **plane** enclosed by two radii and the **arc** (of a **circle**, **ellipse** or other closed curve) between them.
Example

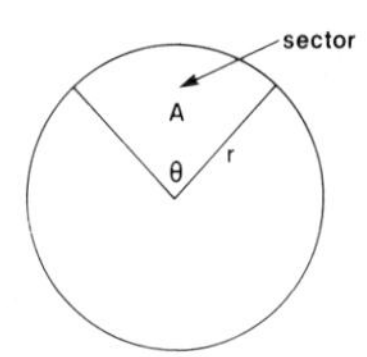

If the **radius** is r, and the angle **subtended** at the centre is θ (in **radians**), then the area of the sector $A = \frac{1}{2}r^2\theta$.
LATIN, *secere*, to cut

segment **1.** (of a **line**) The part of a line between two given points on the line. **2.** (of a **circle**) The part of a plane enclosed by an **arc** of a circle and its **chord**. **3.** (of a **sphere**) The solid formed by formed by two parallel planes cutting through a sphere.
Examples
1.

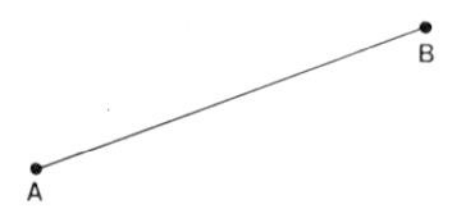

AB is a line segment.

2.

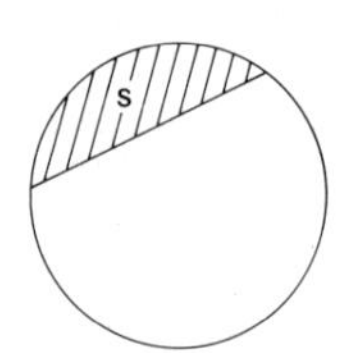

S is a segment of a circle. Segments in a circle may be **major** or **minor**.
3.

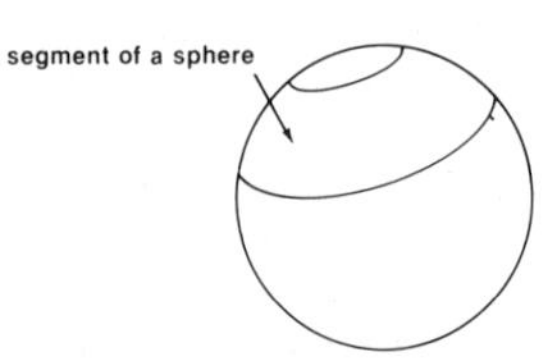

LATIN, *segmentum*, part cut from

semi circle a half of a **circle**. It is the figure enclosed by a **diameter** and the arc of the circle joining its end points.
Example

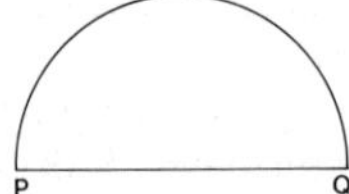

PQ is a diameter
LATIN, *semi*, half; *circus*, a ring

sense direction
Example **Clockwise** motion is in opposite sense to **anti-clockwise**.

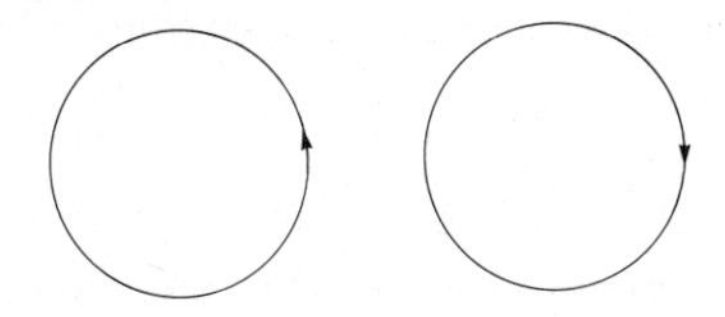

Opposite senses
LATIN, *sentire*, to feel

sequence a **set** of **numbers, terms** etc. placed in a certain order. Symbol used is often $\{U_n\}$.
Examples 2, 4, 6, 8, 10, ... can be written $\{2n\}$ where n is a positive integer. A special sequence where each number is the sum of the previous two terms is called **Fibonacci's sequence**.
LATIN, *sequi*, to follow

series a collection of **numbers** or letters which are separated by + or − signs. Each number or **term** is usually related to the previous number or term by some rule.
A **finite** series means the collection has a definite number of terms. An **infinite** series means there is no end to the number of terms in the collection (*see also* ARITHMETIC, GEOMETRIC PROGRESSION).
Examples
$1 + \frac{1}{2} + \frac{1}{3} + \frac{1}{4} + \frac{1}{5} \ldots$ is an infinite series.
$1 + 2^2 + 3^2 + 4^2 + 5^2 + \ldots + 20^2$ is a finite series.
$1 + x + x^2 + x^3 + x^4 + \ldots$ is an infinite series whose sum can be shown to be equal to $\frac{1}{1-x}$.
LATIN, *serere*, to join

set a collection of things which are classified so that one can tell whether an element belongs to the set or not. The symbol for set is usually { }.
Examples The set of the first 10 letters of the alphabet is {a, b, c, d, e, f, g, h, i, j}
The set of all birds on the island.
The set of all positive whole numbers less than 20 is {1, 2, 3, 4, ..., 18, 19}
Note: A set has members called **elements**, and each element is unique. Two or more sets can **intersect**, or be united (*see* UNION).
LATIN, *secta*, a following (of people)

set builder notation a shorthand way of describing a set of points which obey some rule. The symbol used is $\{(x, y)\text{: RULE}\}$.
Examples $\{(x, y): x = 2\}$ means the set of points where $x = 2$. On a graph on the number plane this looks like the diagram below.

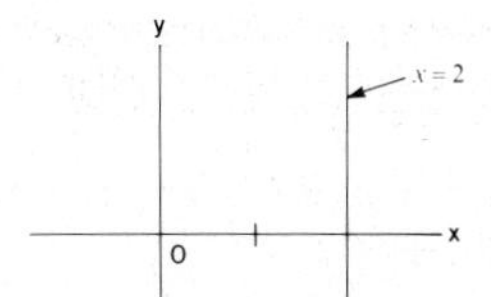

$\{(x, y): y = 2x + 3\}$ means the set of points which obey the rule $y = 2x + 3$. On a graph this looks like:

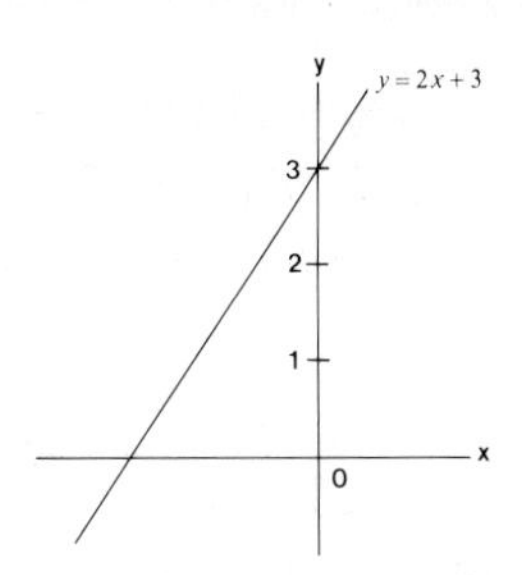

LATIN, *secta*, a following (of people); *nota*, a note; OLD ENGLISH, *byldan*, to build

sex- prefix meaning six.
Example A sextet is a group of six musicians. Sextillion is a 1 followed by 21 zeros (in the United States and France), and one followed by 36 zeros (in Australia, Germany and Great Britain).
LATIN, *sex*, six

shear a **transformation** in which all points of an object slide **parallel** to a base (**line** or **plane**).
Example

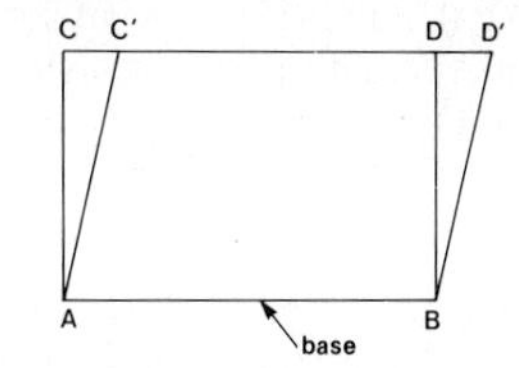

$AC'D'B$ is the result of a shear applied to $ACDB$.
OLD ENGLISH, *sceran*, to cut

sieve of Erastosthenes a way of finding all the prime numbers between 1 and n (n is a positive whole number). After writing down all the numbers from 2 to n one removes those after 2 which are multiples of 2, also removes those after 3 which are multiples of 3, and so on until all multiples of primes are removed. Only the prime numbers are left.
Example Find the prime numbers between 1 and 40: 1, 2, 3, ~~4~~, 5, ~~6~~, 7, ~~8~~, ~~9~~, ~~10~~, 11, ~~12~~, 13, ~~14~~, ~~15~~, ~~16~~, 17, ~~18~~, 19, ~~20~~, ~~21~~, ~~22~~, 23, ~~24~~, ~~25~~, ~~26~~, ~~27~~, ~~28~~, 29, ~~30~~, 31, ~~32~~, ~~33~~, ~~34~~, ~~35~~, ~~36~~, 37, ~~38~~, ~~39~~, ~~40~~. The prime numbers are therefore 2, 3, 5, 7, 11, 13, 17, 19, 23, 29, 31, and 37.
after Erastosthenes, Greek mathematician; OLD ENGLISH, *sife*, filter, sieve

sigma notation a shorthand way of representing a sum of **elements**. The symbol used is Σ (Greek letter sigma).
Example 1 + 2 + 3 + 4 + ... + *n* can be written as

$$\sum_{r=1}^{n} r$$

which is said as: sum the values of *r* going from 1 to *n*.
GREEK, Σ [18th letter in the Greek alphabet]

sign a symbol. It usually represents an **operation** to be performed.
Example +, −, ×, ÷ are the signs for **adding, subtracting, multiplying**, and **dividing**.
LATIN, *signum*, sign, mark, seal

signed numbers the positive and negative numbers.
Example +3, +2, +11, −2, −7, −21.5 are signed numbers.
LATIN, *signum*, sign; *numerus*, number

significant figures the number of digits written to indicate the accuracy of a measured or calculated quantity. Often shortened to sf or sig. figs.
Example 74.309 to four significant figures is 74.31 (we have **rounded** off the fifth figure), and is written 74.31 (4 sig. figs.).
LATIN, *significare*, to make a sign, be meaningful; *figura*, a form, shape

similar having **corresponding** sides **proportional**, and corresponding angles equal.
Example Similar triangles have all angles equal and corresponding sides proportional. Also, similar rectangles and hexagons would have the same properties.

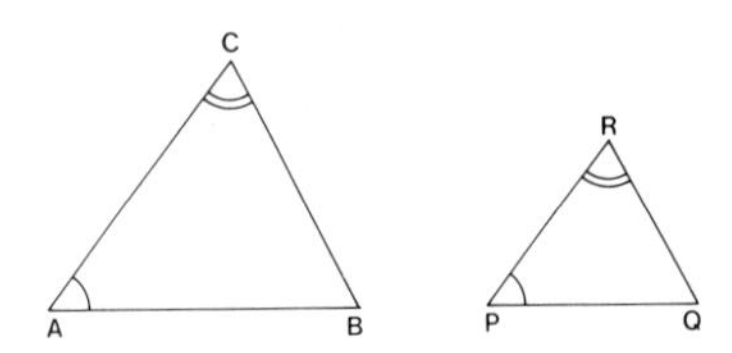

$\frac{AB}{PQ} = \frac{AC}{PR} = \frac{BC}{QR}$ and $\angle A = \angle P$, $\angle C = \angle R$, $\angle B = \angle Q$.
LATIN, *similis*, like

simple consisting of one thing or **element**.
Examples
A simple **closed** curve is a plane curve whose end point joins with its beginning in such a way that no part crosses with another part. One property of a simple curve is that it divides the **region** into two parts, an inside and an outside.

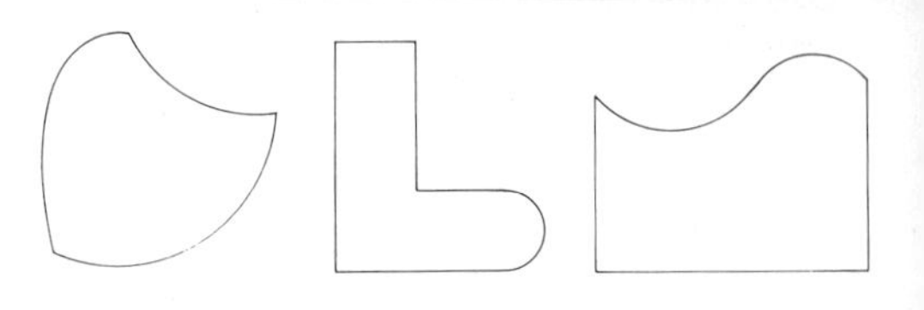

A simple **polyhedron** is a solid which does not contain any tunnels or holes.

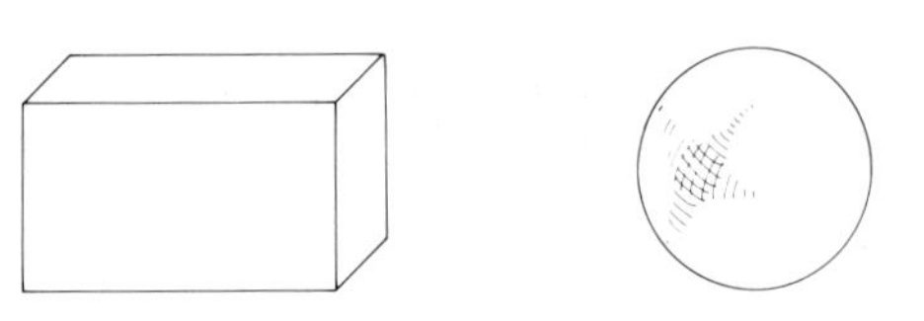

Simple polyhedra

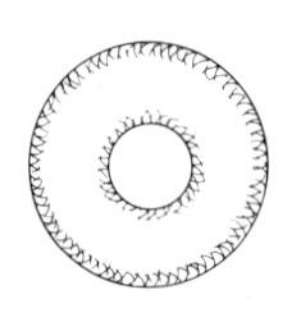

A donut is non-simple
LATIN, *simplus*, simple

simple interest interest paid only on the original sum of money invested (*see also* COMPOUND INTEREST). Interest is usually paid as money. Shorthand notation is S.I. Simple interest is equal to the principal (original sum) multiplied by the **rate** (as a **decimal fraction**) multiplied by the number of years invested.
Example If I invest $200 at 10% simple interest for four years, the simple interest is:

$$\text{S.I.} = \text{P} \times \frac{\text{R}}{100} \times \text{T}$$
$$= 200 \times \frac{10}{100} \times 4$$
$$= \$80$$

LATIN, *simplus*, simple; *interesse*, to matter, be of concern

simple harmonic motion motion of a particle whose acceleration is **proportional** to the **displacement** from a point of origin and in the opposite direction. It is the motion shown by an elastic spring. The **equation** of the motion can be written

$$\frac{d^2x}{dt^2} = -kx$$

(*see* DERIVATIVE, SECOND DERIVATIVE). Sometimes abbreviated S.H.M.

Example

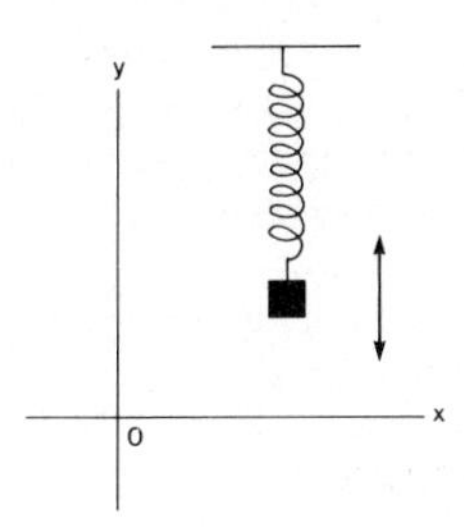

A spring with a weight on it when pulled downwards and released will go through the pattern called S.H.M. The general **solution** to such an **equation**

$$\frac{d^2y}{dt^2} = -ky$$

is $y = a \sin(\omega t + c)$ (ω and c are constants related to the spring and weight) and is used in physics to explain various patterns in nature.
LATIN, *simplus*, simple; *harmonia*, agreement; *motio*, movement

simplify to put into a simpler form.
Example A collection of numbers and letters such as $3 + a + 4 + (-b) + c - 2a$ can be simplified by combining like terms, as follows: $(3 + 4) + (a - 2a) + c - b$, which is equal to $7 - a + c - b$. Instead of six terms, we now have four terms (each symbol between the + or − signs is called a **term**). Also, an algebraic fraction like
$\frac{x^2 - 4}{2x + 4}$ can be simplified as follows:

$$\frac{x^2 - 4}{2x + 4} = \frac{(x - 2)(x + 2)}{2(x + 2)}$$

$$= \frac{x - 2}{2}; \text{ if } x \neq -2.$$

LATIN, *simplus*, simple

Simpson's rule a method used to work out definite **integrals** numerically. It uses the idea of a **parabola** passing through consecutive **sets** of three points.
If the definite integral

$$I = \int_a^b f(x)dx$$

and the interval $b - a$ is divided into $2n$ strips of width h, then,

$$I \doteq \frac{4h}{3}\{f(a+h) + f(a+3h) + f(a+5h) + \ldots + f(a+\overline{2n-1}h)\} + \frac{2h}{3}\{f(a+2h) + f(a+4h) + \ldots + f(a+\overline{2n-2}h)\} + \frac{h}{3}(f(a) + f(a+2nh)$$

i.e. $I \doteq \frac{h}{3}$[first ordinate + last ordinate + 4 (sum of odd ordinates) + 2 (sum of even ordinates)]

Example

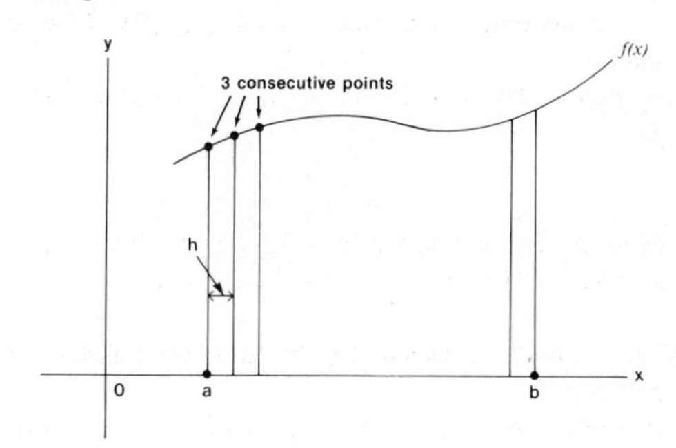

$\int_a^b f(x)\,dx$ is the area under the curve between $x = a$ and $x = b$. $f(a+2nh) = f(b)$.
after Simpson, Thomas, English mathematician;
LATIN, *regula*, rule

simultaneous equations two or more **equations** that have common **solutions**. The equations must be **independent** of each other.
Example The equations $x + y = 1$ and $2x - y = 2$ are a pair of simultaneous **linear** equations, for which the solutions are $x = 1$ and $y = 0$. The solution can be represented on a graph by drawing $x + y = 1$ and $2x - y = 2$ and finding where they cut. The common point is the solution since it lies on both lines. Another pair, $x^2 + y^2 = 4$ and $3x^2 - y^2 = 12$ is a pair of simultaneous equations of degree 2. The solutions are (2,0) and (−2,0).

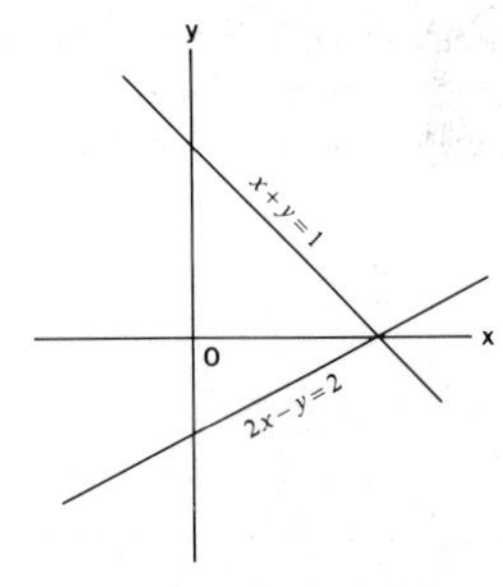

LATIN, *simul*, at the same time; *aequare*, to make equal

sine in **trigonometry**, the **ratio** of the side opposite to the given angle to the longest side of a right angled triangle. The sine of an angle θ is written $\sin \theta$.
Example

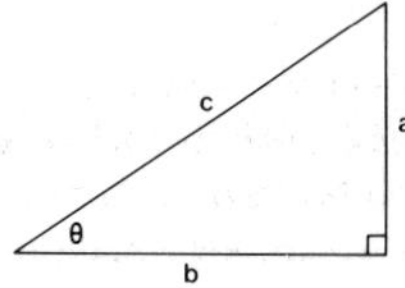

$\sin \theta = \frac{a}{c}$

For an angle greater than 90°, $\sin \theta = \dfrac{y \text{ coordinate}}{OP}$

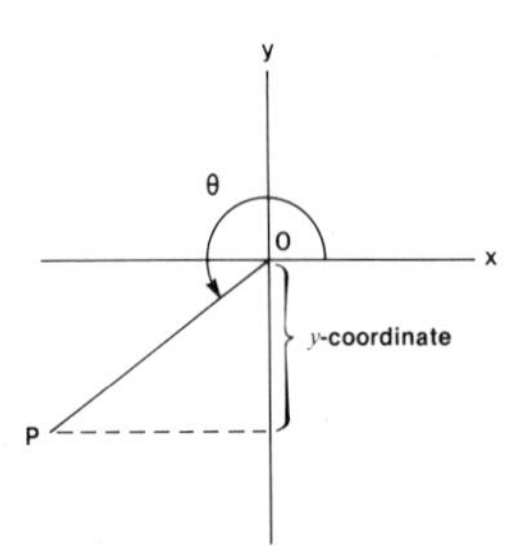

LATIN, *sinus*, bend, curve, fold

sine curve an undulating or wave-like curve representing the graph of $y = \sin x$.
Example

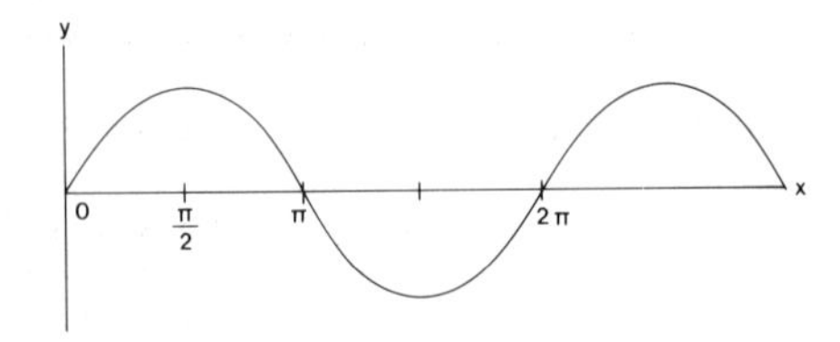

LATIN, *sinus*, bend, curve, fold

sine rule a rule which states that in any triangle the **ratio** of the lengths of any two sides is equal to the ratio of the **sines** of angles opposite those sides.
Example

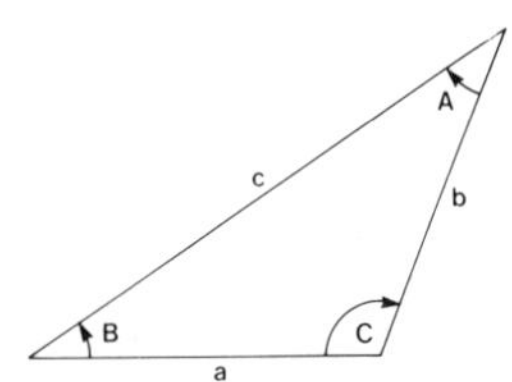

$$\frac{a}{\sin A} = \frac{b}{\sin B} = \frac{c}{\sin C} \text{ or}$$

$$\frac{a}{b} = \frac{\sin A}{\sin B} \text{ and } \frac{b}{c} = \frac{\sin B}{\sin C}$$

LATIN, *sinus*, bend, curve, fold; *regula*, rule

singular matrix a **matrix** whose **determinant** is 0, that is, for which an **inverse** cannot be found.
Examples If $A = \begin{pmatrix} a & b \\ c & d \end{pmatrix}$, then, if $ad - bc = 0$, A is singular. The following matrices are singular.

$$\begin{pmatrix} 1 & 0 \\ 3 & 0 \end{pmatrix} \quad \begin{pmatrix} 2 & 5 \\ 2 & 5 \end{pmatrix}$$

LATIN, *singulus*, single; *matrix*, womb

SI system (Système international) an international system of units, often referred to as the metric system but embracing much more. It is used especially in science and comprises three classes of units: base; derived; and supplementary.
Examples
BASE UNITS: Mass—kilogram (kg); Length—metre (m); time—second (s); electric current—ampere (A); thermodynamic temperature—kelvin (K); luminous intensity—candela (cd); amount of substance—mole (mol).
DERIVED UNITS: Units can be derived from base units by multiplying such a unit by itself, or by combining two or more base units by simple multiplication or division (as in m², m/s, and kg/m/s). Some derived units with special names are shown in the following table:

Quantity	Name of unit	Symbol for unit	Derivation of unit
frequency	hertz	Hz	per second
force	newton	N	$m \cdot kg/s^2$
pressure	pascal	Pa	N/m^2
work; energy; quantity of heat	joule	J	N·m
power; radiant flux	watt	W	J/s
electric charge; quantity of electricity	coulomb	C	A·s
electric potential potential difference	volt	V	W/A
capacitance	farad	F	C/V
(electric) resistance	ohm	Ω	V/A
conductance	siemens	S	A/V
magnetix flux	weber	Wb	V·s
magnetic flux density	testa	T	Wb/m^2
inductance	henry	H	Wb/A
luminous flux	lumen	lm	cd·sr
illuminance	lux	lx	$m^2/cd \cdot sr$
Celsius temperature	degree Celsius	°C	K
absorbed dose; absorbed dose index; kerma; specific energy imparted (radiation)	gray	Gy	J/kg
activity (of a radionuclide)	becquerel	Bq	per second

SUPPLEMENTARY UNITS: unit of plane angle—radian (rad); unit of solid angle—steradian (sr). Reference to the table above will show that the supplementary units can be used to form derived units.
SYMBOLS: When symbols are used for SI units, the symbol stands with no indication of plurality, and is not punctuated. Thus one metre is 1 m and ten metres is 10 m.
NON-SI UNITS IN USE: Eight units that are not part of the SI have been retained in general usage. They are shown in the following table. Note that the preferred symbol for litre in Australia and many other countries is L.

Quantity	Unit	Symbol for unit	Value in SI units
time	minute	min	60 s
	hour	h	3 600 s
	day	d	86 400 s
plane angle	degree	°	$\pi/180$ rad
	minute	′	$\pi/10\ 800$ rad
	second	″	$\pi/648\ 000$ rad
volume	litre	l (or L)	10^{-3} m^3
mass	tonne	t	1 000 kg

PREFIXES: There are a number of prefixes by means of which it is possible to form decimal multiples and submultiples of SI units. There are four prefixes which do not conform to the degree pattern—that is, they are not obtained by successive multiplications of 10^3 or 10^{-3} and they should be avoided in scientific usage. The prefixes are: hecto (h)—10^2; deca (da)—10^1; deci (d)—10^{-1}; centi (c)—10^{-2}.

Factor	Prefix	Symbol for prefix
10^{18}	exa	E
10^{15}	peta	P
10^{12}	tera	T
10^9	giga	G
10^6	mega	M
10^3	kilo	k
10^{-3}	milli	m
10^{-6}	micro	μ
10^{-9}	nano	n
10^{-12}	pico	p
10^{-15}	femto	f
10^{-18}	atto	a

after Système international d'Unités

skew lines lines that are not **parallel** and do not intersect.
Example

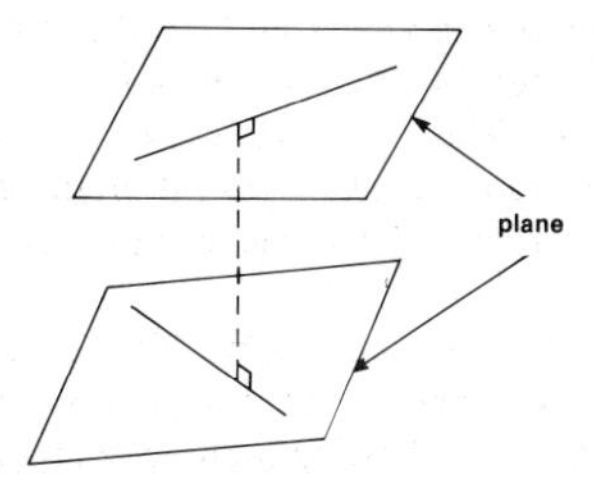

Note: The shortest distance between two skew lines is the line perpendicular to both.
OLD FRENCH, *eskiuer*, to take an oblique course

skewness a measure of how much a **distribution** is **symmetrical** about a **mean**. One way of measuring this is to use the formula S = **(mean − mode)/standard deviation**.
Examples

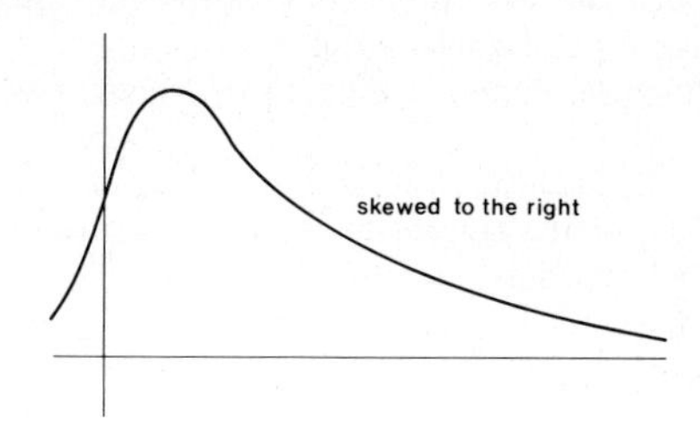

$f(x)$ is a positively skewed distribution, since (mean-mode) is positive. Graphically, this means the frequency curve has a long tail to the right. If (mean-mode) is negative, then $f(x)$ is a negatively skewed distribution.
OLD FRENCH, *eskiuer*, to take an oblique course

slant height **1.** (of a right circular cone) The length of a line from the vertex of the cone to the circle forming the base. **2.** (of a regular pyramid) The length of a line from the top drawn perpendicular to one edge of the base.
Examples
1.

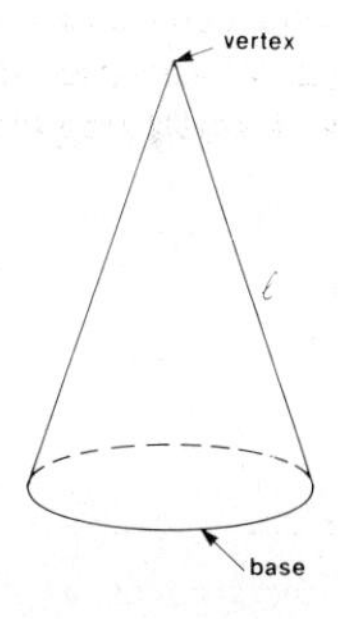

l is the slant height

2.

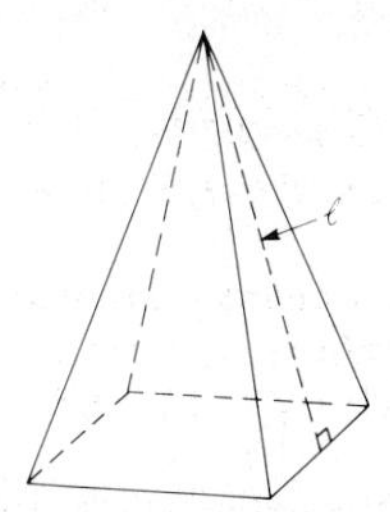

l is the slant height
OLD NORSE, *slenta*, side slip

slide rule a calculating device, consisting mainly of two sliding **logarithmic scales**, which is used to multiply, divide numbers (and other complex **operations**) by adding or subtracting, etc., their powers. It simplifies calculations and prior to the use of electronic **calculators** was extensively used in engineering and similar fields.
OLD ENGLISH, *slidan*, to slide; LATIN, *regula*, rule

slope the **gradient** of a curve at any point. It is the gradient of the **tangent** to the curve at that point. Symbol sometimes used for slope is m.
Examples
1.

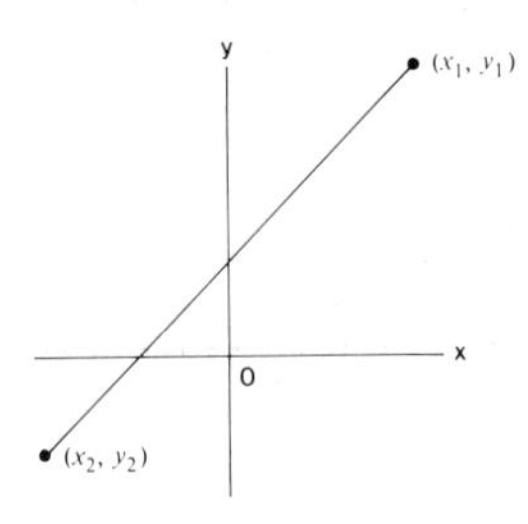

$$\text{slope } (m) = \frac{y_1 - y_2}{x_1 - x_2}$$

2.

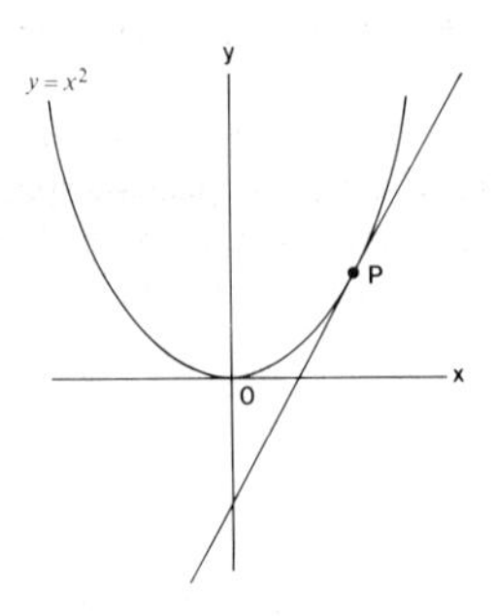

The slope at P is the gradient of the tangent to the curve at that point. It can be calculated by finding the **derivative** of y at the point.
OLD ENGLISH, *aslope*, to slip

solid a figure having length, breadth, and depth.
Example **cube, sphere, prism.**

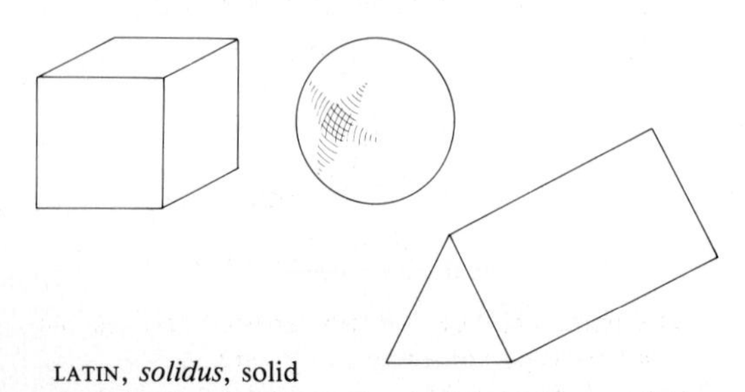

LATIN, *solidus*, solid

solid angle the angle formed at a point by a surface. It is measured by the area formed on the surface of a **sphere** (of unit **radius** which is centred at the point) by joining the **perimeter** of the surface to the point. It is measured in **steradians**.
Examples

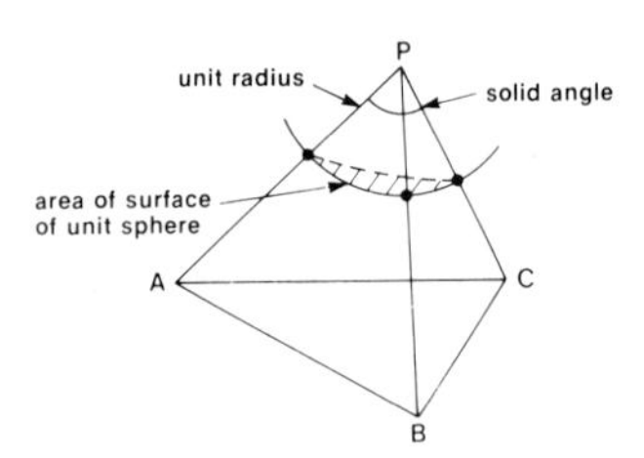

The above example has a solid angle formed by triangle ABC at point P.

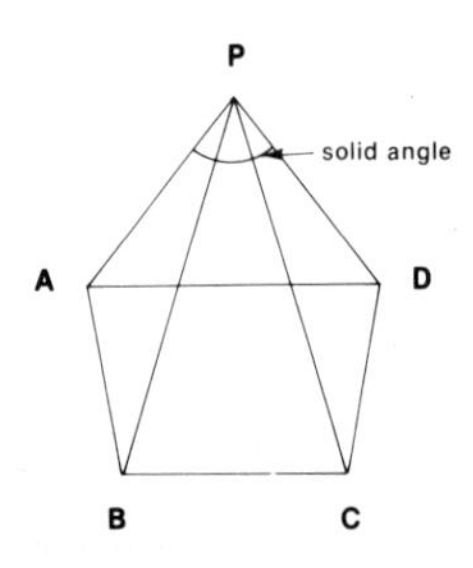

The above example has a solid angle formed by surface $ABCD$ at point P.
LATIN, *solidus*, solid; *angulus*, angle, corner

solid geometry the study of **three-dimensional** figures in space.
Example The properties of **pyramids, tetrahedrons,** etc., can be worked out using solid geometry. Note: Solid **analytical geometry** is the study of these figures using a three axis **coordinate** system.

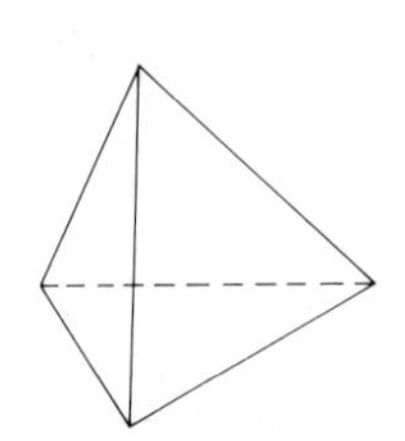

Tetrahedron

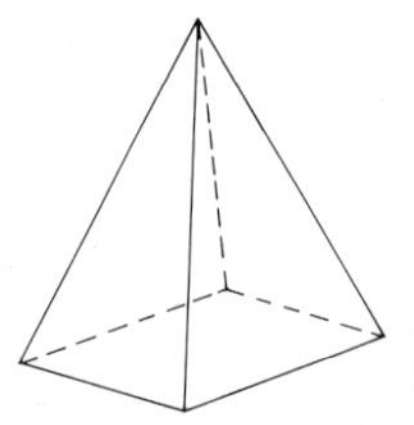

Pyramid

LATIN, *solidus*, solid; *geometria*, measuring land

solid of revolution a **solid** formed by turning a **plane** figure about an **axis**.

Examples

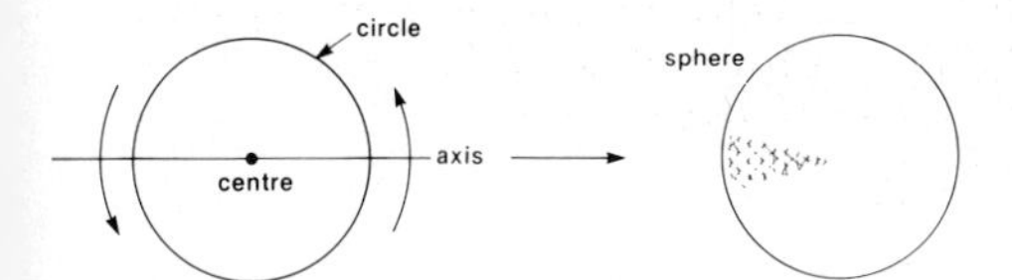

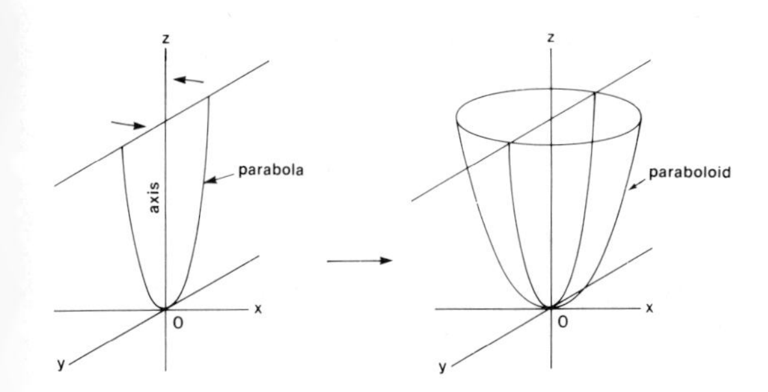

LATIN, *solidus*, solid; *revolvere*, to revolve, roll back

solution 1. (of an equation) The **set** of numbers or values which satisfies an **equation** (i.e. which makes the left hand side [L.H.S.] equal to the right hand side [R.H.S.]. 2. (of an inequality) The set of values which satisfy the **inequality**.
Examples
1. The equation $x + 7 = 11$ has the solution $x = 4$, since L.H.S. $= 4 + 7 = 11 =$ R.H.S.
$y^2 = 4$ has two solutions, $y = 2, -2$, since L.H.S. $= 2^2 = 4 = (-2)^2 = 4 =$ R.H.S.
$x + y = 2$ has an infinite number of solutions, since for every **real** number given to x, a corresponding real number y can be worked out.
2. The solution of the inequality $x + 3 > 0$ is the set of values where $x > -3$.
LATIN, *solvere*, to loosen

solve to work out the **answer** to a question or problem
Examples
1. One can solve the problem of finding the value of x which satisfies the equation $x - 3 = 5$ by adding 3 to both sides.
$x - 3 + 3 = 5 + 3$ or
$x = 8$.
2. Solve the following (give answers relating to x):
$2x + 3 > 11$ (subtract 3 from both sides)
$2x + 3 - 3 > 11 - 3$
$2x > 8$ (divide both sides by 2)
$x > 4$
LATIN, *solvere*, to loosen

space 1. Area or extension. 2. The **set** of all points in a (usually) **three-dimensional** region.
Example The space enclosed by a cube is l^3, where l is the length of one side.

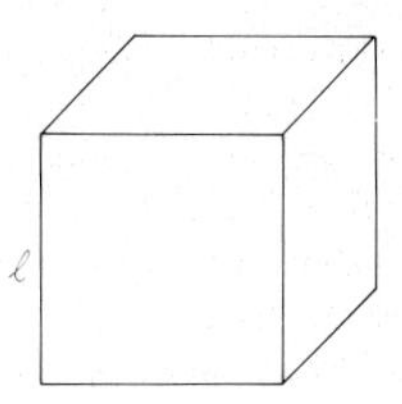

LATIN, *spatium*, space

speed the **rate** of change of distance with time.
Example A car on the freeway can travel up to a speed of 110 km/h. This means in one hour it can cover 110 km.

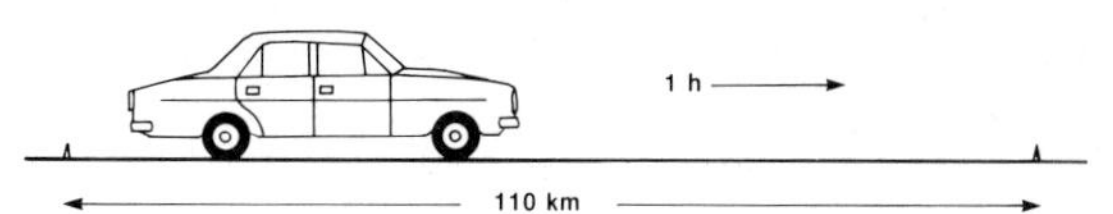

OLD ENGLISH, *spaed*, success, speed

sphere the **set** of points in space which are a fixed distance from a given point. The fixed distance is called the **radius**, and the central point, the centre of the sphere.
Example A perfectly round ball is a sphere, such as a ball bearing. The moon is nearly a sphere, and so too are the sun and the planets. Note: The **volume** of a sphere $= \frac{4}{3}\pi r^3$, and the surface area $= 4\pi r^2$, where r is the radius of the sphere.

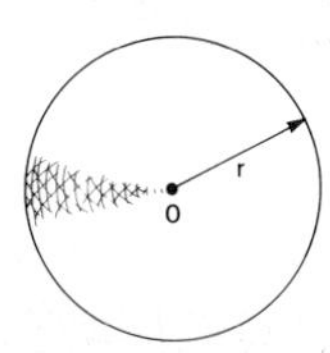

GREEK, *sphaira*, ball or globe

spherical angle the angle on the surface of a sphere formed by two **great circle** arcs intersecting. It is measured by the angle between the two **tangents** to the great circle **arcs** at the point of intersection.
Example

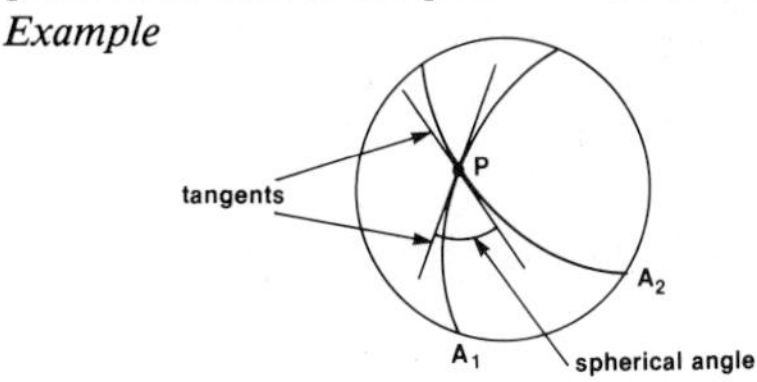

The angle between the two tangents at P is the spherical angle formed by arcs A_1 and A_2.
GREEK, *sphaira*, sphere; LATIN, *angulus*, angle

spherical coordinates a group of three numbers which locate a point in **three-dimensional** space. Symbols are usually (r, θ, ϕ), where r is the distance of the point from the origin or reference point, θ is the angle from one axis, and ϕ the angle from the other.
Example

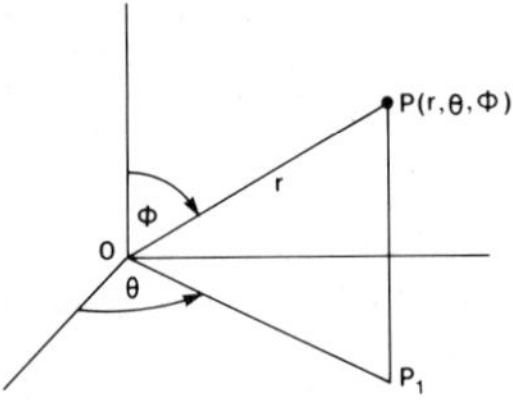

GREEK, *sphaira*, sphere; LATIN, *coordinare*, to set in order, regulate

spheroid a shape which approximates a sphere (*see* OBLATE).

spiral a curve formed by a point moving around a fixed point and constantly moving away from or approaching the fixed point. It can be represented by the polar equation $r = a\theta$ (where r is the distance of the moving point from the fixed point and θ is the angle $\angle XOP$). Also known as **helix** (in three dimensions).
Example

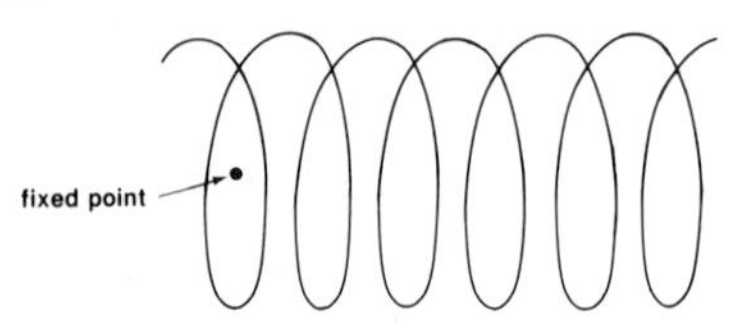

LATIN, *spira*, coil, spire

square **1.** (geometry) A **rectangle** with all sides equal. The properties of a square are those of a rectangle plus the **diagonals** are equal and **bisect** each other at right angles. **2.** (algebra) The second **power** of a number or **term**. **3.** An areal measurement (e.g. square centimetre).
Examples
1.

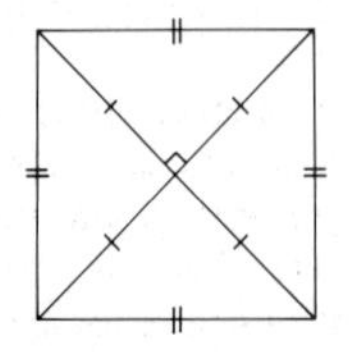

2. 2^2, x^2, $(\frac{1}{7})^2$ are all squares of numbers or terms.
LATIN, *quadra*, a square

square root a **number** which when multiplied by itself gives the given number. It can be represented as the length of one side of a square whose area is given. The symbol or sign is $\sqrt{\ }$ (*see also* ROOT).
Example Square root of 9 = + or − 3, since $+3 \times +3 = 9$, and also $(-3) \times (-3) = 9$.

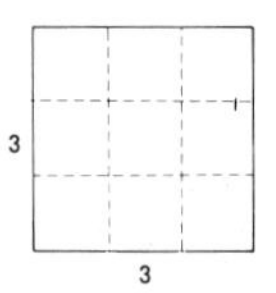

LATIN, *quadra*, a square; OLD NORSE, *rot*, root, branch

standard deviation (statistics) a special measure of how scattered a collection of data is about its **mean**. Also known as the root-mean-square deviation. σ is the symbol for standard deviation.

$$\sigma = \sqrt{\frac{\text{sum of the squares of the differences}}{\text{no. of elements (in the group)}}}$$

Note: The differences are the difference between each value and the mean.
Example

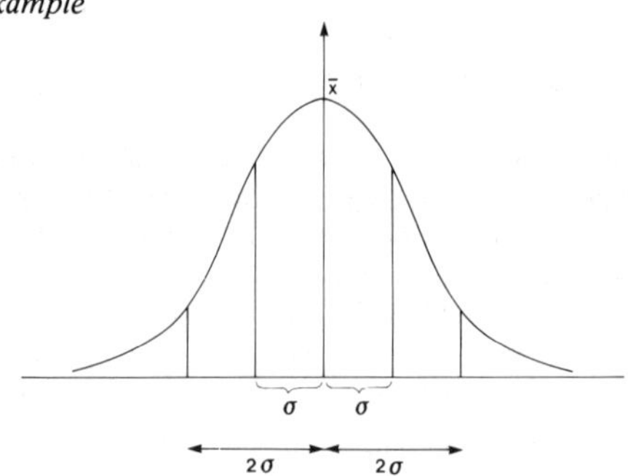

The graph is drawn from a set of observations x, whose mean is $\bar{x}$. The observations are scattered reasonably **symmetrically** about the mean. One standard deviation (σ) from the average either side accounts for over ⅔ of the observations. Two standard deviations account for 95% of the observations.
OLD FRENCH, *estandart*, flag marking rallying point; LATIN, *deviare*, to deviate, turn out of the way

standard error of the mean (statistics) a way of estimating how close to the actual **mean** is the mean of a given **sample**. If σ is the **standard deviation** of the whole group, then $\sigma_n = \dfrac{\sigma}{\sqrt{n}}$ is the standard error of the mean (of a sample of n values), and gives a measure of the degree of uncertainty in a sample of a given size.

Example If the average weight of adult males was 70 kg with a standard deviation of 10 kg, then the standard deviation for the average weight of 100 men would be $\frac{10}{\sqrt{100}} = 1$ kg. This means that we would be 95% confident that the actual average for the 100 men would not be more than two standard deviations (or 2 kg) away from the average found of the whole group. In this case 1 kg is the standard error of the mean.
OLD FRENCH, *estandard*, flag marking rallying point; LATIN, *errare*, to wander; *medius*, middle

standard score the **transformation** $t = (x - u)/\sigma$ which transforms a **normal** curve

$$y = \frac{1}{\sigma\sqrt{2\pi}} e^{-\frac{1}{2}\frac{(x-u)^2}{\sigma^2}}$$

into the standard normal curve

$$z = \frac{1}{\sqrt{2\pi}} e^{-\frac{1}{2}t^2}$$

where $z = \sigma y$. Note: u is the mean, σ is the standard deviation and x is the variable associated with the **probability distribution**.
Example Since a normal curve can be transformed to the standard curve, which is available in table form, then one can work out the probability that any given x will occur within given limits x_1, and x_2.
OLD FRENCH, *estandard*, flag marking rallying point

stationary point(s) a point or points on a curve where the **gradient** of the **tangent** to the curve at that point is **zero**. **Maxima**, **minima**, and horizontal **points of inflection** are all stationary points.
Example $y = x^3 - 3x$ can be drawn as a curve. P and Q are stationary points. P and Q are found by finding where the **gradient** of y is equal to 0, i.e. where $\frac{dy}{dx} = 0$. In the example

$$\frac{dy}{dx} = 3x^2 - 3 \therefore 3x^2 - 3 = 0$$

when $x^2 = 1$, i.e. when $x = 1$ or -1.

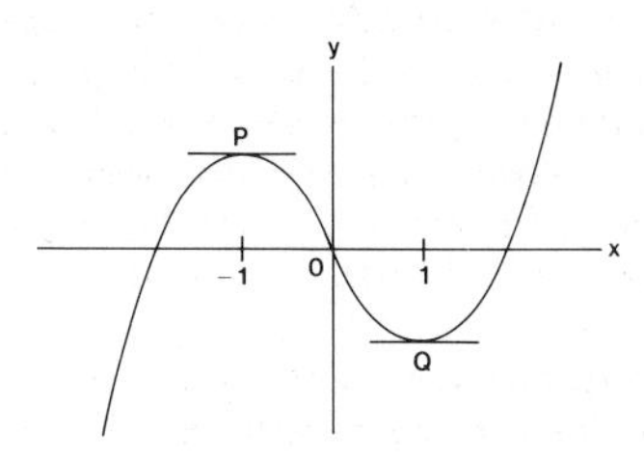

LATIN, *stationarius*, standing still; *punctus*, pricked

statistics **1.** The study of groups of data, their collection, arrangement and analysis. **2.** Numerical facts or data collected and classified.
Examples
1. In statistics we study such things as **averages**, **ranges**, how to group data in a meaningful way, and how to derive other useful information.

Sales	10	12	15	9	8	7	8	10	12	16	19	18
Month	J	F	M	A	M	J	J	A	S	O	N	D

Average Sales for the year = 12
Range of Sales: Lowest = 7, Highest = 19.
Hence, range is 12.
2. Many statistics concerning the habits of the population were gathered by the survey.
LATIN, *statisticus*, of state affairs

step graph a graph consisting of a series of horizontal lines, each separated from the other.
Example

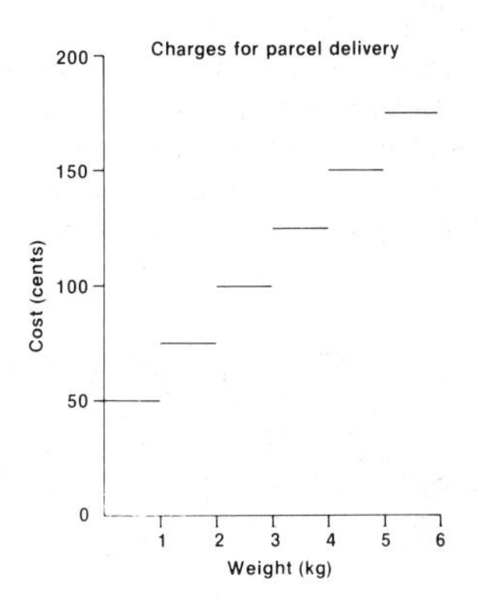

OLD ENGLISH, *staepe*, step; GREEK, *graphein*, to write

steradian, *see* RADIAN

straight angle, *see* FLAT ANGLE

straight line, *see* LINE (STRAIGHT)

string (computing) a set of characters or **numerals** arranged in a line.
Example A computer language often has instructions allowing strings of characters to be defined as a group. In **BASIC**, the instruction, LET A$ = "START OF PROGRAM" associates the set of characters, "START OF PROGRAM" with the variable A$. The instruction PRINT A$ tells the computer to print out "START OF PROGRAM".
OLD ENGLISH, *streng*, cord, string

sub-interval a part of an **interval**.
Example

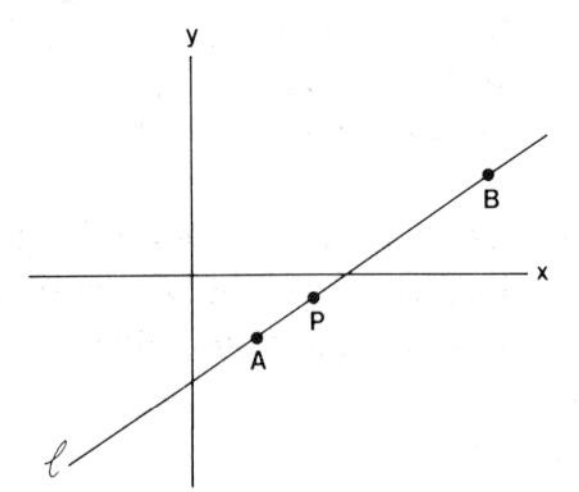

If AB is an interval of the line l, then AP is a sub-interval.
LATIN, *sub*, under; *intervallum*, space between ramparts

subnormal the portion of the **principal** axis of a **parabola** (or general **conic**) cut off by the perpendicular to the **tangent** at the **point of contact** to the parabola and the line from the point of contact of the tangent to the parabola perpendicular to the principal axis.
Example

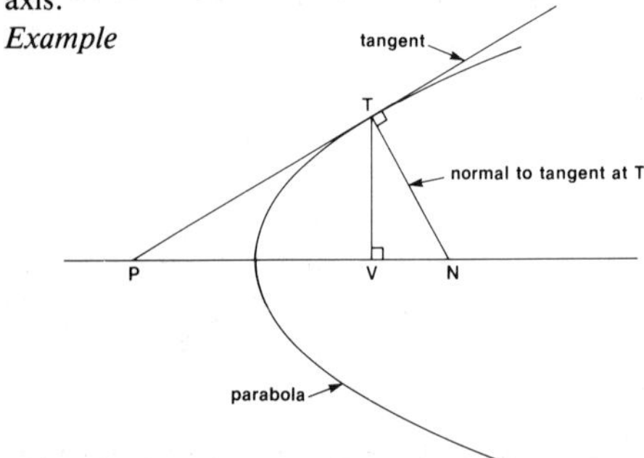

VN is the subnormal of the tangent PT to the parabola shown.
LATIN, *sub*, under; *norma*, a carpenter's square

subset a **set** of **elements** which form part of a larger set. The symbol for subset is $\subset$.
Example The set of positive integers {1, 2, 3, 4, 5, ...} is a subset of the set of the rational numbers $\{\frac{m}{n}\}$, where m, n are whole numbers. Men are a subset of the set of all human beings. This could be written {M} $\subset$ {H}, where {M} stands for the set of all men, and {H} stands for set of all human beings.

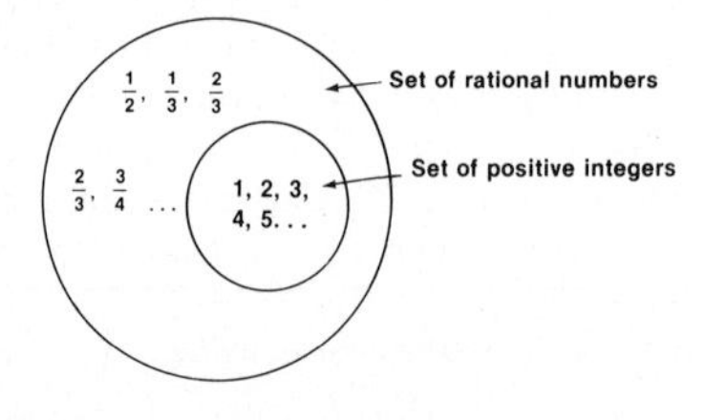

LATIN, *sub*, under; *secta*, a following of people

substitution the act or result of putting one **number** or quantity in place of another.
Examples
Using substitution we can simplify $(x - 3)^2 + (x - 3) - 1 = 0$. Substitute $y = x - 3$ in the above equation, then $y^2 + y - 2 = 0$, and we can solve for y. Having worked out y, $x = y + 3$ will give us x. For example, $y^2 + y - 1 = (y + 2)(y - 1) = 0$, i.e. $y = 1$ or -2. Therefore $x = 4$ or 1.
The perimeter of a rectangle is $2(l + b)$ where l is length, and b is breadth. We can substitute the measure of l and b in the formula when we know them. If $l = 3$ and $b = 1$, then p the perimeter is 8.
LATIN, *sub*, under; *statuere*, to cause to stand

subtend form an angle by joining two lines from two given points.
Example

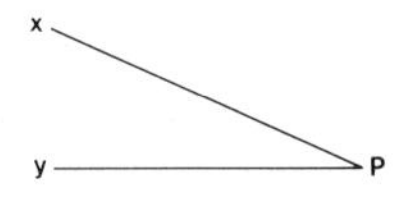

X and Y subtend the angle XPY at P.

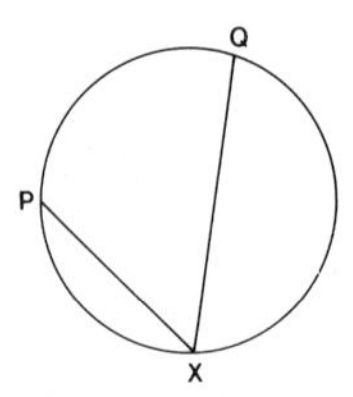

P and Q subtend the angle PXQ at X.
LATIN, *sub*, under; *tendere*, to extend

subtract to take away, or to find the difference between two quantities. The **sign** is $-$. Hence, *subtraction*. Note: Subtract is the **inverse** of **add**.
Example When you subtract 3 from 7 you get 4, which can also be written as $7 - 3 = 4$. The difference between 7 and 3 is 4, found by $7 - 3 = 4$.
LATIN, *subtrahere*, to draw away

sum the result of adding two or more **numbers** or quantities together. Hence, **summation**, the act of adding things up, or the process of totalling.
Examples
The sum of 3, and 5, and 7 is 15. This can be written as $3 + 5 + 7 = 15$.
To find the approximate area under the following curve we can use the summation of small rectangles as shown.

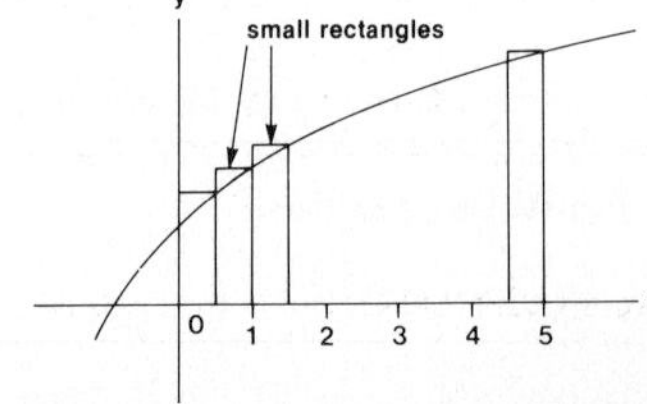

The area under the curve from 0 to 5 $\approx$ sum of each small rectangle as illustrated. As the rectangles are made smaller, so the area becomes more accurate.
LATIN, *summus*, highest, topmast

sum to infinity, *see* LIMITING SUM

supplementary angles angles whose sum is 180°.
Example **Adjacent** angles in a **parallelogram** are supplementary, as are the two angles formed by a line cutting another straight line.

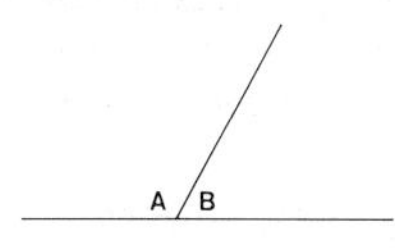

$$A^\circ + B^\circ = 180^\circ$$

The opposite angles of **cyclic quadrilateral** are always supplementary.

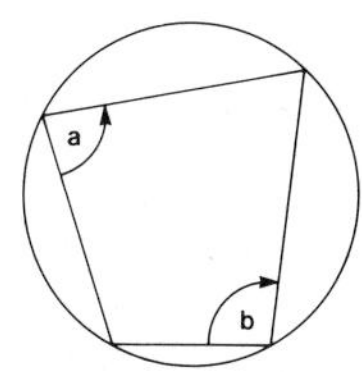

$$a^\circ + b^\circ = 180^\circ$$

LATIN, *supplere*, to fill out, supply; *angulus*, angle

surd an **irrational** number.
Example $\sqrt{3}$, $\sqrt{10} + 3$, $\sqrt[3]{17}$, $\frac{1}{\sqrt{3}}$, $\frac{1}{\sqrt{7} - \sqrt{2}}$
LATIN, *surdus*, deaf, mute

surface a **set** of points which makes up a **space** of two dimensions (length and breadth, but no thickness).
Example A plane surface is one formed from the intersections of two **straight lines.**

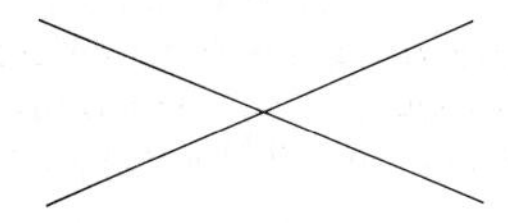

The surface of a **sphere** is the set of points each point of which is always a fixed distance from a given central point.

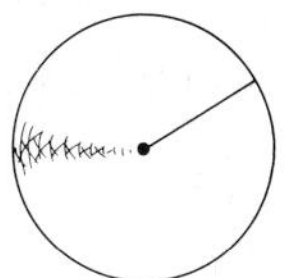

LATIN, *superficies*, surface

symbol something that stands for another. Also called a **sign.**
Example $a, b, 1, 2, 3, +, \times, -, \div, =, \{\ \}, >$, etc.
GREEK, *sumbolon*, mark, token

symmetry exact **correspondence** in position of the several **points** or parts of a figure or body with reference to a dividing **point, line** or **plane.**
Examples

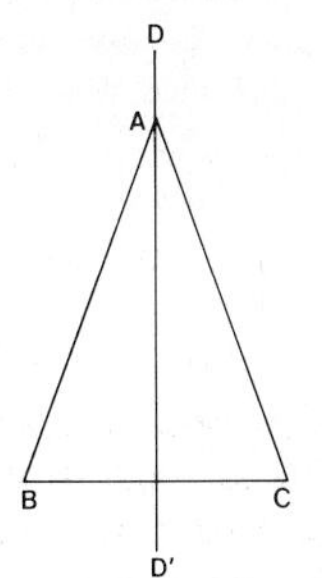

The triangle ABC is symmetrical about the line DD'.

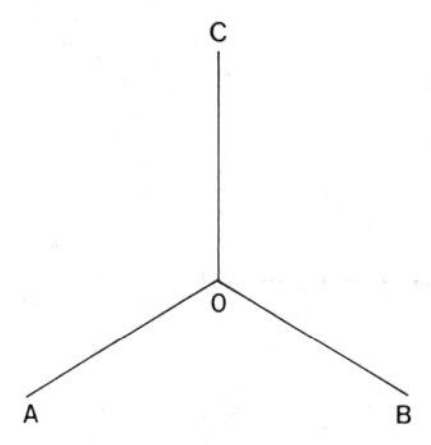

ABC has **rotational** symmetry about the point O (or line through O **perpendicular** to the plane ABC).
GREEK, *summetros*, of like measure

system an organized or connected group of objects or things. *See* MATHEMATICAL SYSTEM, DECIMAL SYSTEM, NUMBER SYSTEM, SI SYSTEM.
GREEK, *sustema*, organized whole

Tt

table(s) a set of related data which is laid out so that given one value one can look up or read out another value or values connected with the first.
Examples
A table of metric measurements for length is as follows:

10 millimetres (mm)	=	1 centimetre (cm)
10 centimetres (cm)	=	1 decimetre (dm)
10 decimetres (dm)	=	1 metre (m)
10 metres (m)	=	1 decametre (dam)
10 decametres (dam)	=	1 hectometre (hm)
10 hectometres (hm)	=	1 kilometre (km)

Logarithmic tables are used to simplify calculations involving complicated numbers.

Number	Logarithm
1.0	0.0
1.1	.0414
.	
.	
2.0	.3010
.	
.	
.	
9.9	.9956

Trigonometric tables are used to find the value of particular trigometric function of that angle

Angle	Sine
0°	0.0
1°	.0175
2°	.0349
.	
.	
.	
30°	.5
.	.
.	.

LATIN, *tabula*, board, list

tally 1. A stick on which marks are made to keep a count or score. 2. The score kept on a piece of wood, score card etc. 3. A mark used in scoring (in groups of five usually) which consists of four vertical lines cancelled by one **diagonal** or horizontal line.
Examples
1.

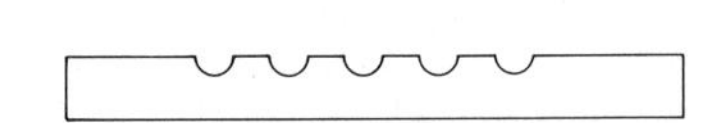

The stick has five notches on it, indicating a tally of 5.
2.

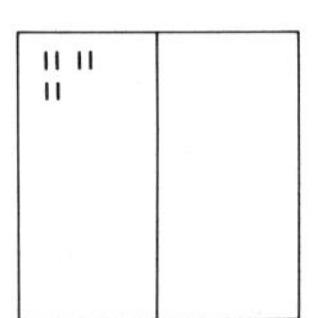

3. 𝍸 𝍸 𝍸 | The tally is 16.
LATIN, *talea*, twig, stick

tangent 1. (to a curve) A straight line which touches (but does not intersect) a curve or surface at a point. 2. (of an angle) The ratio of the side opposite the given angle to the side adjacent in a right angled triangle. The tangent of an angle is written *tan* for short. 3. (coordinate geometry) The **ratio** of the y coordinate to the x coordinate of the point whose radius has moved anticlockwise from the x-axis a given angle to the point. The radius is of unit length.
Examples
1.

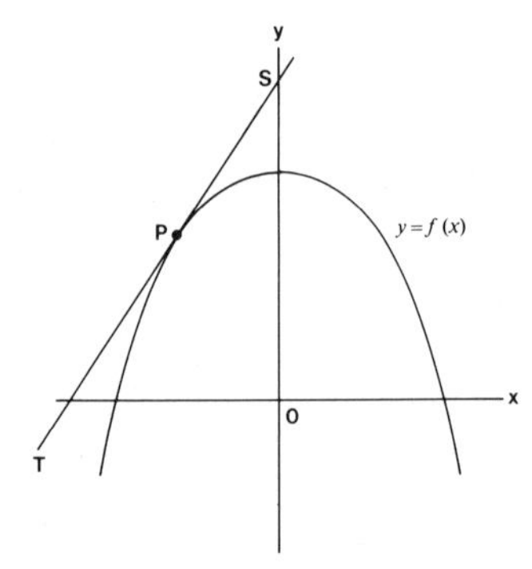

TS is a tangent to the curve $y = f(x)$ at point P.
2.

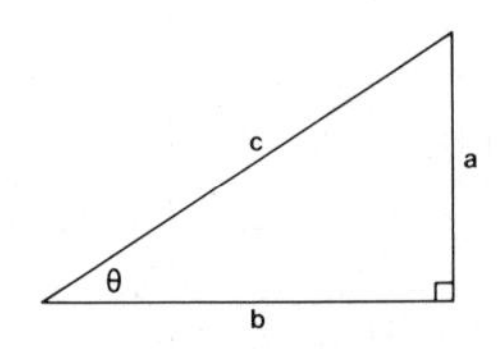

$$\text{Tangent of } \theta = \frac{a}{b}$$

3.

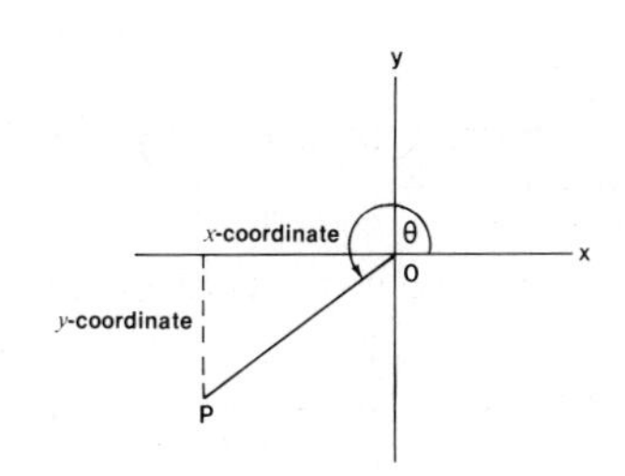

$$\tan\theta = \frac{y \text{ coordinate of } P}{x \text{ coordinate of } P}$$

LATIN, *tangere*, to touch

tangram a Chinese puzzle. It consists of a square cut into seven shapes, which are to be put together into different figures. The seven shapes are five **triangles**, one **square** and a **parallelogram**.
Example A tangram and its pieces put together into the form of a cat.

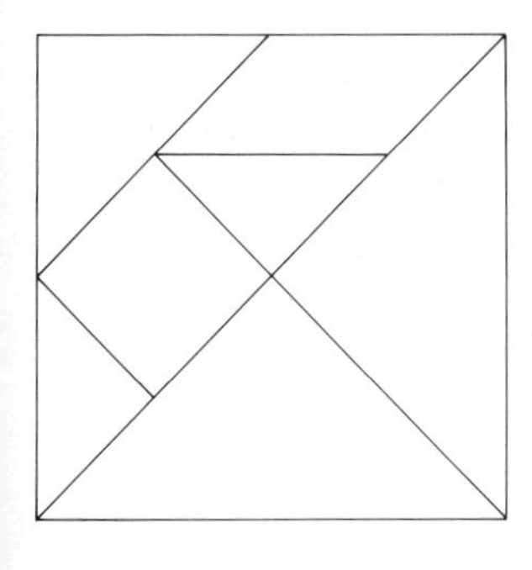

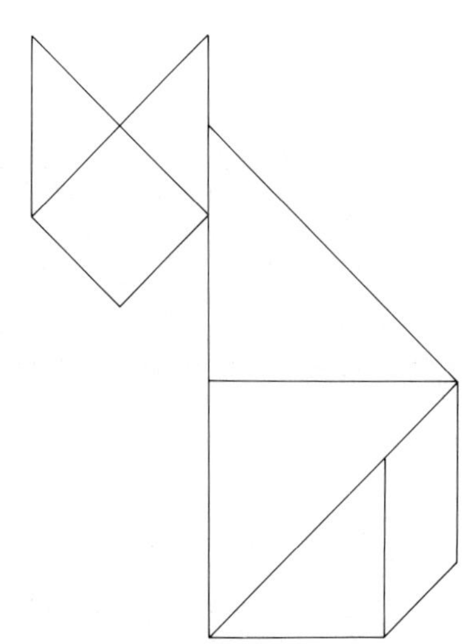

origin obscure

term a number or letter or combination which stands alone in an **expression** or **sequence** or **series**. The combination commonly uses only multiplication or division signs.
Example 3, x, y^2, $3x$, $\frac{4x}{y}$ are terms. In the series $a + ar + ar^2 + ar^3 + \ldots$, a, ar, ar^2, etc. are also terms.
LATIN, *terminus*, boundary line, limit

terminating decimal a **fraction** written as a **decimal**, which does not repeat.
Example $\frac{2}{5} = 0.4$, $\frac{1}{40} = 0.025$ are terminating decimals, but $\frac{1}{3} = 0.33333\ldots$ is not.
LATIN, *terminare*, to limit

terms **1.** In terms of—using (some particular idea or ideas). **2.** In **geometric** (**algebraic**) terms—using geometric (algebraic) symbols.
Examples
1. Express a circle in terms of **coordinate geometry**: $x^2 + y^2 = r^2$ is the equation of a circle centre O, radius r.

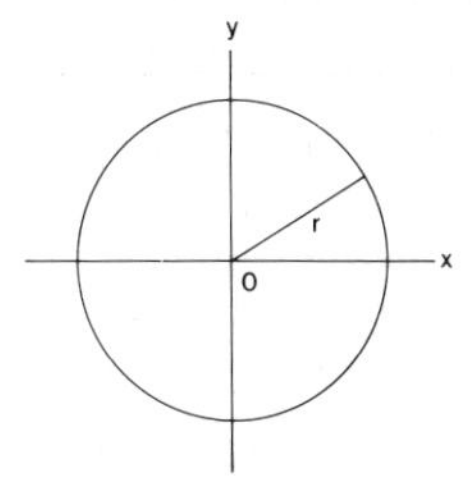

2. Describe in geometric terms the following **equation**: $y = 3x + 1$. (Answer: $y = 3x + 1$ is a straight line, of gradient 3 and y intercept 1.)
LATIN, *terminus*, boundary line, limit

tessellation a pattern made up of **congruent** figures joined together leaving no gaps.
Examples

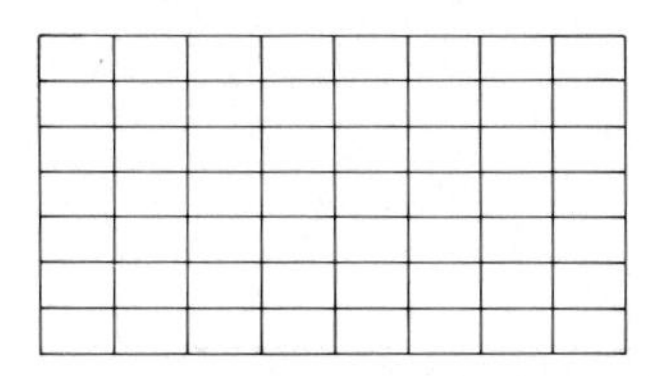

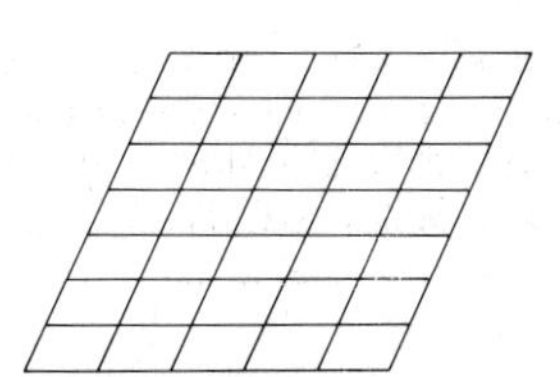

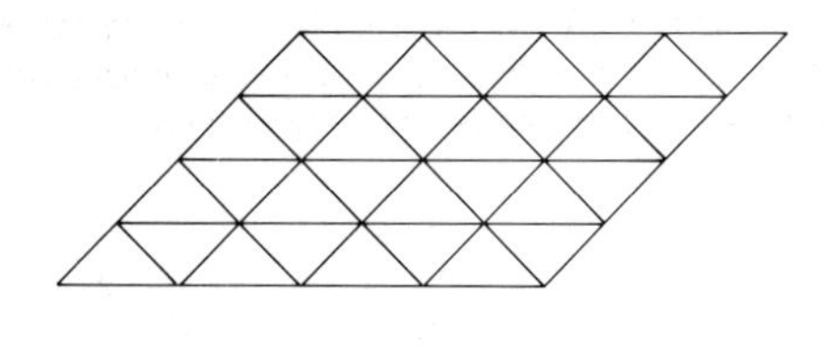

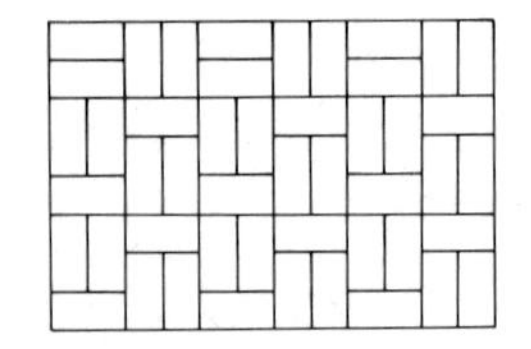

We see tessellations in different styles of tiling in bathrooms etc.
LATIN, *tessella*, a small cube

tetrahedron a four faced solid, where each face is a **triangle**. A regular tetrahedron has all faces **equilateral** triangles.
Example

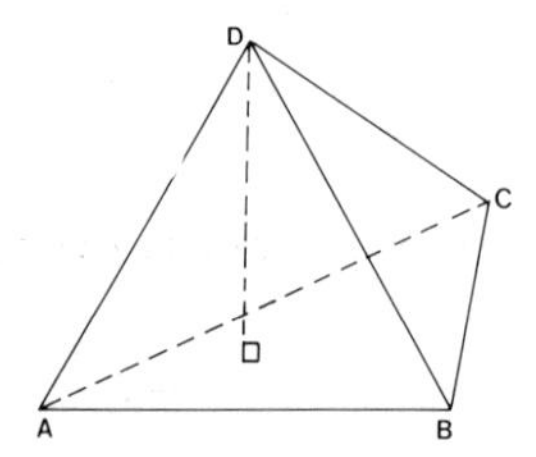

The volume of a tetrahedron is ⅓ × area of base × height.

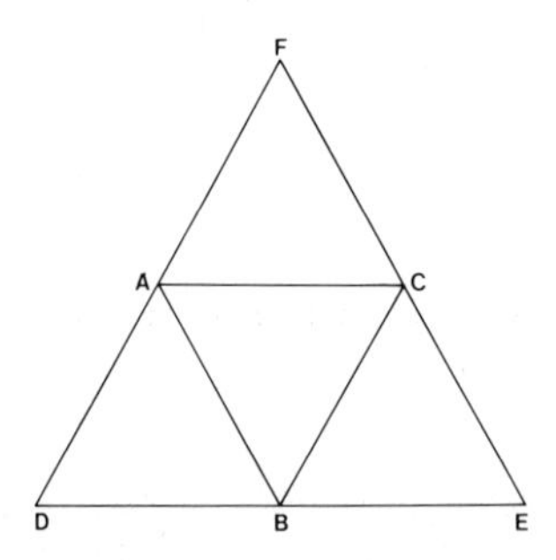

By folding triangles *ABD*, *CBE*, *AFC* about *AB*, *BC*, *AC* respectively, a tetrahedron is formed.
GREEK, *tetraedros*, four faced

theorem a statement that can be proved to be true on the basis of clearly stated assumptions.
Examples The **factor** theorem states that if $P(x)$ is a **polynomial** and if $P(a) = 0$, then $x - a$ is a factor. One of the theorems about a **triangle** is the following: in any triangle the **medians** intersect one another in a common point (called the **centroid**).

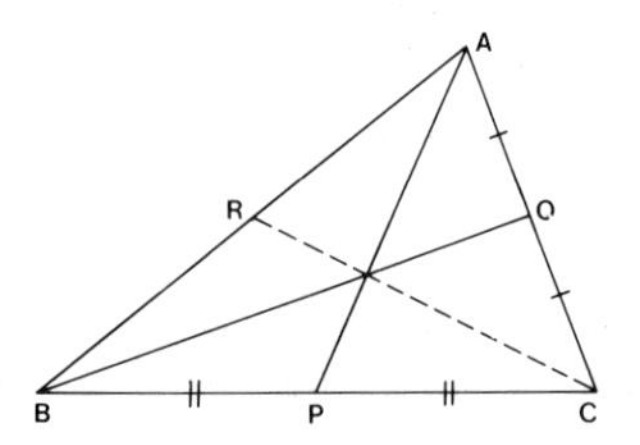

GREEK, *theorema*, speculation, proposition

theory of games, *see* GAMES THEORY

three-dimensional space the space of normal senses, which has length, breadth, and depth (or height). **Points**, **lines**, or **surfaces** can be represented in three dimensions by **algebraic relations** (**equations** etc.) using the methods of **coordinate** or **analytical geometry** (symbol is 3-D).
Examples

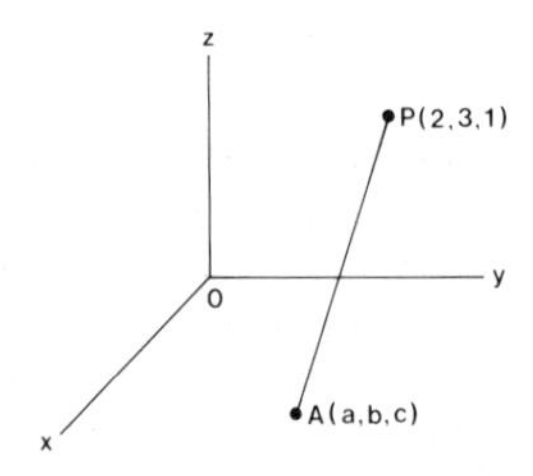

P can be located with respect to three reference axes (all at right angles to each other, where *x* could represent length, *y* could represent breadth, and *z* could represent height) by measuring how far *P* is from each of the three axes. This can be written as, say, $P(2,3,1)$, if *P* was 2 units from the *x* axis, 3 units from the *y* axis and 1 unit from the *z* axis.
A line *AP* can be represented by the following equations:

$$\frac{x-2}{2-a} = \frac{y-3}{3-b} = \frac{z-1}{1-c}, \text{ where } A \text{ is } (a,b,c).$$

A plane can be represented by the equation of the type $ax + by + cz = d$, where a, b, c, d are **constants**.
OLD ENGLISH, *thric*, three; LATIN, *dimensio*, a measuring; *spatium*, space

time series (statistics) a group of recordings of the values of **variable** quantities measured at definite intervals of time.
Example Over a year the following sales of ice cream were recorded.

Sales (tonnes)	21	23	17	14	11	6	5	7	9	10	15	22
Month	J	F	M	A	M	J	J	A	S	O	N	D

This is a time series of the sales of ice cream over time.
OLD ENGLISH, *tima*, time, period; LATIN, *serere*, to join

times multiplied by. Symbols used are × and sometimes •.
Examples
3 times 7 is 21. This can be written $3 \times 7 = 21$, or $3 \cdot 7 = 21$ (not to be confused with 3.7, meaning 3 point 7 or 3 decimal 7).
$3a$ is short for 3 times a, as $2xy$ is short for 2 times x times y.
OLD ENGLISH, *tima*, time, period

tonne a unit of mass equal to 1000 kilograms.
Example A tonne is the unit in which bulk commodities such as wheat, steel, ore and coal are measured.
LATIN, *tunna*, cask, vat

topology the study of the properties of **geometric** figures which remain unchanged even when the figures are distorted. Hence, *topological transformation*, a stretching or twisting change applied to a given shape. Under a topological transformation the following things remained unchanged: (i) the number of **nodes**; (ii) the **order** of each node; (iii) the number of **arcs**; (iv) the **sequence** of points along each arc; (v) the number of **regions**. When objects are *topologically equivalent*, they can be transformed topologically one to another or others.
Examples
1. If a square is drawn on a rubber sheet, and the sheet distorted, the square can be pulled into a **circle** or **ellipse** or any simple **closed curve** shape. Topology studies what things about shapes remain unchanged when the shapes are distorted, and is sometimes called "rubber sheet" geometry.

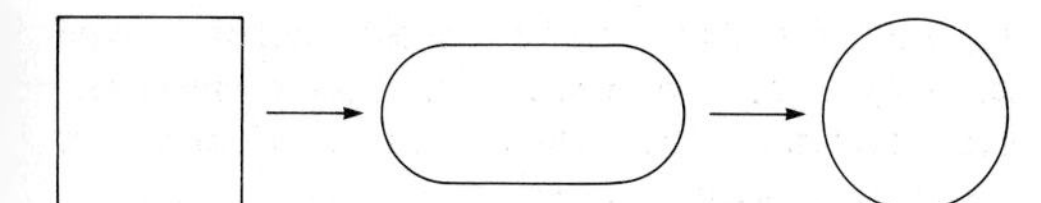

2. A diagram showing the train routes and stations along them in a large city is a topological transformation of the actual map of the routes.

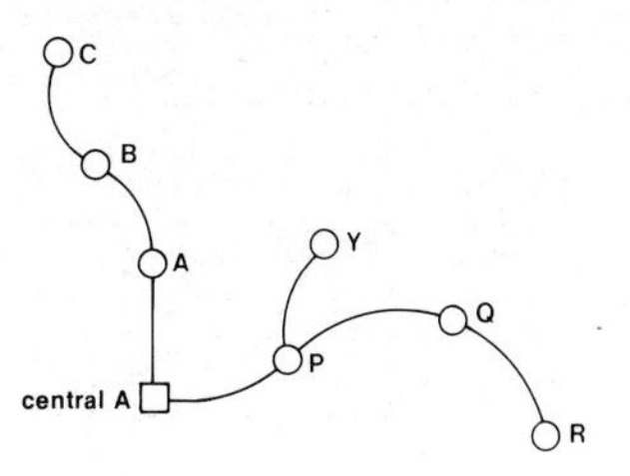

Actual map

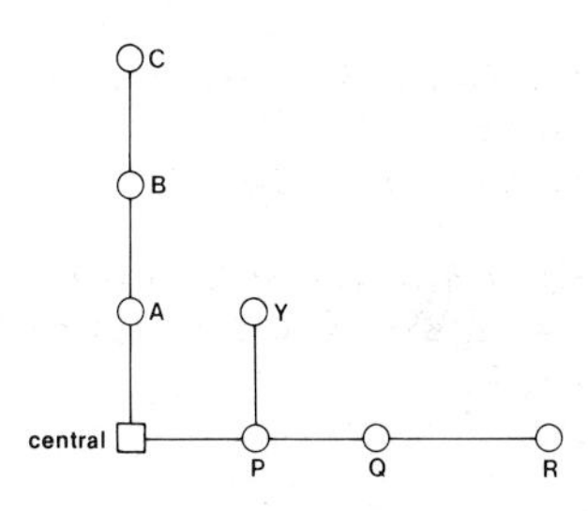

Topological map

3. The following shapes are topologically equivalent.

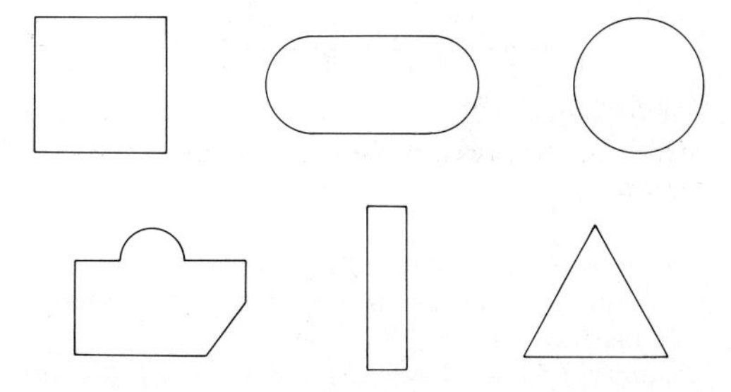

Also a tea cup and a hoop are topologically equivalent.

GREEK, *topos*, place; *logos*, study of

torus a donut shaped solid. It is made by rotating a closed curve (usually a **circle**) about an axis in its own plane.
Example

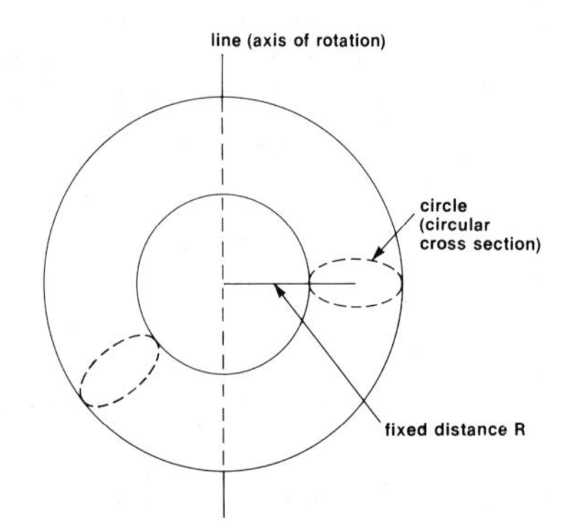

The **area** of a torus $= 4\pi^2Rr$ where R is the distance from the line to the centre of the circular cross-section (circle) and r is the radius of the circular cross-section. The **volume** of a torus $= 2\pi^2Rr^2$.
LATIN, *torus*, protuberance, round swelling

total the amount one gets by adding up. Also called **sum**.
Examples

$$\begin{array}{r} 4 \\ 13 \\ 11 \\ \underline{7} \\ \text{Total} = 35 \end{array}$$

Find the total of $3x$, $-2x$, $7x$. Total $= 3x - 2x + 7x = 8x$.
LATIN, *totus*, whole

transcendental number a **real number** which is not a **root** of any **algebraic equation** with **rational coefficients**.
Example π (= 3.14159...) and e (= 2.71828...) are transcendental numbers.
LATIN, *transcendere*, climb over; *numerus*, number

transformation the **relation** between a point and a corresponding point called an **image** point. Also known as a **mapping**.
Example **Rotations, translations, reflections,** and **enlargements** of objects are different kinds of transformations. Transformations can be grouped in several classes, depending on the extent of the change applied. A transformation which results in each corresponding point occupying the original point is called an **identity** transformation. A transformation which preserves the lengths and angles of a **geometric** figure is called an **isometric** transformation. There are also **similarity, affine,** and **topological** transformations.

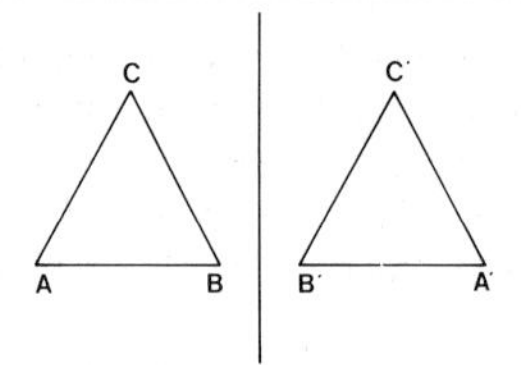

Reflection tranformation

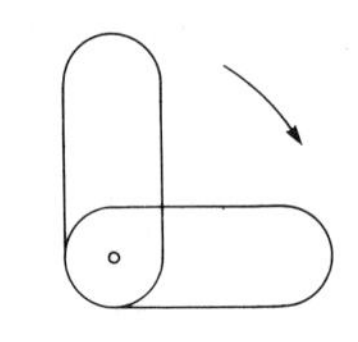

Rotational transformation

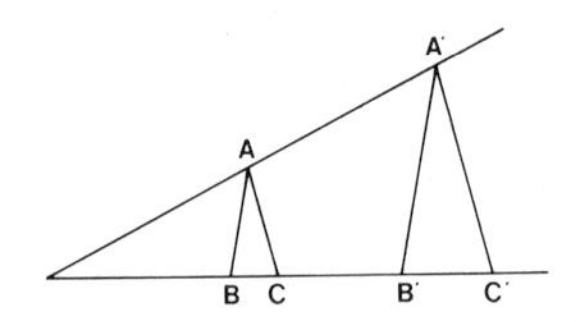

Enlargement tranformation

LATIN, *trans*, across; *forma*, form, shape

transitive (of a relation) characterising a **relation** such that: If A is related to B in the same way as B is related to C, then A is also related to C.
Examples
1. If $3 < 7$, and $7 < 10$, then $3 < 10$. This is a transitive relation.
2. If Joe is taller than Bill, and Bill is taller than Peter, then it follows that Joe is taller than Peter, and the relation is transitive.
LATIN, *transitus*, gone across

translation a **transformation** in which every point moves the same distance and in the same direction.
Example

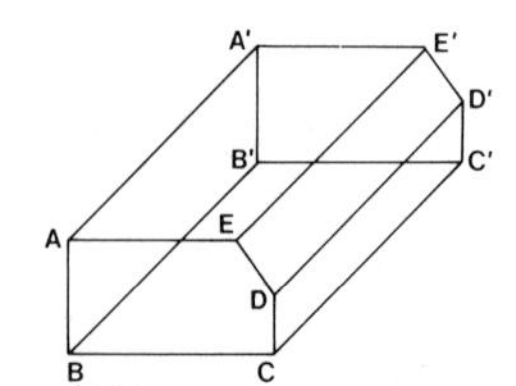

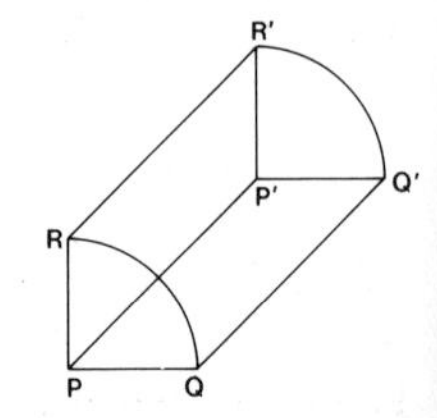

The shape $ABCDE$ is translated to position $A'B'C'D'E'$ and the quadrant PQR translated to the $P'Q'R'$.
LATIN, *translatus*, carried across

transpose 1. (algebra) To transfer a quantity or **term** from one side of an **equation** to the other. 2. (matrices) The rearrangement of the **rows** and **columns** of a matrix so that the rows become columns and columns becomes rows. It is written $'$.

Examples

1. If $x + 3 = 7$, by transposing 3 from the left hand side to the right hand side we get $x = 7 - 3 = 4$ and so solve the equation.
2. If $M = \begin{pmatrix} 1 & 2 & 3 \\ 0 & -1 & 4 \end{pmatrix}$, when M' (the transpose of M is

$$\begin{pmatrix} 1 & 0 \\ 2 & -1 \\ 3 & 4 \end{pmatrix}$$

OLD FRENCH, *transposer*, move from one place to another

transversal a line that cuts across two or more lines. In general, when a transversal cuts a pair of **parallel** lines, the: (i) **alternate** angles formed are equal; (ii) **corresponding** angles formed are equal; (iii) **cointerior** or **allied** angles are **supplementary**.

Example

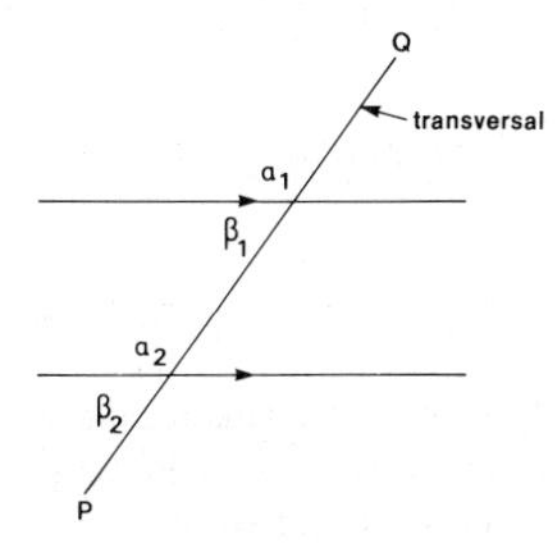

PQ is a transversal. Note: There are several properties relating to the angles formed by a transversal cutting two parallel lines. For example $\alpha_1 = \alpha_2$, $\beta_1 = \beta_2$ and $\alpha_2 + \beta_1 = 180°$.

LATIN, *transversus*, lying, turned across

trapezium (also **trapezoid**) a four sided figure with one pair of sides **parallel**. The area of a trapezium is equal to half the sum of the parallel sides times the distance between them: Area $= \frac{1}{2}(AD + BC) \times h$.

Example

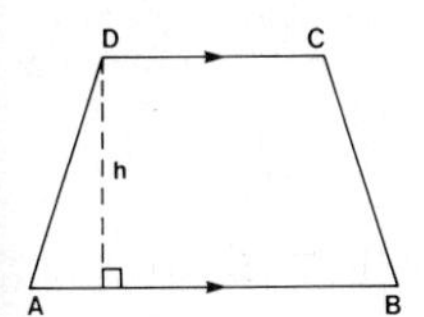

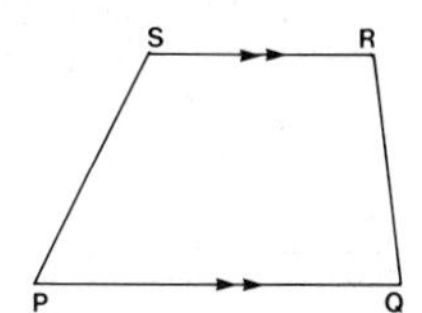

ABCD and *PQRS* are both trapeziums. If $AD = BC$ then *ABCD* is known as an **isoceles** trapezium.

GREEK, *trapezion*, small table

trapezoidal rule a way of finding the approximate value of a definite **integral** using **trapezoids**. If $A = \int_a^b f(x)dx$, then A can be approximated by dividing the interval a into equal subdivisions ($a = x_0, x_1, x_2, \ldots x_n = b$) and then approximating the curve $y = f(x)$ by joining $f(x_0)$ and $f(x_1)$ by a straight line and so on until $f(x_n)$ is reached. The value of A is then approximated by

$$(b - a)/n\{\tfrac{1}{2}(y_0 + y_n) + \sum_{i=1}^{n-1} y_i\}$$

Example If $y = x^2 + 2x + 1$, then $\int_0^1 ydx$ can be found approximately as follows. Divide the interval (0,1) into four parts, then, applying the trapezoidal rule, we get:

$$\begin{aligned}\int_0^1 ydx &\approx \frac{1-0}{4}[\tfrac{1}{2}(1 + 4) + 1\tfrac{9}{16} + 2\tfrac{1}{4} + 3\tfrac{1}{16}] \\ &\approx \tfrac{1}{4}(\tfrac{5}{2} + \tfrac{9}{4} + \tfrac{37}{8}) \\ &\approx \tfrac{75}{32} \\ &\approx 2\tfrac{11}{32}\end{aligned}$$

$$\text{By actual integration, } \int_0^1 ydx = [\frac{x^3}{3} + x^2 + x]_0^1 = 2\tfrac{1}{3}$$

GREEK, *trapezion*, small table

traversable referring to a **network** which can be drawn without going over any line twice.

Example

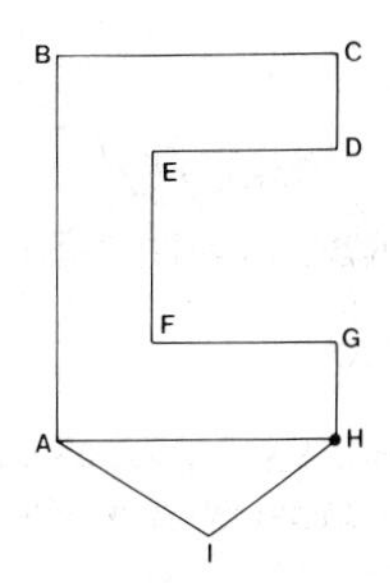

The network *ABCDEFGHI* can be traced out without going over any line twice.

LATIN, *transversus*, transverse

tree diagram a diagram used to represent possible outcomes (represented as "branches") as a result of a **sequence** of **events** taking place, each event having itself a number of possible results. It is used in **probability** situations.

Example What is the probability of a one turning up if a die is tossed three times? The various outcomes can be represented by a probability tree diagram as shown below. Y denotes a one turning up; N denotes any other number turning up. For a one to turn up three times in a row (YYY).

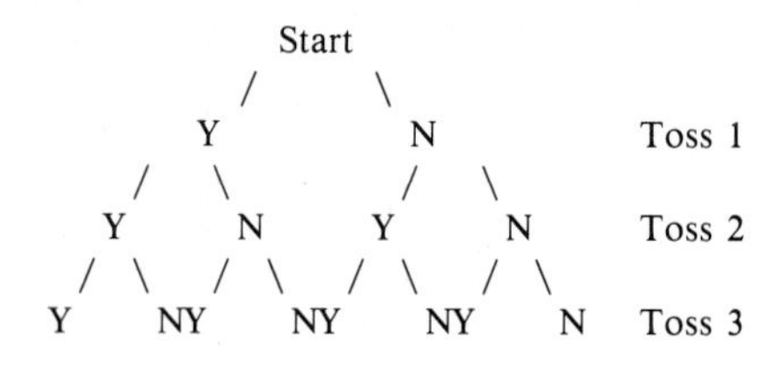

Probability $= \frac{1}{6} \times \frac{1}{6} \times \frac{1}{6} = \frac{1}{216}$
For a one not to turn up at all in the three tosses (NNN). Probability $= \frac{5}{6} \times \frac{5}{6} \times \frac{5}{6} = \frac{125}{216}$
Other combinations can be worked out. Another way of looking at this is as a **binomial expansion** of $(\frac{1}{6} + \frac{5}{6})^3$.
OLD ENGLISH, *treo*, tree; GREEK, *diagramma*, marked out by lines

triangle a plane three sided figure. The area of any triangle = ½ base × height (the **perpendicular** distance from the **vertex** to the base). The sum of the three interior angles of a triangle is two right angles (180°).
Examples
This is an **isoceles** triangle (two sides are equal).

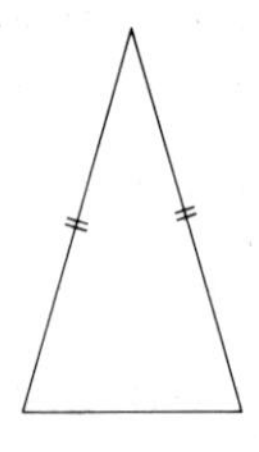

This is an **equilateral** triangle (all sides are equal).

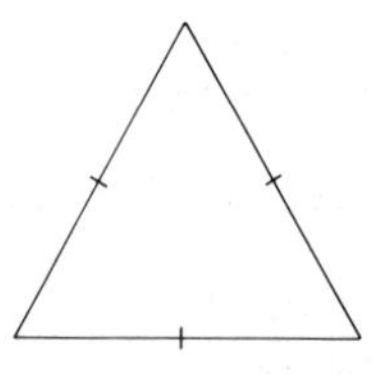

This is a **right angled** triangle (one angle = 90°).

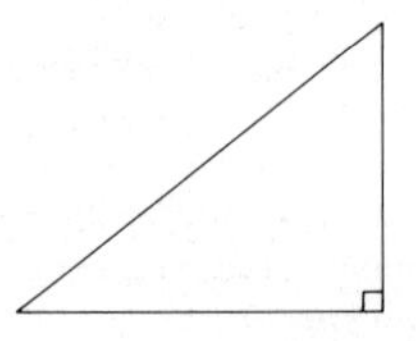

This is a **scalene** triangle (all sides unequal).

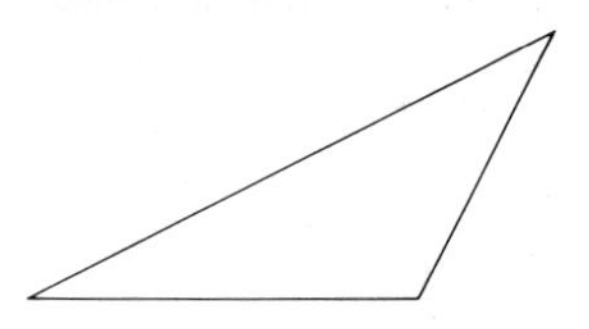

LATIN, *triangulus*, three angled

triangular number a number which can be represented by a triangle of dots, where each dot is regularly spaced.
Examples

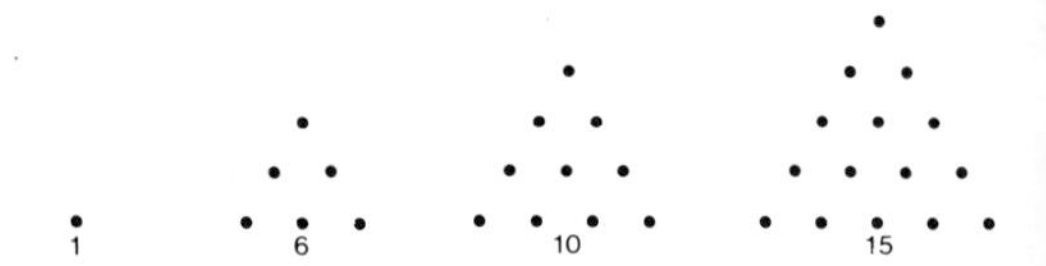

Notice that each triangle has on each side an equal number of dots. Triangular numbers are examples of number **patterns**.
LATIN, *triangulus*, three angles; *numerus*, number

trigonometric formulae various rules relating two or more trigonometric **functions**. Also called **trigonometric identities**.
Examples For any angle θ,
$\sin^2\theta + \cos^2\theta = 1$
$\tan\theta = \dfrac{\sin\theta}{\cos\theta}$
$\tan^2\theta + 1 = \sec^2\theta$.
$\sin 2\theta = 2 \sin\theta \cos\theta$, etc.
GREEK, *trigonon*, triangle; *metron*, measure; LATIN, *forma*, form, shape

trigonometric function a function of any angle defined as follows: It is the **ratio of sides** (each function has its own ratio of sides) of a right angled triangle containing the angle (*see* TRIGONOMETRY).
Examples
The **sine** (abbreviated as sin) of an angle is a trigonometric function defined as follows: $\sin\theta = \dfrac{a}{c}$.

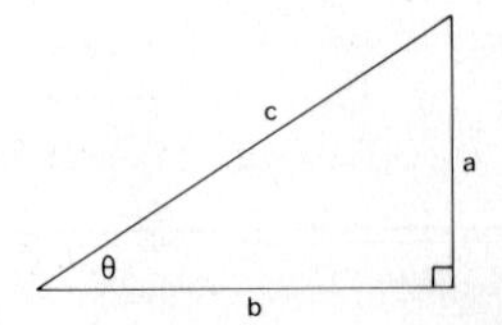

Similarly $\cos\theta$ (short for cosine) $= \frac{c}{b}$, and $\tan\theta = \frac{a}{b}$. In general $\sin\theta$ can be defined by the ratio $\frac{a}{c}$ (the side opposite to the angle, divided by the longest side), and the tangent is $\frac{a}{b}$ (side adjacent divided by the longest side).

In the graph below, the formula $y = \sin\theta$ has been drawn.

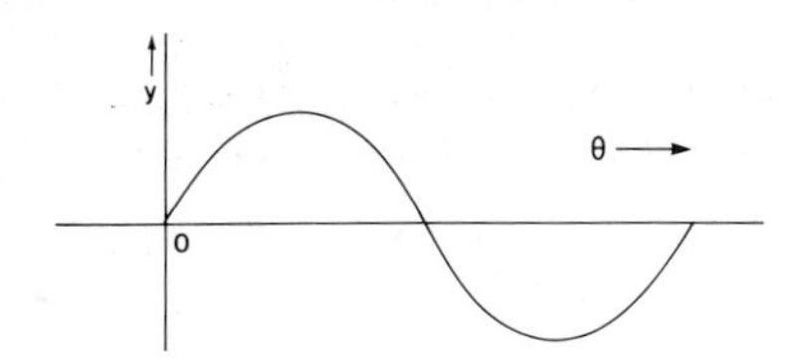

GREEK, *trigonon*, triangle; *metron*, measure; LATIN, *fungere*, to perform

trigonometric ratios the **ratios** of the appropriate sides of a right angled triangle (or the ratios of the appropriate **coordinates** generally) which define the various trigonometric **functions** of an angle.
Examples

$$\sin\theta = \frac{a}{c} \qquad \operatorname{cosec}\theta = \frac{c}{a}$$

$$\cos\theta = \frac{b}{c} \qquad \sec\theta = \frac{c}{b}$$

$$\tan\theta = \frac{a}{b} \qquad \cot\theta = \frac{b}{a}$$

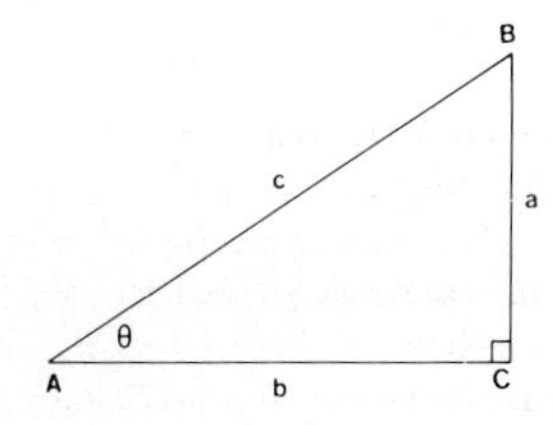

GREEK, *trigonon*, triangle; *metron*, measure; LATIN, *ratio*, reason

trigonometry the study of the **relations** between **angles** and sides of **triangles**, and various **functions** of these relations.
Example In the right angled triangle *ABC* the following relations between the angle θ and the sides exist: $\sin\theta = \frac{BC}{AC}$, $\cos\theta = \frac{AB}{AC}$, $\tan\theta = \frac{BC}{AB}$.

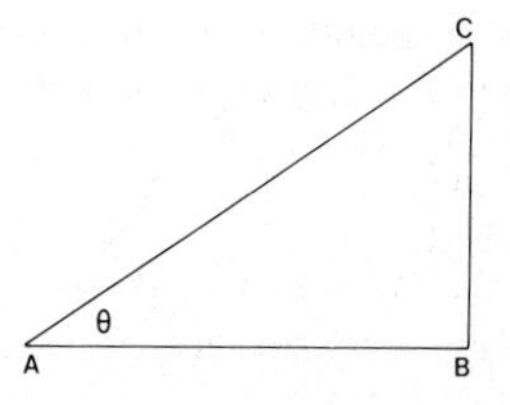

In any triangle the following **ratios** can be shown to be true: $\frac{a}{\sin A} = \frac{b}{\sin B} = \frac{c}{\sin C}$ (*see* SINE RULE). Also it can be shown that $c^2 = a^2 + b^2 - 2ab\cos C$ (*see* COSINE RULE).

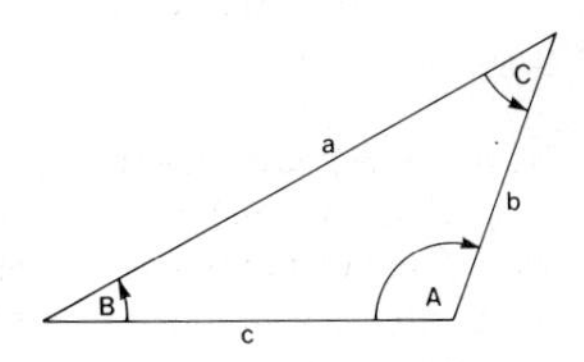

GREEK, *trigonon*, triangle; *metron*, measure

trinomial an **algebraic expression** made up of three **terms** combined together by either + or − signs.
Examples
$3x + 2y - z$ is a trinomial.
$x + 5x + 4$ is another.
GREEK, *trinomos*, three parts

trivial ordinary or commonplace. Requiring no real depth of understanding.
Example Showing that $a^2 - b^2 = (a + b)(a - b)$ is trivial. It is done by simply multiplying $(a + b)$ and $(a - b)$ out. Is $\frac{3}{4}$ the same as $\frac{9}{12}$? The answer is "yes" and the proof is trivial (simply divide numerator and denominator of $\frac{9}{12}$ by 3).
LATIN, *trivialis*, commonplace

true **1.** Correct, not false. **2.** Satisfying some rule or pattern.
Examples
1. It is true that $7 > 4$.
2. What values of x make $x^2 < 5$ true?
OLD ENGLISH, *treowe*, loyal, trusty

truth set a collection of quantities or things which obey a given rule or pattern.
Example The numbers 1, 2, 3, form the truth set for the relation $n + 1 < 5$, where n is a positive whole number.
OLD ENGLISH, *treowthe*, firmness, faith, truth; LATIN, *secta*, a following (of people)

turning point a **point** on a **curve** where the **gradient** or **slope** of the curve just before it and just after it are opposite.
Examples

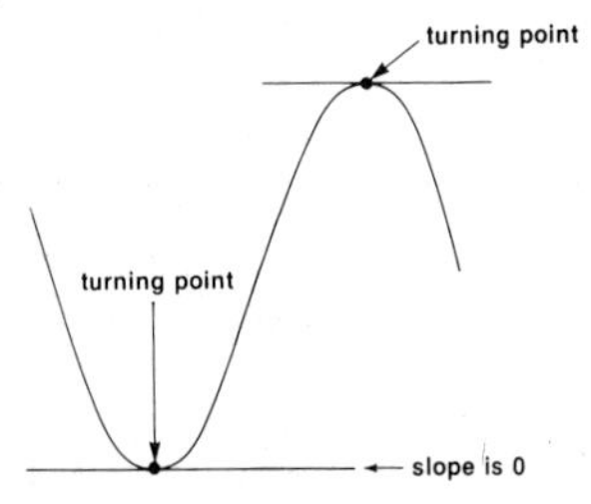

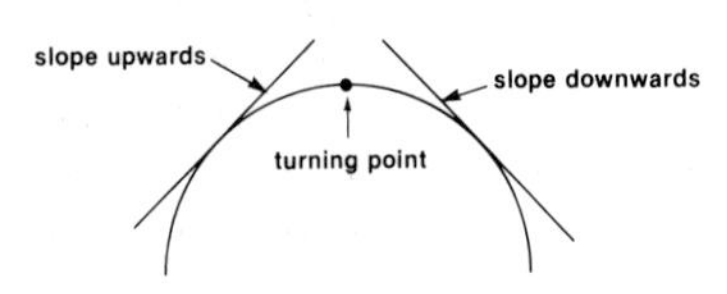

If the curve is smooth, then the slope of the curve at the turning point is 0 (that is, neither up nor down).
See MAXIMUM, MINIMUM, POINT OF INFLECTION.
LATIN, *tornare*, to turn in a lathe, round off

Uu

union the joining of (two **sets**). The union of two sets is the set that contains all the **elements** of the two given sets, and no other elements. The symbol for union is ∪ (pronounced "cup", or "union").
Example If $A = \{2, 5, 9, 16, 17\}$ and $B = \{3, 5, 7, 9, 11\}$ then $A \cup B$ (the union of A and B) is

$$\{2, 3, 5, 7, 9, 11, 16, 17\}$$

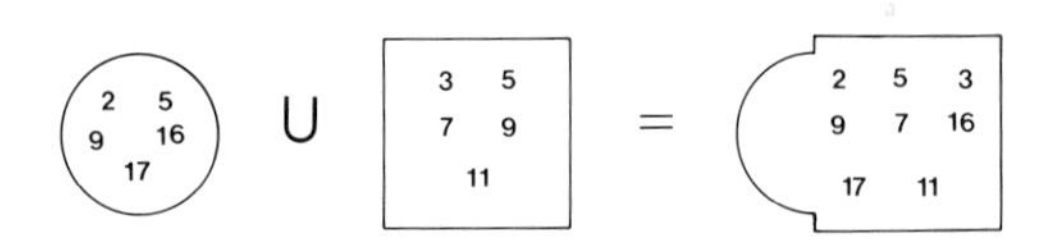

LATIN, *unus*, one

unitary method a way of working out problems dealing with different quantities by finding out the value of one quantity first.
Example If six eggs cost 96 cents, what would 11 eggs cost? This problem can be worked out by finding the cost of *one* egg first, then finding out what 11 eggs cost. For example, six eggs cost 96¢. One egg costs 96 ÷ 6 (or 16¢). Therefore, 11 eggs cost 16 × 11 = $1.76. This shows how the unitary method is used to solve problems of this kind.
LATIN, *unus*, one; GREEK, *methodos*, a going after, pursuit

unit **1.** A single thing or number. **2.** A standard of measurement. **3.** A quantity whose value is 1. **4.** In a system with a **binary operation** ⊕, an **element** e for which $e \oplus a = a \oplus e = a$ for every element in the set.
Examples
1. I have completed the first unit of my mathematics studies.
2. A metre is a standard unit of length.
3. If a car is regarded as a unit, then the car park's size is the same as 50 units.
4. 1 is the unit for multiplication (⊕ = ×), 0 is the unit for addition (⊕ = +).
LATIN, *unus*, one

unity one.
Example Any number divided by itself is unity, that is, 6 ÷ 6 = 1.
LATIN, *unus*, one

universal extending over or including the whole of something.

Example The universal **set** is the set of all **elements** of a given **class**. The symbol used is E or $\mathcal{E}$.. If we are dealing with positive **integers** then E = {1, 2, 3, 4, 5, 6, ...} is the universal set.
LATIN, *universus*, whole, entire, "turned into one"

unknown a **symbol** (usually a letter) used to represent a value yet to be found.
Example In $\frac{x + 3}{10} = 1$, x is called an unknown, or unknown quantity.
OLD ENGLISH, *ungecnawan*, not known

variable a quantity which, represented by some **symbol** (such as a letter), can take on different values. The opposite of a variable is a **constant**.
Example In the equation $y = 4x + 3$, y and x are variables, 4 and 3 are constants.
LATIN, *varius*, changeable

variance (statistics) a **measure** of how much variation there is from the **average** of a **set** of values. It is worked out as the average (**mean**) of the squares of the variations from the average (mean) of all the values.
Example The following results from a mathematics test marked out of 50 were recorded: 30, 32, 35, 37, 37, 38, 39, 39, 40, 40, 40, 41, 42, 42, 43, 43, 44, 44, 45, 46, 47, 47, 48, 49, 50. The average is

$$\frac{30 + 32 + 35 + \ldots + 50}{25} = 41.5$$

The average of the squares of the variations is

$$\frac{(30-41\tfrac{1}{2})^2 + (32-41\tfrac{1}{2})^2 + \ldots + (50-41\tfrac{1}{2})^2}{25}$$

This is the variance.
LATIN, *variare*, to vary

vary **1.** Vary as directly as: is **proportional** to. **2.** Vary inversely as: is inversely proportional to. **3.** Vary jointly as: is proportional to the product of.
Examples
1. If A varies directly as B, then $A \propto B$, or $A = kB$, where k is a constant.
2. If A varies inversely as B, then $A \propto \frac{1}{B}$, or $A = \frac{k}{B}$, where k is a constant.
3. If A varies jointly as B and C, then $A \propto BC$, or $A = kBC$.
LATIN, *variare*, to vary

vector a quantity which has magnitude (size) and direction. It is written in several different ways, such as v, $\vec{v}$, $\mathbf{v}$, and as a column of numbers $\begin{pmatrix} a \\ b \end{pmatrix}$ where a and b are any **real** numbers.
Examples

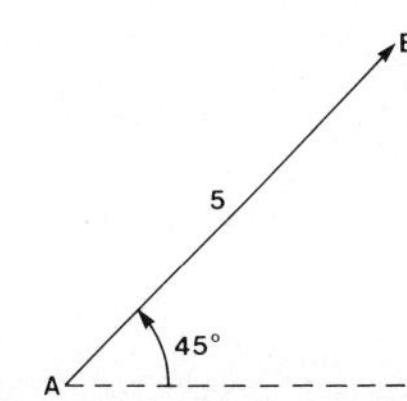

AB is a vector. It is 5 units long and is pointing in the direction approximately 45° from the horizontal.

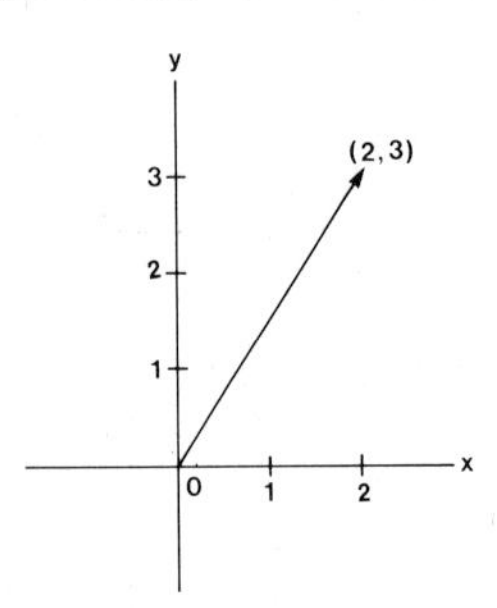

$\binom{2}{3}$ represents the vector.

Note: Vectors can be combined using a triangle. $\vec{C}$ is the resultant or combined result of adding two vectors $\vec{A}$ and $\vec{B}$. Vectors can represent physical quantities like forces, or describe translations.

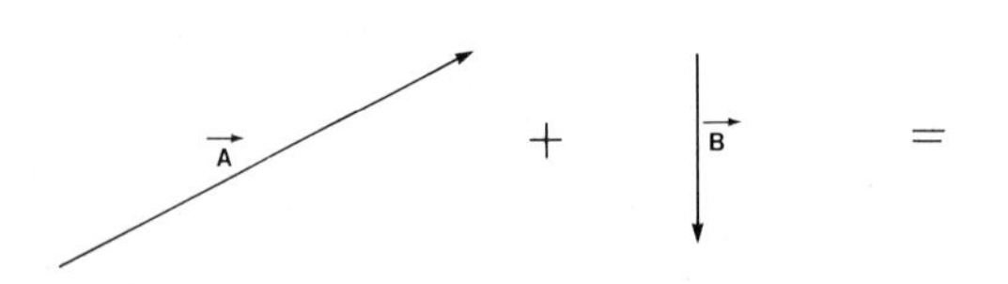

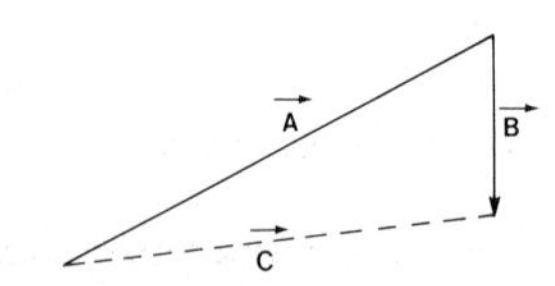

LATIN, *vector*, carrier

vector product (also **cross product**) the result of combining two **vectors** so that the resulting vector has a magnitude equal to the product of the lengths of the two vectors multiplied by the **sine** of the angle between them, and is directed at right angles to the **plane** of the first two. $\vec{A} \times \vec{B} = |\vec{A}||\vec{B}| \sin\theta \vec{C}$ where $\vec{C}$ indicates a unit vector (in the direction which corresponds to the z-axis if $\vec{A}$ or $\vec{B}$ correspond to the x or y-axis directions).
Example

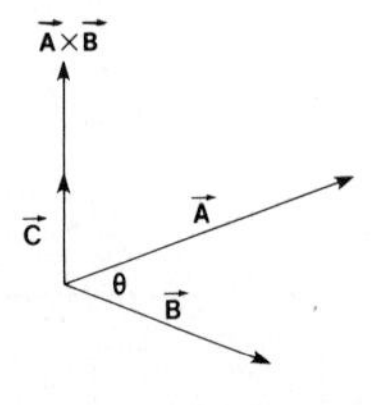

LATIN, *vector*, carrier; *productus*, that produced

velocity **1.** **Speed** in a particular direction. **2.** The **rate** of change of **displacement** with time. It is a **vector** quantity.
Examples
1.

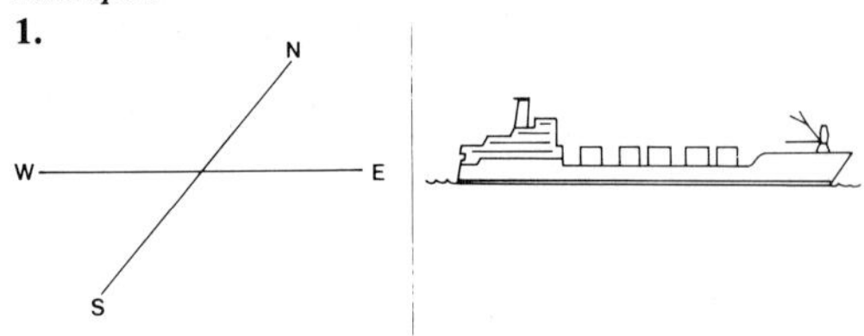

The velocity of the ship is 15 knots due E.
2. A ball thrown from a building changes velocity as it falls for two reasons: a. Velocity = acceleration multiplied by time ($v = gt$, where g is the acceleration due to gravity); and b. its direction is changing all the time.

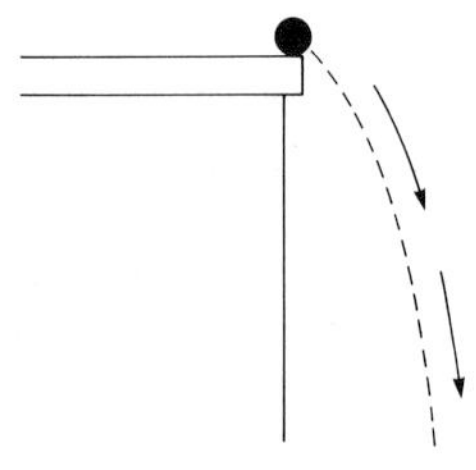

LATIN, *velox*, fast

venn diagram a diagram using circles (or oval shapes etc.) to show how **sets** are related.
Example

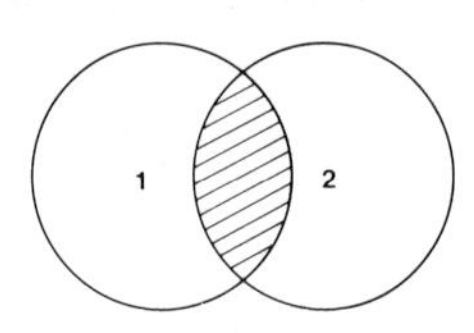

$A \cap B$ (**intersection** of A and B) is the shaded portion. $A \cup B$ is the **union** of A and B, consisting of Area 1, 2, and the shaded portion.
after Venn, John, English mathematician

vertex (*pl.* **vertices**) **1.** The **point** opposite to the **base** of a **plane** or solid figure. **2.** An angular point. **3.** The point on a **parabola** which lies at the point of intersection of the parabola with its axis.
Examples
1. The vertex of a **cone** is the topmost point.

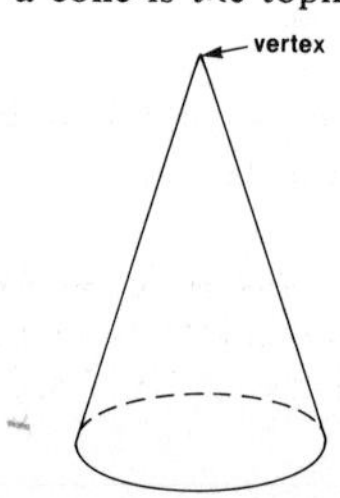

2. A **polygon** has many vertices.

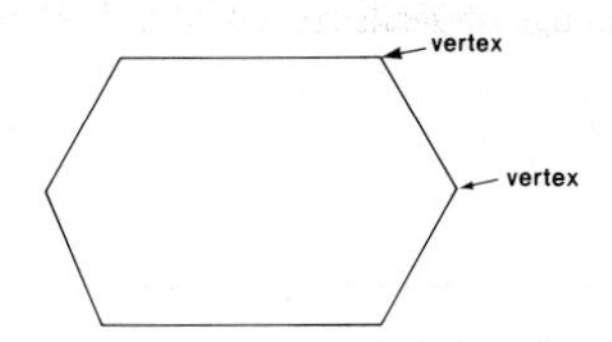

3.

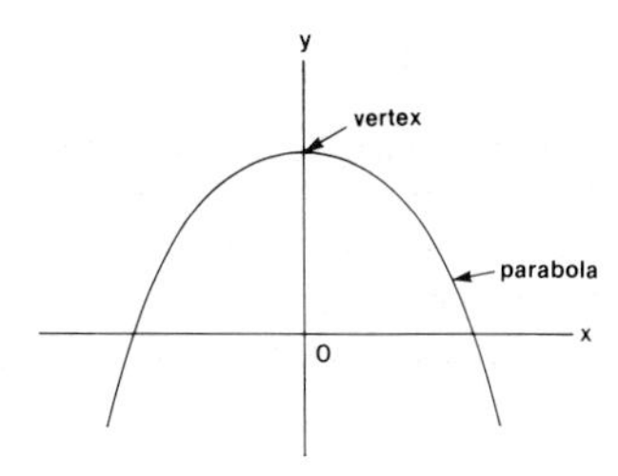

LATIN, *vertex*, highest point

vertical at right angles to the **horizontal**. A vertical **displacement** is **vector** movement of an object etc. at right angles to the horizontal. In a **triangle**, a vertical **angle** is that one opposite the base. A vertical angle (or vertically opposite angle) is also the name given to either of the two angles opposite each other when two lines intersect.
Example
The line AH is the vertical line to the base of the $\triangle ABC$.

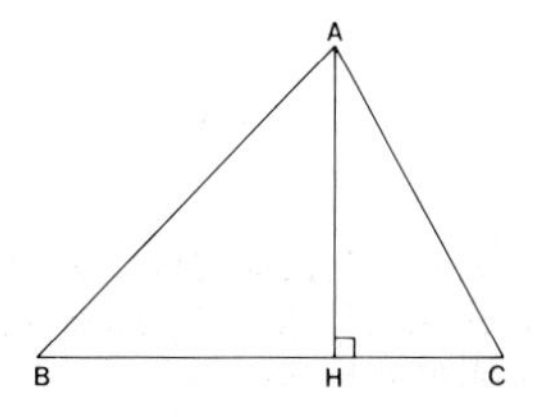

LATIN, *vertex*, highest point

vinculum a straight line drawn over two or more quantities to show they are grouped together. An alternative symbol for **brackets**.
Example $\overline{a + b} = (a + b)$, $3 \times \overline{a + b} = 3(a + b)$.
LATIN, *vincire*, to tie

volume the amount of space an object occupies. It is measured in **cubic units**.
Examples
The volume of a cube is found by multiplying the length of one side by itself twice, i.e.

$$\text{volume} = l \times l \times l.$$

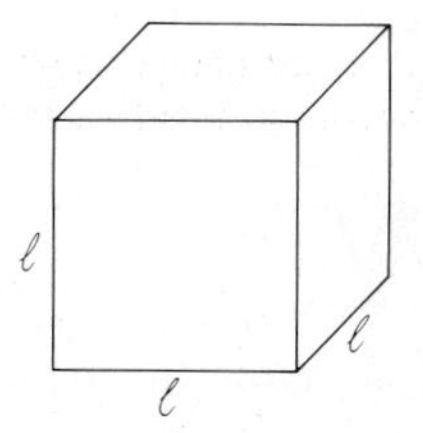

The volume of a box, or **prism** (in general) is the area of the base multiplied by the height or length, i.e. volume $= A \times h$.

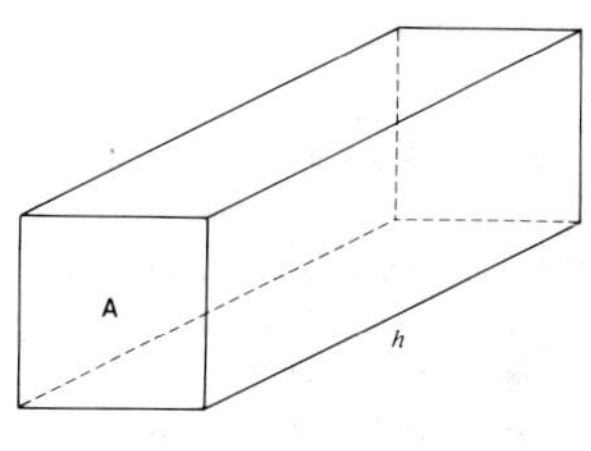

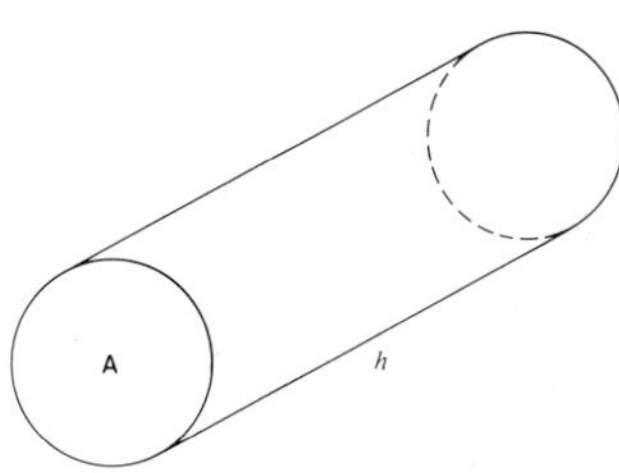

LATIN, *volumen*, roll of parchment

vulgar fraction common **fraction** (one whose **numerator** is smaller than its **denominator**).
Example $\frac{3}{4}$, $\frac{5}{7}$, $\frac{31}{64}$ are all vulgar fractions. $\frac{15}{7}$, $\frac{91}{18}$ are not vulgar fractions (they are **improper fractions**).
LATIN, *vulgus*, the common people

Ww

weighted average (also **weighted mean**) the average or mean of a set of values or quantities calculated as follows: Each of the values or quantities is multiplied by one of a group of numbers (to indicate each value's contribution to the average) and is then averaged. If $q_1, q_2, q_3, \ldots q_n$ is a set of n values and $w_1, w_2, w_3, \ldots w_n$ is a set of n numbers to indicate relative importance then W (weighted average) =

$$\frac{q_1\, w_1 + q_2\, w_2 + q_3\, w_3 + \ldots q_n\, w_n}{w_1 + w_2 + w_3 \ldots w_n}$$

Example A group of workers was put into five categories as far as wages were concerned.

Wages earned per week	$150	$200	$250	$300	$350
No. of people	10	15	20	16	19

$$W = \frac{150 \times 10 + 200 \times 15 + 250 \times 20 + 300 \times 16 + 350 \times 19}{10 + 15 + 20 + 16 + 19}$$

$$= \frac{20950}{80}$$

= $262 (to the nearest dollar).

OLD ENGLISH, *wight*, weight; FRENCH, *avarie*, damage to ship or cargo

whole numbers a **number** without any **fractions** in it.
Example −2, −73, 0, 4, 52, 10738, etc.
OLD ENGLISH, *hal*, whole, sound, unharmed; LATIN, *numerus*, number

yield **1.** An amount produced. **2.** The **profit** obtained from an investment.
Examples
1. The wheat crop this year gave a yield of 50 tonnes.
2. If I invest $1000 in shares which are paying 10% per annum, what is my yield after one year? (My yield is $1000 × 10% = $100.00.)
OLD ENGLISH, *gieldan*, to yield, pay

Zz

Zeno (495–480 BC) Greek mathematician famous in particular for his four paradoxes.

zero **1.** The **number** value of a **set** that has no members in it; nought. **2.** A whole number which lies between 1 and -1. **3.** The number which, when added to any other number, leaves the latter number unchanged. The symbol for zero is 0.
Examples
1. The number of members in the set $S = \{ \}$ is 0. In the symbol 309, 0 means no tens.
2.

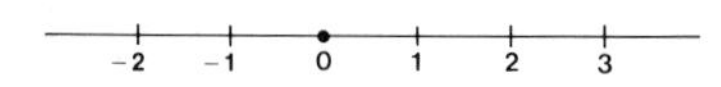

3. $6 + 0 = 6$; $a + 0 = a$.
ARABIC, *cifr*, zero (symbol for)

zero(s) of a polynomial **solutions** (or **roots**) of a **polynomial equation**. If $P(x)$ is a polynomial and when $x = a$, $P(a) = 0$, then $x = a$ is a zero of that polynomial.
Examples
$x + 3$ is a polynomial of degree 1. $x + 3 = 0$ is the polynomial equation. $x = -3$ is a zero of the polynomial.
$x^2 - 2x - 3$ is a polynomial of degree 2.
$x^2 - 2x - 3 = 0$ is the polynomial equation.
$(x + 1)(x - 3) = 0$ hence $x = -1,3$ are zeros of that polynomial.
ARABIC, *cifr* zero (symbol for)

FORMULAE

I. Plane figures

Circle

Circumference (C) $= \pi d$

$= 2\pi r$

Area (A) $= \pi r^2$

$= \dfrac{\pi d^2}{4}$

Parallelogram

Perimeter (P) $= 2\ (a + b)$

Area (A) $= bh$ (base × height)

Pythagoras' theorem

$a^2 + b^2 = c^2$

(where c is the length of the hypotenuse, and a and b are the lengths of the other two sides).

Rectangle

Perimeter (P) $= 2\ (b + h)$

Area(A) $= bh$ (base × height)

Square

Perimeter (P) $= 4s$ (where s is the length of a side)

Area (A) $= s^2$

Trapezium

Perimeter (P) $= a + b + c + d$

Area (A) $= \frac{1}{2}\ h\ (b + d)$

Triangle

Perimeter (P) $= a + b + c$

Area (A) $= \frac{1}{2}\ bh$ ($\frac{1}{2}$ base × height)

$= \sqrt{s(s-a)\ (s-b)\ (s-c)}$ [Hero's formula]

where $s = \frac{1}{2}$P

$= \frac{1}{2}(a + b + c)$

For an equilateral triangle, where all sides are equal:

Perimeter (P) $= 3a$

(where a is the length of a side)

Area (A) $= \dfrac{\sqrt{3}}{4}\, a^2$

II. Solid figures

Cube

Volume (V) $= a^3$

(where a is the length of a side)

Surface area (A) $= 6a^2$

Cuboid (rectangular right prism)

Volume (V) $= abc$

Surface area (A) $= 2(ab + bc + ac)$

Pyramid

Volume (V) = $\frac{1}{3}$ Bh (area of base × height)

Surface area (A) = area of base + area of triangular sides

Right circular cone

Volume (V) = $\frac{1}{3}\pi r^2h$

($\frac{1}{3}$ area of base × height)

Area (of curved surface) = πrl

Right circular cylinder

Volume (V) = πr^2h

(area of base × height)

Surface area (A) = area of curved surface + area of ends

= $2\pi rh + 2\pi r^2$

= $2\pi r(r + h)$

Right prism

Volume (V) = Bh (base area × height)

Area (A) = 2B + Ph

(2 × base area + perimeter × height)

Sphere

Volume (V) = $\frac{4}{3}\pi r^3$

Area (A) = $4\pi r^2$

III. General

Average (mean)

A = average; T = total; N = number of items

$$A = \frac{T}{N}$$

Cost

C = total cost; N = number of items; c = cost of each item

C = Nc

Distance

D = distance; V = velocity; t = time; g = acceleration due to gravity

D = Vt (for constant velocity)

D = $\frac{1}{2}gt^2$ (for falling objects)

Interest (simple)

I = simple interest; P = principle; R = rate; T = time; A = amount

$$I = \frac{PRT}{100}$$

A = P + I

Profit
P = profit; S = selling price; C = cost price
P = S − C

For profit percentage of cost price : $\frac{P}{C} \times \frac{100}{1}$

For profit percentage of selling price: $\frac{P}{S} \times \frac{100}{1}$

Selling price
S = C + P
(see abbreviations under *Profit*)

IV. Other formulae

Cosine rule
$a^2 = b^2 + c^2 - 2bc \cos A$

Sine rule

$$\frac{a}{\sin A} = \frac{b}{\sin B} = \frac{c}{\sin C}$$

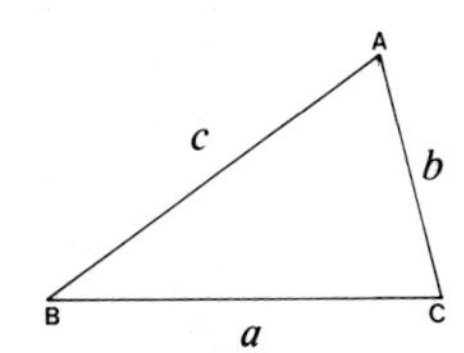

V. Derivatives

The right hand column represents the derivative of the function in the left hand column.

$y\ (=f(x))$	$\frac{dy}{dx}\ (=f'(x))$
x	1
ax	a (a is a constant)
ax^n	anx^{n-1} (a is a constant)
$\log_e x$	$\frac{1}{x}$
$\log_b x$	$\frac{1}{x} \log_e b$ (b is a constant)
$\cos x$	$-\sin x$
$\sin x$	$\cos x$
$\sec x$	$\tan x \,.\, \sec x$
$\operatorname{cosec} x$	$-\cot x \,.\, \operatorname{cosec} x$
$\cos u$	$-\sin u \frac{du}{dx}$ (u is a function of x)
$\sin u$	$\cos u \frac{du}{dx}$ (u is a function of x)
$\tan u$	$\sec^2 u \frac{du}{dx}$ (u is a function of x)
$\log_e u$	$\frac{1}{u} \frac{du}{dx}$
$\sin^{-1} \frac{x}{a}$	$\frac{1}{\sqrt{a^2 - x^2}}$
$\cos^{-1} \frac{x}{a}$	$-\frac{1}{\sqrt{a^2 - x^2}}$
$\tan^{-1} \frac{x}{a}$	$\frac{1}{\sqrt{a^2 + x^2}}$

IV. Integrals

The right hand column represents the integral of the left hand column.

$y\ (=f(x))$	$\int y\,dx = \int f(x)\,dx$
x^n	$\dfrac{x^{n+1}}{n+1} + c$ (where c is a constant)
$\dfrac{1}{x}$	$\log_e x + c$
e^{ax}	$\dfrac{e^{ax}}{a} + c$
$\log_e ax$	$x\log_e ax - x + c$
$\cos x$	$\sin x + c$
$\sin x$	$-\cos x + c$
$\tan x$	$\log_e (\cos x) + c$

8 9 0 1 2 3 4 5
B C D E F G H I J